In Situ Aeration and Aerobic Remediation

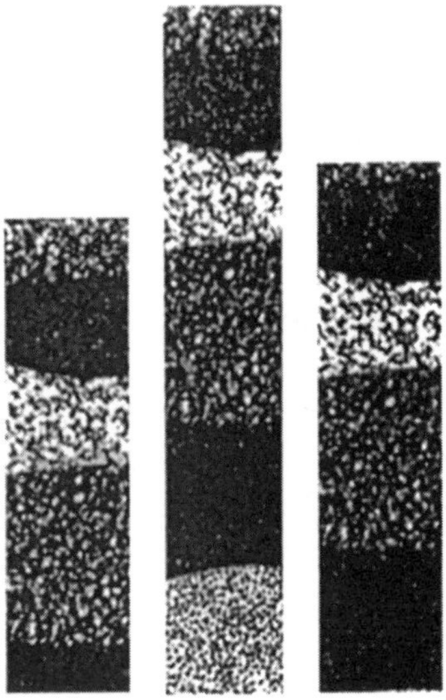

Editors

Andrea Leeson, Paul C. Johnson, Robert E. Hinchee, Lewis Semprini, and Victor S. Magar

The Sixth International In Situ and On-Site Bioremediation Symposium

San Diego, California, June 4–7, 2001

BATTELLE PRESS
Columbus • Richland

Library of Congress Cataloging-in-Publication Data

International In Situ and On-Site Bioremediation Symposium (6th : 2001 : San Diego, Calif.)
In situ aeration and aerobic remediation : the Sixth International In Situ and On-Site Bioremediation Symposium : San Diego, California, June 4-7, 2001 / editors, A. Leeson … [et al.].
p. cm. -- (The Sixth International In Situ and On-Site Bioremediation Symposium ; 10)
Includes bibliographical references and index.
ISBN 1-57477-120-5 (hc. : alk. paper)
1. In situ bioremediation--Congresses. 2. Hydrocarbons--Biodegradation--Congresses. 3. Bioremediation--Congresses. I. Leeson, Andrea, 1962- . II. Title. III. Series: International In Situ and On-Site Bioremediation Symposium (6th : 2001 : San Diego, Calif.). Sixth International In Situ and On-Site Bioremediation Symposium ; 10.
TD192.5.I56 2001 vol. 10
628.5 s--dc21
[628.5]

2001037942

Printed in the United States of America

Battelle Press
505 King Avenue
Columbus, Ohio 43201, USA
614-424-6393 or 1-800-451-3543
Fax: 1-614-424-3819
Internet: press@battelle.org
Website: www.battelle.org/bookstore

For information on future environmental conferences, write to:
Battelle
Environmental Restoration Department, Room 10-123B
505 King Avenue
Columbus, Ohio 43201-2693
Phone: 614-424-7604
Fax: 614-424-3667
Website: www.battelle.org/conferences

CONTENTS

Aerobic Cometabolism

Multiphase Extraction

Groundwater Circulation Well Operation and Monitoring

Physical or Chemical Pretreatment Approaches

Remediation of Oil Spills in Aquatic Environments

FOREWORD

The papers in this volume correspond to presentations made at the Sixth International In Situ and On-Site Bioremediation Symposium (San Diego, California, June 4-7 2001). The program included approximately 600 presentations in 50 sessions on a variety of bioremediation and supporting technologies used for a wide range of contaminants.

This volume, *In Situ Aeration and Aerobic Remediation*, focuses primarily on enhanced aerobic bioremediation for removal of petroleum compounds, and aerobic cometabolism for removal of various chlorinated hydrocarbons, including TCE, methylene chloride, and ethylene dibromide. Other topics include extraction strategies for simultaneous recovery of multiple contaminant phases, groundwater circulating wells, and physical and chemical pretreatment methods to enhance aerobic biodegradation.

The author of each presentation accepted for the symposium program was invited to prepare an eight-page paper. According to its topic, each paper received was tentatively assigned to one of ten volumes and subsequently was reviewed by the editors of that volume and by the Symposium chairs. We appreciate the significant commitment of time by the volume editors, each of whom reviewed as many as 40 papers. The result of the review was that 352 papers were accepted for publication and assembled into the following ten volumes:

Bioremediation of MTBE, Alcohols, and Ethers — 6(1). Eds: Victor S. Magar, James T. Gibbs, Kirk T. O'Reilly, Michael R. Hyman, and Andrea Leeson.

Natural Attenuation of Environmental Contaminants — 6(2). Eds: Andrea Leeson, Mark E. Kelley, Hanadi S. Rifai, and Victor S. Magar.

Bioremediation of Energetics, Phenolics, and Polycyclic Aromatic Hydrocarbons — 6(3). Eds: Victor S. Magar, Glenn Johnson, Say Kee Ong, and Andrea Leeson.

Innovative Methods in Support of Bioremediation — 6(4). Eds: Victor S. Magar, Timothy M. Vogel, C. Marjorie Aelion, and Andrea Leeson.

Phytoremediation, Wetlands, and Sediments — 6(5). Eds: Andrea Leeson, Eric A. Foote, M. Katherine Banks, and Victor S. Magar.

Ex Situ Biological Treatment Technologies — 6(6). Eds: Victor S. Magar, F. Michael von Fahnestock, and Andrea Leeson.

Anaerobic Degradation of Chlorinated Solvents— 6(7). Eds: Victor S. Magar, Donna E. Fennell, Jeffrey J. Morse, Bruce C. Alleman, and Andrea Leeson.

Bioaugmentation, Biobarriers, and Biogeochemistry — 6(8). Eds: Andrea Leeson, Bruce C. Alleman, Pedro J. Alvarez, and Victor S. Magar.

Bioremediation of Inorganic Compounds — 6(9). Eds: Andrea Leeson, Brent M. Peyton, Jeffrey L. Means, and Victor S. Magar.
In Situ Aeration and Aerobic Remediation — 6(10). Eds: Andrea Leeson, Paul C. Johnson, Robert E. Hinchee, Lewis Semprini, and Victor S. Magar.

In addition to the volume editors, we would like to thank the Battelle staff who assembled the ten volumes and prepared them for printing: Lori Helsel, Carol Young, Loretta Bahn, Regina Lynch, and Gina Melaragno. Joseph Sheldrick, manager of Battelle Press, provided valuable production-planning advice and coordinated with the printer; he and Gar Dingess designed the covers.

The Bioremediation Symposium is sponsored and organized by Battelle Memorial Institute, with the assistance of a number of environmental remediation organizations. In 2001, the following co-sponsors made financial contributions toward the Symposium:

Geomatrix Consultants, Inc.
The IT Group, Inc.
Parsons
Regenesis
U.S. Air Force Center for Environmental Excellence (AFCEE)
U.S. Naval Facilities Engineering Command (NAVFAC)

Additional participating organizations assisted with distribution of information about the Symposium:

Ajou University, College of Engineering
American Petroleum Institute
Asian Institute of Technology
National Center for Integrated Bioremediation Research & Development (University of Michigan)
U.S. Air Force Research Laboratory, Air Expeditionary Forces Technologies Division
U.S. Environmental Protection Agency
Western Region Hazardous Substance Research Center (Stanford University and Oregon State University)

Although the technical review provided guidance to the authors to help clarify their presentations, the materials in these volumes ultimately represent the authors' results and interpretations. The support provided to the Symposium by Battelle, the co-sponsors, and the participating organizations should not be construed as their endorsement of the content of these volumes.

Andrea Leeson & Victor Magar, Battelle
2001 Bioremediation Symposium Co-Chairs

AIR SPARGING LIFE CYCLE DESIGN FOR AN MTBE AND BTEX SITE

Keith A. Fields (Battelle, Columbus, Ohio)
Thomas C. Zwick, Andrea Leeson, and Godage B. Wickramanayake (Battelle, Columbus, Ohio); Herb Doughty (Southwest Division Naval Facilities Engineering Command, San Diego, California); Tracy Sahagun (Assistant Chief of Staff Environmental Security, Camp Pendleton, California)

ABSTRACT: This study presents the life-cycle design, operation, and exit strategy that Battelle implemented at a gas station site (Site 43286) at Marine Corps Base (MCB) Camp Pendleton. This site required remediation due to elevated concentrations of benzene, toluene, ethylbenzene, and xylenes (BTEX) and methyl tert-butyl ether (MTBE) in groundwater, and due to its proximity to an ecologically sensitive area. The remedial approach consisted of operating an in situ air sparging/soil vapor extraction (IAS/SVE) system and then implementing natural attenuation. The full-scale IAS/SVE system, which operated from February 1999 to August 2000, consisted of 18 sparge wells and 4 vapor extraction wells. Life-cycle design approaches were implemented during operation of the IAS/SVE system, taking advantage of the most cost-effective methods of off-gas treatment and negotiating shutdown when system operation was no longer cost effective. After approximately 18 months of operation, over 90% mass removal of BTEX and over 75% mass removal of MTBE was achieved and contaminant removal reached asymptotic conditions. Although final cleanup levels had not been met, continued operation was no longer cost-effective. Therefore, after discussions with the regulators, it was agreed that the system would be shut down, and that quarterly monitoring would continue at the site to evaluate rebound for 1 year. If significant rebound occurs, system operation will recommence; otherwise, monitored natural attenuation will be implemented, as necessary, until final cleanup goals are achieved.

INTRODUCTION

Site 43286, an active gas station constructed in 1958 and remodeled in 1996, was impacted by a release of gasoline from underground storage tanks (USTs) used to contain fuel at the facility. The site is located near the center of Marine Corps Base (MCB) Camp Pendleton within the Las Flores Groundwater Basin, which is considered to be suitable for municipal or domestic water usage. The site is bounded to the east by sensitive ecological habit around Las Flores Creek (see Figure 1). Due to the groundwater basin designation and the sensitive ecological habitat, final cleanup goals for the site are maximum contaminant levels (MCLs) for benzene, toluene, ethylbenzene, and xylenes (BTEX) and methyl tert-butyl ether (MTBE). A feasibility analysis was performed to determine the most cost-effective remedial approach for the site. In situ air sparging/soil vapor extraction (IAS/SVE) was selected as the preferred remedy.

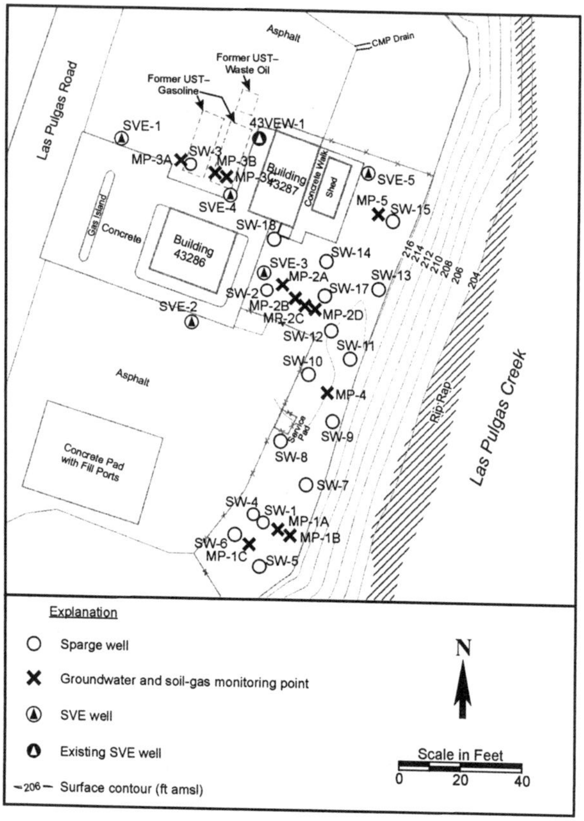

FIGURE 1. Site map showing sparge well locations.

IAS is an innovative in situ treatment technology that uses injected air to remove volatile or biodegradable contaminants from the saturated zone. The most common application of air sparging entails the injection of air directly into the saturated subsurface in order to remove volatile contaminants, such as BTEX, from the dissolved or residual phase to the vapor phase through air stripping. The stripped compounds are then biodegraded and/or removed via SVE in the vadose zone.

In a case study of 49 air sparging sites, Bass et al. (2000) reported that groundwater volatile organic compound (VOC) concentration reductions ranged from 27% to almost 100%. The overall success of the systems varied due to differing designs, construction techniques, and operational conditions, which indicates the need for a correct design, installation, and operation. However, even successful systems with greater than 90% concentration reductions may not achieve stringent cleanup criteria (U.S. Army Corps of Engineers, 1998; Hughes and Dacyk, 1998). Therefore, cost-effective design, installation, and operation must include realistic performance objectives tied to the exit strategy.

DESIGN

Major aspects of IAS/SVE system design included treatment area definition, pilot-scale testing, well-spacing, and injection/extraction flowrates. When selecting the treatment area it is recommended that the source zone be targeted (Leeson et al., 2001). At Site 43286, 18 sparge wells were installed in areas with the highest contaminant concentrations, near the former USTs, and throughout the plume area (see Figure 1). A relatively dense sparge well spacing of approximately 4.5 m (15 feet) was used at the site (Leeson et al. 2001). The study by Bass et al. (2000) showed that the most successful air sparging systems consisted of multiple wells spaced less than 30 feet apart. Air was injected into each well at a flowrate of 2.4 L/s (5 cfm) based on the results of pilot-scale testing. Results of pilot-scale testing and the resulting design decisions are provided in Table 1.

TABLE 1. Pilot-scale testing results and subsequent design decision.

Aspect of Study	Pilot-Scale Testing Results	Design Decision
Soil-Gas R_I Testing	R_I approximately 6 m (20 feet) at a vapor extraction rate of 14 L/s (30 cfm)	Vapor extraction rate of 14 L/s sufficient to control vapor migration.
Dissolved Oxygen Monitoring	Dissolved oxygen concentrations increased in monitoring points greater than 3 m from sparge wells.	Well spacing of 4.5 m appropriate.
Water Level Monitoring	Groundwater elevations in monitoring wells increased at startup of the IAS and returned to pre-injection conditions within 2 hours.	Air is not trapped in aquifer (i.e., not forced to move laterally, which could potentially impact creek).
Soil-Gas O_2, CO_2, TPH, and vacuum monitoring	• Average O_2 concentrations increased from 2% to 19% • Average CO_2 concentrations decreased from 9% to 1% • Initial TPH concentrations were high (typically >2,000 ppmv) • Soil-gas vacuum readings were negative near Building 43286	Biological degradation in the vadose zone will be enhanced by vapor extraction and air injection. Also, the SVE system was able to maintain a negative pressure around the building (reducing possibility of vapor migration)
Extraction Flowrates and Pressures	14 L/s/well extracted at 20 kPa	Adequate extraction rates can be achieved under reasonable vacuum.
Injection Flowrates and Pressures	Up to 3.8 L/s per well injected at approximately 69 kPa	Adequate flowrates can be achieved at relatively low injection pressure. Selected blower instead of compressor for air injection.

Prior to system operation, an air discharge permit had to be obtained. The air discharge permit required the SVE system to be operated whenever the IAS system was operating. In order to accommodate reductions in the soil vapor concentrations with time, the air discharge permit was developed using a life-cycle approach. The air discharge permit was approved for 4 modes of off-gas treatment: thermal oxidation, catalytic oxidation, activated carbon, and direct discharge. The approved concentration range for each process is provided in Table 2.

TABLE 2. Off-gas treatment.

Process	Inlet VOC Concentration Range (ppmv)	Allowable VOC Discharge Concentration (ppmv)
Thermal Oxidation	>2,000	350
Catalytic Oxidation	>100 and <2,000	100
GAC	>35 and <100	35
Direct Discharge	<35	35

OPERATION

Sparge wells were operated continuously during the 18 months of operation. Contaminant concentrations in samples collected quartely from groundwater wells indicated significant reductions ranging from 72% to 99%. Figure 2 shows the MTBE and BTEX concentration reductions observed in a typical, source area groundwater monitoring well after commencing operation of the system. Total dissolved contaminant mass reduction observed during operation of the system was estimated at 78% for MTBE and 94% for BTEX. Figure 3 shows the MTBE plume prior to IAS/SVE operation, after 14 months of operation, and 9 months after system shutdown.

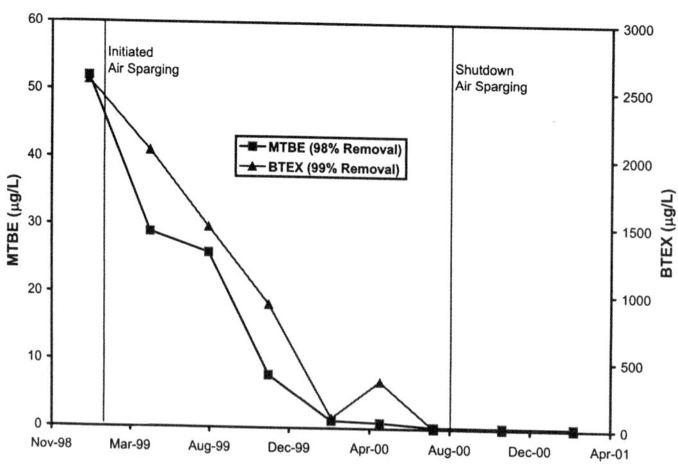

FIGURE 2. Typical contaminant concentration response observed in groundwater monitoring wells.

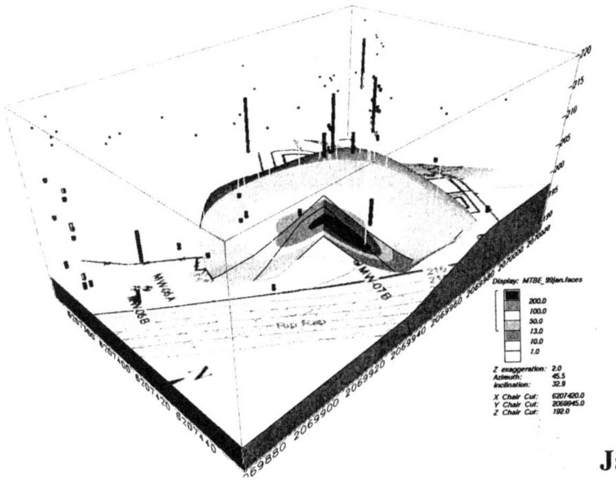

January 1999

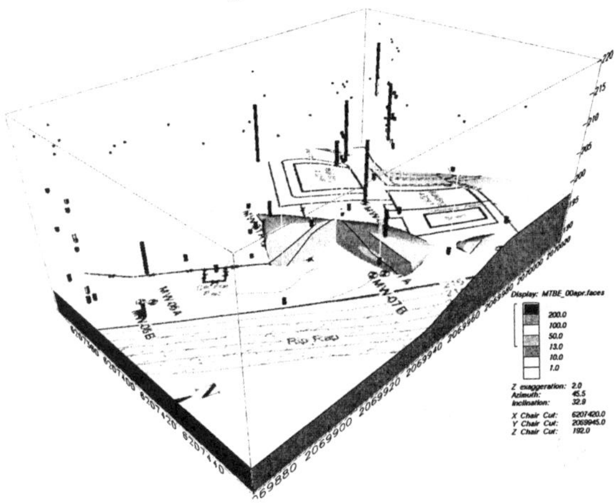

April 2000

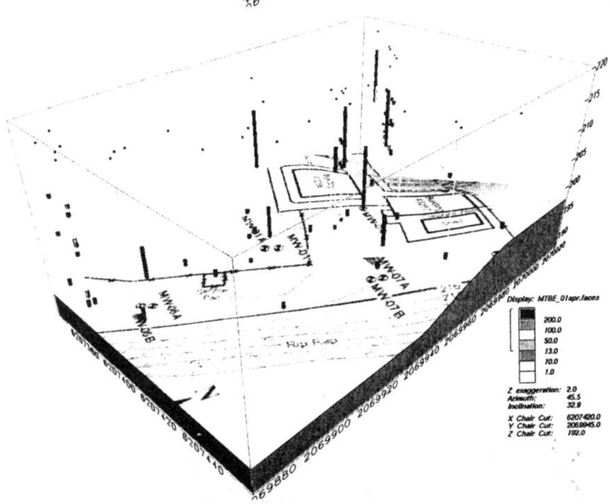

April 2001

FIGURE 3. MTBE plumes prior to IAS/SVE, after 14 months of operation, and 9 months following system shutdown.

Approximately 570 lb of petroleum hydrocarbons were removed by the SVE system and treated using thermal oxidation or GAC. As indicated on Figure 3, influent VOC concentrations decreased significantly during the first three months of operation. Therefore, in order to make use of the most cost-effective treatment method, Battelle removed the thermal oxidizer and replaced it with a GAC system in early April 1999. SVE operations costs were optimized to the extent practical using these alternate off-gas treatment methods. As a result of using the most cost-effective treatment method, operation costs were cut significantly, up to 40%.

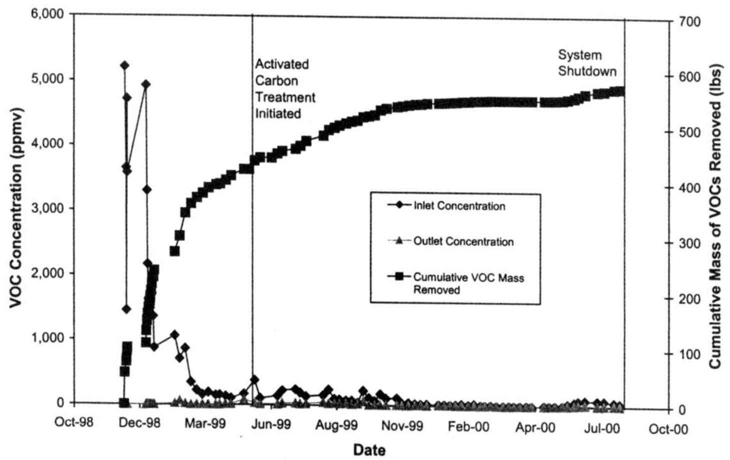

FIGURE 4. Cumulative mass removal by the SVE system.

EXIT STRATEGY

The basis of the exit strategy for Site 43286 was established during the design phase of this project as part of the Remediation Work Plan. The exit strategy consisted of evaluating contaminant concentrations and mass removal on a quarterly basis and shutting down the system when an asymptotic level of removal had been achieved. A contingency was added to the exit strategy to address contaminant rebound. Under this contingency, the system would be left on "on standby" following system shutdown during a transition period of approximately 1 year, so that the system could be easily reinitiated if rebound of contaminant concentrations occurred. If significant rebound does not occur, monitored natural attenuation will be implemented until final cleanup goals are achieved. The recommendation for system shutdown was supported as follows:

- **Asymptotic levels achieved by SVE.** Through August 2000, approximately 572 lb of petroleum hydrocarbons were removed by the SVE system (Figure 4). However, only 4 lb of hydrocarbons were removed between December 1999 and May 2000. These data indicate that the IAS/SVE system was effective in removing the majority of VOCs and continued operation would not be cost-effective.

- **Asymptotic levels achieved by IAS.** Groundwater monitoring results indicated that the contaminant concentrations had reached asymptotic levels (see Figure 2) and mass removal of BTEX and MTBE was significant, 94% and 78% respectively. Also, samples from only two wells contained groundwater concentrations above the cleanup goals: benzene in MW-04 and MTBE in MW-07B. These results showed that the system had remediated the majority of groundwater VOCs. Continued operation to meet cleanup goals was not cost effective or necessary considering the risk to potential receptors.

- **Reduced Risks to Receptors.** The primary risks necessitating active remediation at Site 43286 was the sensitive ecological area surrounding Las Flores Creek and the protection of a potential drinking water source. Data showed that the plume was shrinking and VOCs were not impacting Las Flores Creek.

CONCLUSIONS

Nine months have passed since shutting down the system at Site 43286 and rebound has not been observed. Recommendations and lessons learned on this project are as follows:

- Develop exit strategy criteria and negotiate with the regulators as early in the process as possible.

- Negotiate air discharge permits to utilize multiple treatment technologies in order to take advantage of the most cost-effective off-gas treatment technology as concentrations decrease.

- Consider pulsed air injection. Although not done at this site, pulsing would have lowered costs of the compressor because the total air flowrate would have been reduced by only injecting air into a subset of the total wells at the site. Also, studies suggest that mass removal can be increased by 20% to 30% through pulsed operation (Bruce et al., 2001; Kirtland and Aelion, 2000).

- Use a dense sparge well spacing. At Site 43286, a sparge well spacing of approximately 4.5 m was effective in permanently reducing contaminant concentrations.

- Focus on cost-effectiveness over the life-cycle of the project and utilize cost considerations as part of the exit strategy.

REFERENCES

Bass, D.H., N.A. Hastings, and R.A. Brown. 2000. Performance of Air Sparging Systems: A Review of Case Studies. *Journal of Hazardous Materials*, (72)101-119.

Bruce, C.L., P.C. Johnson, R.L. Johnson, and A. Leeson. 2001. Assessment of the In Situ Air Sparging Design Paradigm at a Gasoline-Spill Site. Submitted to *Bioremediation Journal.*

Hughes, W.D., and P.I. Dacyk. 1998. Air Sparging of Chlorinated Volatile Organic Compounds in a Layered Geologic Setting. Proceedings from The First International Conference on Remediation of Chlorinated and Recalcitrant Compounds. Monterey, CA, May 18-21. *Physical, Chemical, and Thermal Technologies: Remediation of Chlorinated and Recalcitrant Compounds*. Pp. 279-284.

Leeson, A., P.C. Johnson, R.L. Johnson, R.E. Hinchee, D.B. McWhorter, and C. Coonfare. 2001. *Air Sparging Design Paradigm.* Prepared for ESTCP, Arlington, VA. December 31.

Kirtland, B.C., and C.M. Aelion, 2000. Petroleum Mass Removal from Low Permeability Sediment Using Air Sparging/Soil Vapor Extraction: Impact of Continuous or Pulsed Operation. *Journal of Contaminant Hydrology*, (41)367-383.

U.S. Army Corps of Engineers. 1998. Air Sparging and Soil Vapor Extraction at Landfill 4, Ft Lewis Washington. *Cost and Performance Report*. September.

ACCELERATING AEROBIC BIODEGRADATION OF GRO/DRO: A FIVE YEAR STUDY

Herminio Muniz (Hart Crowser, Anchorage, AK, USA)
Stephen Koenigsberg, (Regenesis, Inc., San Clemente, California, USA)
Kevin Lapus (Regenesis, Inc., San Clemente, California, USA)

ABSTRACT: Gasoline Range Organics (GRO) and Diesel Range Organics (DRO) contaminated the aquifer underlying a warehouse/equipment storage facility in Homer, AK. This contamination originated from leaks in underground storage tanks located in the northern section of the site. Groundwater flow at the site is controlled in large part by a finger drain running through the facility. This finger drain, cut four feet into the ground and filled with coarse gravel, is responsible for the majority of contaminant migration. This paper presents the results of a 5-year treatment program conducted by Hart Crowser using Oxygen Release Compound (ORC®). The goal of the treatment program was to remediate the entire range of hydrocarbon contamination (GRO and DRO). Treatment was focused along the length of the finger drain. Initially, Hart Crowser conducted a five-month pilot study to determine the ability of ORC to accelerate the aerobic biodegradation of GRO and DRO. After a successful pilot program, Hart Crowser went on to a full-scale application. At the beginning of the full-scale treatment, DRO concentrations averaged 87 mg/L and GRO concentrations averaged 14.5 mg/L. Benzene, toluene, ethylbenzene, and xylene (BTEX) concentrations averaged 4.4 mg/L. After approximately 5 years of treatment, DRO concentrations decreased by 90%, to an average of 8.3 mg/L. GRO concentrations decreased by 91%, to an average of 1.3 mg/L, and BTEX concentrations decreased by 89%, to an average of 0.5 mg/L.

INTRODUCTION

The Homer Electric Association (HEA) warehouse site is located at 3977 Lake Street, Homer, Alaska. The facility has been in operation for over 40 years as a power generation station, utility support facility, and for materials storage (current use).

In September 1990, two 2,000-gallon diesel underground storage tanks (USTs), one 4,000-gallon gasoline UST, and associated dispensers and lines were removed and closed (DOWL, 1990). Water and soil sampling results indicated that a release of petroleum hydrocarbons had occurred. Concentrations of BTEX in groundwater in the excavation suggested that the release was in part due to gasoline. Elevated total petroleum hydrocarbon (TPH) concentrations were also detected in the soil samples.

A subsequent release investigation was performed at the site in April 1991 (DOWL, 1991). This investigation showed elevated BTEX concentrations in the groundwater and elevated BTEX, volatile (or gasoline) range petroleum hydrocarbons (VPH), and extractable (or diesel) petroleum hydrocarbon (EPH) concentrations in the soil at the site. In general, samples collected from 5 feet below grade exhibited the highest concentrations of petroleum hydrocarbons. The investigation also showed that the site geology consists primarily of coarse fill material underlain with silt. Groundwater level, at the time of the investigation, was approximately 2 to 3 feet below grade. According to slug tests, hydraulic conductivity ranges from 11.25 to 45 ft/day. The gradient at this site is 0.04 ft/ft. Groundwater flow at this site is to the south with an effective porosity of 30% and a seepage velocity of 2 to 6 ft/day.

Prior to October 1991, three wells were installed downgradient of the Lake Street site on Tract 3A of the Waddell Subdivision. Three groundwater sampling events were performed in October and November 1991, and January 1992 (DOWL, 1992). Although the results were variable, benzene concentrations in excess of the state maximum contaminant level (MCL) of 0.005 mg/L were observed.

In June 1993, Hart Crowser performed a remedial site investigation at the site. This investigation included groundwater measurements and groundwater and soil sampling (Hart Crowser, 1993). A biotreatability study was performed to assess the suitability of this site to biological remediation.

RESULTS

Pilot Study: ORC Versus Air Sparging. One finding of the remedial investigation was the presence of elevated concentrations of iron (average 56.5 mg/L) and manganese (average 3.1 mg/L). This combined with the low oxygen concentrations measured within the monitoring wells suggested that significant metal precipitation could occur with oxygenation of the groundwater. This could result in plugging of both the remediation system equipment and aquifer materials. Therefore, a pilot study was conducted to assess the possible consequences of aquifer oxygenation on both groundwater BTEX concentrations and dissolved metals concentrations.

Two methods of groundwater oxygenation were explored as part of the pilot study. On the southern portion of the site, a 15-foot sparging line was installed along with a downgradient well point. On the north side of the warehouse building, a well (MP) was installed and a magnesium peroxide based ORC was placed in the well. Six well points were placed approximately 6.5 feet downgradient from the sparging line, and six well points were installed downgradient from the ORC well. A control well (MP-Control) was also placed cross gradient from the MP well.

Prior to start-up of the sparging line and placement of the ORC in the MP well, dissolved oxygen readings were made, and samples for BTEX, dissolved iron and manganese, and nutrients were collected from selected wells and well points. Initial testing and system operations began in December 1993. The pilot study was then conducted for 5 months (January through May).

Dissolved oxygen readings were taken in all well points and in MP-Control after 1 month, 2, 3, and 5 months. BTEX samples were collected from selected well points after 1 month and 2 months, and all well points after 3 and 5 months. Dissolved metals and nutrient analyses were collected after 3 and 5 months. Results of dissolved oxygen readings at the ORC area showed that the dissolved oxygen concentration in the MP-Control well remained very low throughout the pilot test period. In contrast, over a 2-month period, the dissolved oxygen concentrations in the down gradient well points rose from essentially non-detectable to several mg/L. At the sparging array the dissolved oxygen concentration increased to about 5 mg/L after 3 months, and 6 mg/L after 5 months.

Initial benzene and total BTEX concentrations at the ORC sites were 0.13 mg/L and 0.35 mg/L. After 2 months, the benzene concentration at WP-3 was 0.005 mg/L, and the total BTEX was 0.009 mg/L. After 3 months with falling groundwater conditions, BTEX and GRO concentrations in the ORC well points ranged from 0.008 mg/L to 0.035 mg/L. However, this was similar to the concentration detected in the control well. After 5 months of operation with rising groundwater conditions, the benzene concentration within the control well was 0.98 mg/L. In the well points, however, benzene concentrations ranged from 0.003 mg/L to 0.043 mg/L.

In the sparging area, the initial benzene concentration was 0.2 mg/L. After 3 months of operation, no benzene was detected in the well point (<0.0025 mg/L), and after 5 months of operation, 0.019 mg/L was detected.

The results of metals analyses showed that little change occurred in the groundwater of the ORC site. However, at the sparging site, dissolved iron decreased from an initial concentration of 38 mg/L in December 1993 to 0.46 mg/L in March 1994 and to 0.05 mg/L in May 1994, indicating significant reduction in the concentration of dissolved iron in this zone.

Both systems proved effective at increasing dissolved oxygen concentrations in the groundwater and reducing BTEX concentrations in the groundwater. However, a three order of magnitude reduction in dissolved iron concentration at the sparging site, whereas no significant change in dissolved metal concentration was observed at the ORC site. Because of the concern for bio-fouling within this high iron environment (upwards of 100 mg/L), and because there was evidence of channeling in the sparging area, ORC was chosen as the treatment technology for full-scale application.

Full-Scale Application of ORC. During excavation activities in October 1994, a finger drain was discovered at a depth of 3 to 4 feet below grade constructed of large, well-sorted, rounded gravel (Hart Crowser, 1997). The finger drain was found to have a width of 6 to 10 ft, and a vertical thickness of 3 to 4 ft. A total of 2150 yd^3 of soil were removed. In December 1994, drill rigs were used to install 38 wells, oriented along the length of the finger drain, to serve as ORC application points (see Figure 1). Each well is 4 to 8 feet deep, with groundwater varying from 0.5 to 4 ft below the ground surface, depending on location and season. ORC was applied to the aquifer via 4-inch filter socks, each sock approximately 1 ft in length. Three to six socks were placed in each well, totaling 175 socks for this initial application. Since this initial application, socks have been removed and replaced three times annually.

A second phase of excavation activities occurred between November and December 1997 to remove impacted soils at the site (Hart Crowser, 1997). Soil was removed from three areas: the former warehouse (removed in October 1997), the north finger drain, and the southern area. The excavation locations are illustrated in Figure 1. Soil from each area was excavated down to the silt layer. An additional 1,600 to 1,700 yd^3 of impacted soil was removed. At the conclusion of each excavation, ORC powder was mixed the groundwater to enhance the aerobic degradation of dissolved-phase hydrocarbon contamination. A total of 1,000 lb of ORC was applied: approximately 600 lb into the warehouse excavation, approximately 300 lb into the north finger drain excavation, and approximately 100 lb into the southern area excavation.

In May 1999, an additional 28 wells were installed along the southern edge of the HEA property. ORC filter socks were applied to these wells to enhance degradation and to prevent off-site migration of dissolved-phase petroleum hydrocarbons. Similar to the wells running along the length of the finger drain, three to six socks were placed into each well.

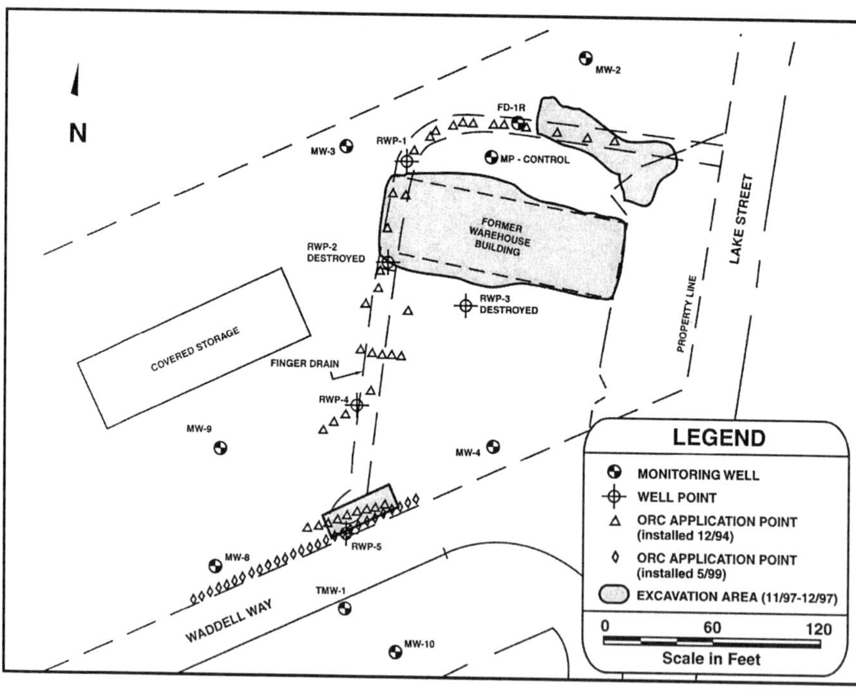

FIGURE 1. Full-Scale Treatment Approach.

MONITORING RESULTS

The monitoring program at this site includes total BTEX, GRO, and DRO. Total BTEX was monitored via EPA Method 5030/8020, GRO via EPA method

5030/8015M, and DRO via EPA Method 3550/8100M. Because the site is located in Alaska, monitoring between December and April was not possible due to the freezing over of the wells. In general, wells were monitored bi-monthly between May and November of every year. Dissolved oxygen levels were also regularly monitored.

At the beginning of the full-scale application, DRO concentrations averaged 87 mg/L, GRO concentrations averaged 14.5 mg/L, and BTEX concentrations averaged 4.4 mg/L. By the second post-monitoring event (approximately 8 months after the initial application), DRO concentrations decreased by 27% (to 63.9 mg/L), GRO concentrations decreased by 43% (to 8.3 mg/L), and BTEX decreased by 34% (to 2.9 mg/L).

Through the last monitoring event (approximately 5 ½ years after the initial application), DRO concentrations have decreased by a total of 90% (to 8.3 mg/L), GRO by a total of 91% (to 1.3 mg/L), and BTEX by a total of 88% (to 0.5 mg/L). Figure 2 presents the results of the full monitoring program for VOCs.

Dissolved oxygen levels fluctuated throughout the treatment program, going as low as 0.2 mg/L and rising as high as 12 mg/L. Although the release of oxygen from ORC is steady, the utilization of this oxygen by in-situ microbial populations can vary over the course of the lifespan of ORC.

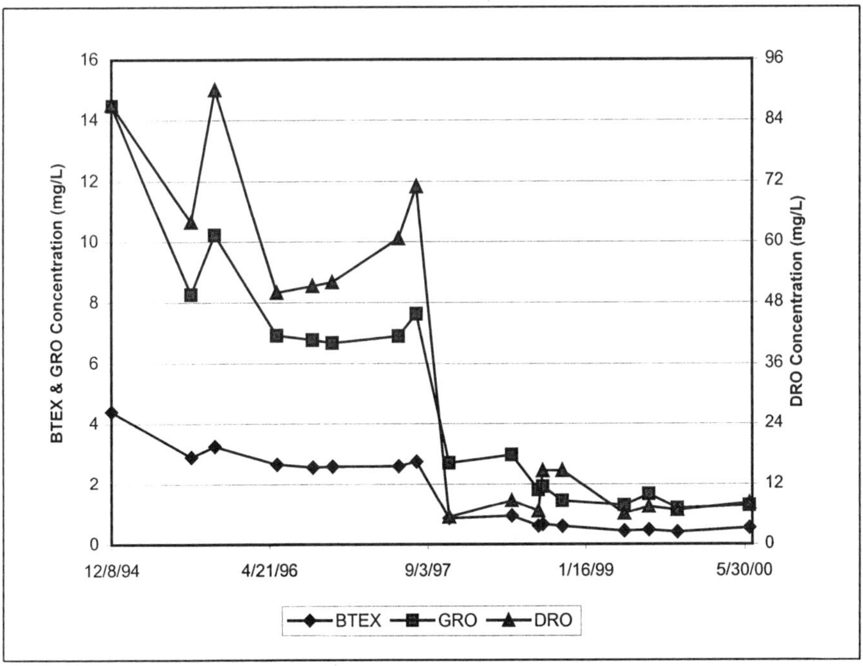

FIGURE 2. VOC Concentration Graph.

CONCLUSION

The goal of the full-scale ORC application at this site was to remediate the entire range of petroleum hydrocarbon contamination (GRO and DRO). Although some monitoring wells contain benzene at levels above an MCL of 0.005 mg/L, ORC has, except for minor exceptions, been able to bring GRO and DRO concentrations to or below the regulatory standards of 1.3 mg/L and 1.5 mg/L, respectively. The treatment program ended June 2000, when ORC filter socks were last applied, and the site is on a monitoring only status.

REFERENCES

DOWL Engineers Inc., 1990. *Summary of Findings Underground Storage Tank Closure Program.*

DOWL Engineers Inc., 1991. *Spill/Release Investigation Report for 3977 Lake Street.*

DOWL Engineers Inc., 1992. *Summary of Findings For Tract 3A, Waddell Subdivision, Homer, Alaska.*

Hart Crowser, 1993. *Remedial Site Investigation, HEA Lake Street Facility, Homer, Alaska, Letter Report.*

Hart Crowser, 1997. *Soil Excavation Report, HEA Lake Street Facility, Homer, Alaska.*

BTEX/TVPH REMEDIATION USING AN OXYGEN RELEASE COMPOUND

Nick M. Fischer (Aquifer Technology, Englewood, CO)
Tom Reed, (URS-Greiner, Wheat Ridge, CO)
Doug Camrud, (Terracon, Wheat Ridge, CO)
Clint Madsen, (Terracon, Wheat Ridge, CO)

ABSTRACT: The project site was a former gasoline station. Underground storage tanks (USTs) and petroleum-contaminated soils were removed from the site in May of 1996. Following the completion of assessment activities a Corrective Action Plan (CAP) was submitted to Oil Inspection Section (OIS) in August of 1996. The CAP recommended the use of a soil vapor extraction (SVE) and an air sparging system to remediate soil and groundwater contamination at the subject site. A CAP Addendum was completed to reassess the remedial technologies. The emergence of a solid peroxygen technology (Oxygen Release Compound - ORC®) designed to deliver oxygen in a passive, low-cost, time-release manner, appeared as a promising option to promote the in-situ bio-remediation of the groundwater. The design of the groundwater remediation program with ORC assumed a treatment area of approximately 8,000 square feet and an average total volatile petroleum hydrocarbon (TVPH) concentration of 5.4 mg/L. Based on these parameters, 1,242 pounds of ORC was applied in May 1998. The material was injected on approximately 20-foot centers to cover the proposed treatment area. Direct-push technology (DPT) was used to deliver the ORC slurry through 34 injection points using an injection pressure of approximately 62 psi. In addition to the ORC application, 148 gallons of microbe slurry, and 116 pounds of nutrient mix were injected to increase the effectiveness of the bio-remediation process. Quarterly groundwater samples collected after ORC application indicated that benzene-toluene-ethylbenzene-xylenes (BTEX) and TVPH concentrations were reduced significantly for about a third the cost of competing technologies. Without partial reapplication, which was not warranted by regulators given that mass and risk were sufficiently reduced, the BTEX and TVPH concentrations remained at the same approximate level of reduction.

SITE INFORMATION AND REMEDIAL DESIGN EVALUATION

Site Conditions. A piezometric surface diagram for groundwater elevations observed on May 27, 1997 is shown in Figure 1 and was prepared with Surfer™ (Golden Software) based on the groundwater table elevation measurements. The Groundwater flow has appeared to be in a generally north-northwesterly direction at 50 ft/yr. With reference to the Figure, groundwater has been observed at depths ranging from 4 to 8 feet bgs in wells MW-1 through MW-13. Free-phase petroleum product has not been observed in the any of the monitoring wells during the groundwater measurement events.

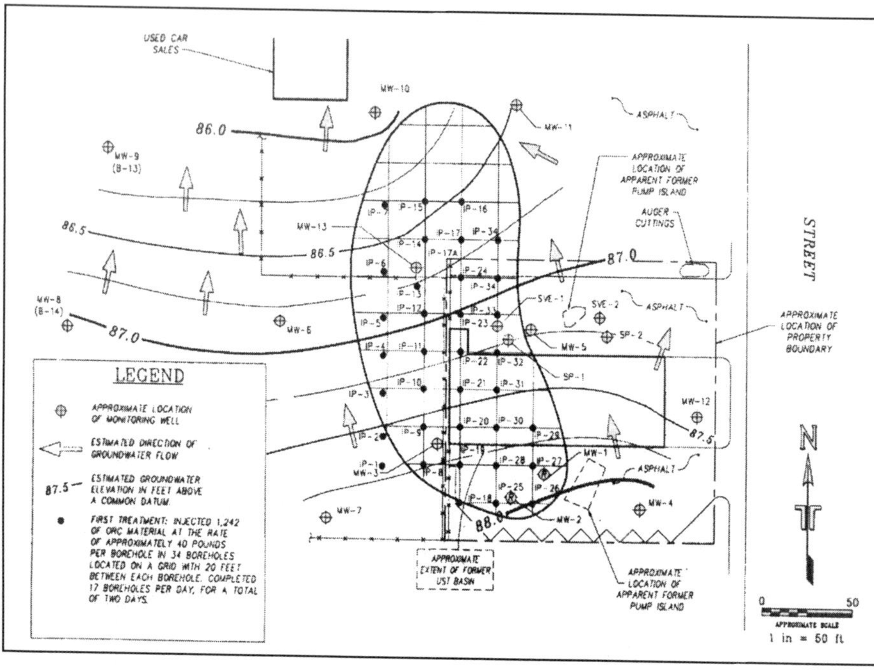

FIGURE 1: Site Map showing potentiometric surface, locations monitoring wells, and the location of 34 boreholes on a 20 foot grid in which ORC application was conducted using DPT.

Remedial and Design Evaluation. Groundwater remedial alternatives were evaluated during this process. Both in-situ and ex-situ conceptual designs and cost estimates were prepared for various systems and clean up methodologies, based on parameters estimated from field-testing and assessment activities. The remedial alternatives were considered for the site are presented in Table 1; all costs include groundwater monitoring.

TABLE 1 - COST CAPARISONS FOR REMEDIAL TECHNOLOGIES

	Remedial Technology	Estimated Cost
1	In-situ SVE with Horizontal Wells, In-situ Air-Sparging/Vertical Wells	\$247,000 to \$284,500
2	In-situ SVE with Horizontal Wells, In-Situ Air-Sparging with Horizontal Wells	\$318,000 to \$378,000
3	Excavation (Ex-situ), In-situ SVE with Horizontal Wells, In-Situ Air-Sparging with Vertical Wells	\$353,500 to \$380,500
4	In-situ SVE with Horizontal Wells, Groundwater Recovery (Ex-situ), with Vertical Wells, Low-profile Aeration/Disposal to Sanitary Sewer	\$360,000 to \$436,000
5	ORC Injection via DPT	\$60,000 to \$120,000

METHODS

Oxygen Release Compound (ORC). The ORC application promotes bioremediation, by enhancing the supply of oxygen in groundwater to the naturally occurring microbes that metabolically transform toxic organic compounds into harmless by-products. The ORC material is a proprietary magnesium peroxide compound that is activated by moisture to release oxygen. The product releases approximately ten percent of the available oxygen in the first few weeks, and the balance over a period of six months to a year. Nutrients and microbes can also be added to groundwater during the ORC injection process to improve the bio-degradation process.

General Evaluation Criteria for Remediation with ORC. A list of criteria used to evaluate whether ORC was feasible for this site are present below. The criteria are based on a literature review and practical experience:

- viable microbial bacteria must be present in the subsurface, e.g. soil (unsaturated) and groundwater (saturated),

- a substrate, e.g. the substance or food on which bacteria live and grow, must be present.

- nutrients must be present so that bacteria growth can be sustained, and

- in saturated conditions, certain water-quality conditions must be present in order to sustain bacterial growth.

Specific Evaluation of Water-Quality Parameters. The following is a list of water-quality parameters that were considered in the ORC feasibility-evaluation process:

- Total Volatile Petroleum Hydrocarbons (TVPH): Groundwater with a concentration of lower than 100 parts per million (ppm) TVPH or lower is generally recommended; higher values may contraindicate ORC application based on cost, relative to the next available technology and its cost.

- The hydrogen ion concentration (pH): A range of from 6-8 is optimal.

- Temperature: Rates of bioremediation are attenuated as a function of temperature. They are essentially halved with incremental decreases of 10° C.

- Reduction-Oxidation Potential (Eh): Reducing environments (low redox) are less desirable. It should be noted that ORC can increase the Eh potential if current conditions are reduced.

- Dissolved oxygen (DO): A concentration of 2 mg/l or greater is desirable in order to sustain significant microbial activity.

- Heterotrophic bacteria: A positive plate count (presence) is required for biodegradation to occur. (Addition of microbes with the injection process may be required in order to increase the initial biological activity.)

- Nutrients: Nutrients should be present in the of 100:10:1 of carbon: nitrogen: phosphorus, respectively. The following nutrient concentrations are required; nitrogen: 1/10 of carbon mass is required for microbial growth (in the form of nitrate, nitrite, or ammonia), phosphorus: 1/100 of carbon mass is required for microbial growth (in the form of ortho-phosphate). It should be noted that nutrients can be added to the subsurface media if nutrient availability is not sufficient to sustain optimal biodegradation activity.

- Chemical oxygen demand (COD) to biological oxygen demand (BOD) concentrations and ratio: Environments with high concentrations of COD/BOD with a ratio of more than 4:1 may not be suitable for bioremediation. It should be noted that the COD/BOD ratio can be decreased by placing ORC into the groundwater.

- Iron: High concentrations of iron in groundwater, on the order of 100 ppm may lead to fouling of the aquifer pore space when oxidized.

- Anions, Cations, and Total Dissolved Solids (TDS): High concentrations of anion, cation or TDS may be toxic to the bacteria and/or interfere with growth.

Specific Assessment of Water Quality Parameters. Groundwater bio-assessment sampling was conducted on April 21, 1998. Three (3) monitoring wells were sampled to review site characterization and ORC feasibility. These wells were MW-6, MW-7, and MW-13. For each groundwater monitoring well sampled, the water level was measured and recorded using a portable water level indicator, each well was purged at least three (3) wellbore volumes until pH, conductivity, and temperature were stabilized, and then each well was sampled for the required bio-assessment analytes. Field measurements conducted and recorded were water levels, DO, pH, temperature, reduction-oxidation potential, and conductivity.

TVPH concentrations for MW-13 were 12 milligrams per liter (mg/L). This concentration satisfies as a substrate for bioremediation and being below the 100 mg/l guidelines recommended for ORC application.

Field Measurements are shown below in Table 2. pH was observed to range from 7.1 to 7.5 and in the optimal range. Temperature was in the acceptable range and observed at 11 degrees centigrade. DO concentrations ranged from 0.6 to 1.7 mg/L. In the plume center (MW-13), DO concentrations were not sufficient to sustain significant microbial growth. ORC application was employed to release oxygen to sustain significant microbial growth.

TABLE 2 - SUMMARY OF FIELD MEASUREMENTS

Well	Date	DO (mg/L)	Temperature (°C)	pH
MW-6	4/21/98	1.6	11	7.2
MW-7	4/21/98	1.7	11	7.5
MW-13	4/21/98	0.6	11	7.2
MW-13	6/19/98	0.7		
MW-13	11/10/98	0.3		

Note: ORC/bacteria/nutrients injected in 34 boreholes on May 27 and 28, 1998.

Heterotrophic bacteria colonies are summarized in Table 3. To increase the effectiveness of the bio-remediation activity, 148 gallons of microbe slurry was added in the injection program.

TABLE 3 - SUMMARY OF HETERTROPHIC BACTERIA COLONIES

Monitoring Well	Date	Aerobic (cfu/mL)
Before ORC/Bacteria/ Nutrient Injection		
MW-6	4/21/98	9,800
MW-7	4/21/98	13,000
MW-13	4/21/98	160,000
	Average =	61,000
After ORC/Bacteria/ Nutrient Injection		
MW-5	11/12/98	400,000
MW-6	11/12/98	190,000
MW-7	11/12/98	55,000
MW-13	11/12/98	3,200,000
	Average =	961,000
MW-5	2/9/99	57,000
MW-6	2/9/99	68,000
MW-13	2/9/99	310,000

Note: ORC/bacteria/nutrients injected in 34 boreholes on May 27 and 28, 1998.

Nutrients may not have been available in sufficient quantities near the plume center (MW#13). To increase the effectiveness of the bio-remediation process 116 pounds of nutrient mix was added to the ORC media to sustain optimal biodegradation activity. A table of nutrient concentrations before ORC application is summarized in Table 4.

TABLE 4 - SUMMARY OF NUTRIENT CONCENTRATIONS

Well	Date	Nitrate (mg/L)	Nitrite (mg/L)	O-Phosphate (mg/L)
MW-6	4/21/98	10.0	<0.25	<0.25
MW-7	4/21/98	21.5	<0.25	<0.25
MW-13	4/21/98	0.3	<0.25	<0.25

COD was determined to range from 15 to 78 mgO2/L. ORC application rates were not adjusted since existing COD concentrations were not excessive.

BOD was determined to range from less than 10 to 17 mgO2/L. ORC application rates were not adjusted since existing BOD concentrations were not excessive.

Concentrations of iron were found to be 0.1 to 0.4 mg/L, thus iron fouling of the host media pore space was not anticipated to be a potential problem.

As summarized in Tables 5 and 6, high concentrations of major anions and cations were not present in the groundwater, thus the concentrations of ions should were not expected to be toxic to the bacteria and/or interfere with growth.

TABLE 5 - SUMMARY OF MAJOR ANIONS

Well	Date	Chloride (mg/L)	Nitrate (mg/L)	Sulfate (mg/L)	BICARBONATE as ($CaCO_3$) (mg/L)
MW-6	4/21/98	211	10	170	1110
MW-7	4/21/98	489	22	136	687
MW-13	4/21/98	114	0.3	21	530

Note: Nitrites, Sulfites were not observed above laboratory detection limits.

TABLE 6 - SUMMARY OF MAJOR CATIONS

Well	Date	Sodium (mg/L)	Potassium (mg/L)	Calcium (mg/L)	Magnesium (mg/L)
MW-6	4/21/98	115	6	139	91
MW-7	4/21/98	190	5	220	82
MW-13	4/21/98	105	3	74	47

Corrective Action Plan Implementation. OIS Remediation Action Category (RAC) II guidelines were approved for this site. Petroleum-impacted soil was excavated and transported to a landfill, remaining soil contaminant concentrations are below RAC II levels, and additional soil remediation was not warranted at the site. The groundwater contaminant levels identified at the site were relatively low in most areas except the general area at monitoring well MW-13, while the contaminant plume was located over a relatively large area. The remediation strategy utilizing ORC to promote the in-situ biodegradation of the dissolved contaminant plume was developed by using a spreadsheet computer program to calculate the required oxygen (to be released from the ORC application) for bioremediation. Using this program an injection scheme was developed and implemented.

The ORC application utilized direct-push technology (DPT). Injection consisted of 3 treatments: 50% of the total quantity of ORC product during the first, and 25% of the total quantity for the second and third treatments.

The first and only ORC application was conducted on May 26 and 27, 1998. The ORC material was injected on approximately 20-foot centers to cover the proposed treatment area as shown. The application area included the area north of the former UST basin toward the downgradient well, MW-10. 1,242 pounds of ORC was injected via direct-push technology (DPT) at an injection pressure of approximately 62 psi, at 34 injection points. In addition to the ORC application, 148 gallons of microbe slurry, and 116 pounds of nutrient mix were injected to increase the effectiveness of the bio-remediation process. 3- injections were completed at various depths for each injection hole. Radian International Mobile Field Services (URS) provided personnel and field equipment to inject. ORC powder was added to water in a mixing bin, forming a slurry. A positive displacement pump injected the slurry via DPT. From conducting DPT soil sampling it appeared that the radius of injection ranged from 5 to 15 feet depending on placement of injection and the amount of ORC slurry injected.

Prior to implementation of the groundwater remediation program, baseline DO levels were measured, and bacterial plate count analyses were performed on groundwater samples collected from the site, and a bio-assessment and ORC feasibility study was conducted. Quarterly groundwater monitoring samples have continued to be collected at the site for chemical and bio-assessment analysis, as well as field DO measurements. This monitoring information was conducted to be periodically reviewed to evaluate if additional injections of ORC should be recommended at the site.

It was planned that DPT soil and groundwater sampling would be performed between applications to assess remediation performance and strategically select subsequent application areas (Since it was possible that the first, or first and second planned treatments could reduce the groundwater contamination such that additional treatments would not be necessary, or would require less ORC). However, since this project was initiated, the OIS changed cleanup guidelines from RAC guidelines to risk based corrective action (RBCA) guidelines. Due to the fact that no offsite properties were impacted, and the risk to human health and environment were not above RBCA screening levels, the OIS concluded that any additional ORC treatment would not be required.

RESULTS AND CONCLUSIONS

Dissolved Oxygen Levels. As shown in Table 2, groundwater samples were collected in both impacted areas and background areas for DO levels before and after ORC application. It appears that the rate of oxygen release was approximately equal to the uptake of oxygen by the metabolic activity of the bacteria.

Heterotrophic Aerobic Plate Counts. As shown in Table 3, the addition of ORC/bacteria/nutrients increased the heterotrophic aerobic plate count in the area in which ORC injection was conducted by a factor of 16 in six months. Aerobic bacteria counts increased from an average of 61,000 to 960,000 cfu/ml. The aerobic heterotrophic plate count increased at monitoring well MW-13 (the highest petroleum-impacted area) from 160,000 to 3,200,000 cfu/ml in the same six month period.

BTEX and TVPH. Due to site construction and the abandonment of site impacted groundwater monitoring wells, monitoring well MW-13 was the only remaining well sufficiently impacted by contamination that merited analysis. Groundwater samples were collected on a quarterly basis. The results are presented in Table 7. BTEX concentrations were reduced by approximately 81%, from 1,387 to 270 μg/L three months after application. Through approximately two years of monitoring TVPH concentrations were reduced by approximately 89%, from 12 mg/L at baseline to 1.4 mg/L.

TABLE 7 - SUMMARY OF BTEX/TVPH AT MW-13

			MW-13			
Date Sampled	Benzene (μg/L)	Toluene (μg/L)	Ethylbenzene (μg/L)	Xylenes (μg/L)	BTEX (μg/L)	TVPH (mg/L)
6/13/96	379	239	1,290	4,608	6,516	19
5/27/97	284	151	728	2,487	3,650	15
4/21/98	110	57	410	810	1,387	12
8/3/98	20	7.9	2.4	240	270	NA
11/10/98	44	15	150	260	469	4.6
2/9/99	62	31	190	200	483	4.3
6/3/99	66	24	190	230	510	8.8
9/16/99	51	8.1	230	77	366.1	4.3
12/16/99	51.2	4.6	151	62.3	269.1	3.18
3/15/00	49.2	6.7	194	89.4	339.3	NA
6/15/00	31.3	5.3	60.2	153	249.8	1.4
9/20/00	27.8	4.2	134	57.6	223.6	NA

Note: ORC/bacteria/nutrients injected in 34 boreholes on May 27&28, 1998. NA = not analyzed.

It appears that ORC application was cost effective and technically feasible, with little long-term site disturbance, and was readily implementable and effective. Only the first of three ORC applications was conducted as originally proposed since the contaminant mass and risk were sufficiently reduced from the first application. Groundwater monitoring at the site currently remains ongoing for closure activities.

A SYSTEMS' APPROACH TO IN-SITU BIOREMEDIATION: FULL SCALE APPLICATION

Michael Scalzi (Innovative Environmental Technologies, Inc., Mt. Laurel, New Jersey)
Xandra Turner, P.E. (InterBio, The Woodlands, Texas)
Eric Andrews (Environmental Management, Inc., Springfield, Illinois)

ABSTRACT: An 18-month, in-situ biological program, specifically aimed at accelerating the current attenuation process such that the owner can sell the property began May 5, 2000. The remedial approach focused on bacterial population management, nutrient augmentation, terminal electron acceptor enhancement, and an efficient subsurface delivery of the enhancement liquids. Specifically, a novel delivery system, utilizing direct push advancement equipment coupled with high-pressure air and liquid systems allowed for uniform lateral distribution of the materials to the impacted capillary fringe zone. The treatment zone has extended from 9 to 11 feet below grade across the 20,000 square foot property. Preliminary analytical results from soils and groundwater at the site indicate that the targeted 18-month remediation time-line will be met. Samples from the groundwater and the soils 300 days into the eighteen-month program indicate total BTEX concentrations in the soils have been reduced an average of 76% with a 92% decrease in toluene. Groundwater BTEX levels have also seen a dramatic decrease, approaching targeted treatment levels in several of the wells within ten months.

INTRODUCTION

There are few communities across the country that have been spared the issue of contamination to their community and individual drinking water supplies. Most often this threat comes from leaking underground storage tanks (USTs) at local gas stations.

When faced with reporting and remedial guidelines from regulatory agencies, the station owner must rely on an environmental consultant to represent the owner's best interests while meeting the reporting requirements. As the process moves from reporting the incident to the investigation phase, the owner must rely on the capabilities of the consultant. In the event that an investigation leads to a remedial action plan and eventual site remediation, the environmental consultant must thoroughly evaluate, recommend and implement a wide variety of remedial technologies. Often, the data available to the consultant, for biologically based remediation focuses on individual elements necessary for effective biological processes. As a consequence, when faced with the conflicting field data and the limited pool of qualified remedial contractors, gas station owners and their consultants are hesitant to recommend biological remediation.

By focusing on a systematic and managed approach to bioremediation in which all the known elements of the biological process are addressed, the feasibility of applying in-situ biological solutions becomes more realistic. These elements include bacterial population enumeration, nutrient evaluation, dissolved oxygen concentration monitoring, contaminant concentrations in both the soils and groundwater, effective delivery mechanisms and understandable data interpretation. To date, most field applications of bioremediation have not focused on all the parameters necessary to successfully establish and sustain a robust and productive consortia of bacteria. A full-scale project focused on monitoring and maintaining a supportive environment for the biological mineralization of petroleum hydrocarbons, more specifically benzene, toluene, ethyl benzene and xylenes (BTEX compounds), would go a long way in advancing biological solutions in the general UST market. Results from a managed project are presented including the issues associated with data interpretation and field experiences.

Objective. The objective of the project was to demonstrate the efficacy of in-situ bioremediation when it is applied as a system. The goal for the customer is to remediate the soils and groundwater to a level at which his environmental consultant may petition for a "No Further Action" letter from the state regulatory agency. To meet this goal the objectives listed in Table 1 will need to be met.

TABLE 1. The soil and groundwater cleanup objectives for Georges 66 Station, Princeton, Illinois.

BTEX	Soil Cleanup Objective	GW Cleanup Objective
Benzene	0.17 ppm	0.025 ppm
Toluene	29 ppm	2.5 ppm
Ethyl benzene	19 ppm	1 ppm
Total Xylenes	150 ppm	10 ppm

Site Description. On May 1, 1998 six gasoline USTs and one waste oil UST at a facility commonly known as George's 66 in Princeton, Illinois were removed. The combined capacity of the gasoline USTs was 13,000 gallons and the waste oil tank was 500 gallons. The facility had been in operation for more than 50 years prior to its closing in 1998. During the tank removal activities, it was discovered that three of the six gasoline tanks and the waste oil tank had leaked. Subsequent to the tank removal activities, approximately 500 cubic yards of soil were removed from the source area and replaced with clean backfill material. An early action report was filed with the Illinois Environmental Protection Agency (IEPA) followed by a series of on-site and off-site investigations to better define and delineate the impacted soils and groundwater. These investigations revealed the presence of petroleum hydrocarbons on two adjacent residential properties. A corrective action plan and budget to address the cleanup needs of the site and the adjacent properties was filed with the IEPA on February 1, 2000. The budget was approved on April 3, 2000 and remedial activities were initiated on May 4, 2000.

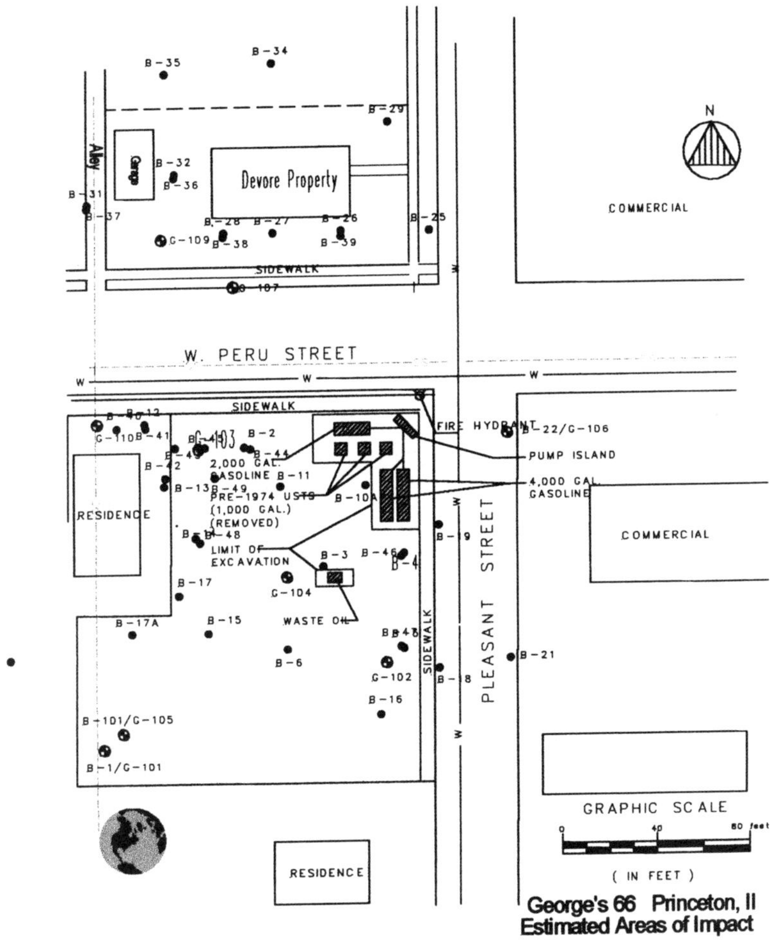

FIGURE 1. George's 66 site, Princeton Illinois

Groundwater depth at the site varies seasonally from between 4 and 10 feet bgs. The soils are generally a low to medium plasticity silty clay with a trace of sand with a calculated hydraulic conductivity of $1.049X10^{5}$ cm/sec. The groundwater gradient trends in a generally northwesterly direction.

MATERIALS AND METHODS

The implementation of the remedial action plan (RAP) consisted of first establishing a 10 foot by 10 foot grid across the site. Throughout the grid oxygen release compound (ORC^{TM}), vegetative heterotrophic bacteria cultures, nutrients, nitrification inhibitors and low levels of hydrogen peroxide were injected via a patented high pressure injection process into the saturated and capillary fringe area, using direct-push drilling equipment. Injections occurred every 100 days, with the second and third injections focusing on those areas exhibiting elevated BTEX levels.

Across the site grid, bio-injection points consisting of concentrated cultures and bioslurry were over-laid on an ORC™ injection grid. The bioslurry consisted of 700 ppm hydrogen peroxide, nutrients composed of a proprietary blend of bioavailable slow-release nitrogen, ortho-phosphate and nitrification inhibitors. Liquid heterotrophic cultures were vegetative *Pseudomonas* consortia with a cell count of no less than $3X10^9$ Colony Forming Units//mL.

Groundwater and soil samples were collected periodically and evaluated for BTEX compounds via EPA Method 8260, ortho phosphate via EPA Method 365.2 in acidified and non-acidified samples, ammonia nitrogen via EPA Method 350.3 in acidified and non-acidified samples, total calcium via Method SM3500B in acidified and non-acidified samples and iron via Method SM3111D in acidified and non-acidified samples. Total and selective heterotrophic evaluations were also performed on groundwater samples.

Total heterotrophic enumeration was performed by obtaining samples in sterile bottles. Dilution occurred as quickly as possible after having been received into the laboratory, samples were shaken and the foam allowed to settle, 1mL of sample was transferred from the sample into 99 mL of sterile phosphate dilution water to give a 10^{-2} dilution. Samples were shaken and allowed to settle. Dilution was continued using either 99 mL or 9 mL dilution water bottles until the required dilution level was reached. Samples were then plated (into sterile petri dishes) at three dilution levels in duplicate e.g. 10^{-5}, 10^{-6}, 10^{-7}, in the expected range. Pour plates were prepared with molten PCA agar (temperature approx. 45°C). A control plate was prepared to check for sterility. Plates were swirled gently to mix inoculums and agar. Plates were allowed to set. Plates were inverted, stacked and incubated at 30°C +/- 2°C for 48 hours. Plates were checked after 24 hours for spreading colonies. If present, a quick count was performed and the sample was re-incubated for another 24 hours. Plates with between 30 and 300 colonies were counted and results recorded.

Selective enumeration of the samples utilized both MacCokey Agar and PIA agar. Using the fixed volume 100 μL micro-pipet, 0.1 mL of the undiluted sample was pipetted onto the center of a MacCokey Agar plate. With a sterile glass rod the sample was carefully spread over the surface of the plate by spinning it on a turntable. Plates were incubated at 35°C for 24-48 hrs. The plates were examined for growth and the results recorded.

RESULTS AND DISCUSSION

In November 1999 a series of soil borings were advanced across the site to better delineate the extent of impacted soils. The results of these soils borings, in combination with the groundwater monitoring data, have served as verification of the methods employed to date at the site.

TABLE 2A: Source Area – – Soil Sample Results at $T_{0\ days}$ and $T_{300\ days}$

Boring Number	B-2	B-44	Change
Depth of Sample (ft Bgs)	6	8.5	
Date of Sample	11/12/98	2/28/2001	
BTEX Total (ppb)	328780	134687.4	-59%
Benzene (ppb)	4380	729.4	-83%
Toluene (ppb)	44200	15363	-65%
Ethylbenzene (ppb)	53200	35445	-33%
Xylenes Total (ppb)	227000	83150	-63%

TABLE 2B: Perimeter of Soil Excavation – Soil Sample Results at $T_{0\ days}$ and $T_{300\ days}$

Boring Number	B-4	B-46
Depth of Sample (ft Bgs)	10.5'	11'
Date of Sample	11/12/98	2/28/01
BTEX Total (ppb)	31292	ND
Benzene (ppb)	452	ND
Toluene (ppb)	9940	ND
Ethylbenzene (ppb)	3700	ND
Xylenes Total (ppb)	17200	ND

TABLE 2C: Cross Gradient – Soil Sample Results at $T_{0\ days}$ and $T_{300\ days}$

Boring Number	B-14	B-48	Change
Depth of Sample (ft Bgs)	11'	10'	
Date of Sample	11/10/99	2/28/01	
BTEX Total (ppb)	472690	2048.5	-99.57%
Benzene (ppb)	4400	67.5	-98.47%
Toluene (ppb)	3690	1	-99.97%
Ethylbenzene (ppb)	108000	1980	-98.17%
Xylenes Total (ppb)	356600	0	-100.00%

Overall, where comparisons are available of soil borings from similar locations and depths, an average decrease in BTEX compounds of 76% is observed. Further, in those areas where soils have been exposed to the petroleum hydrocarbons for greater lengths of time, as in Table 2A, longer remediation times area seen. For those area up-gradient or cross gradient, such as the results seen in Tables 2B and 2C complete remediation of the soils is accomplished within 300 days. Observed remediation times are a direct function of the absorptive characteristics of the soil and the robustness of the established biofilm. Across the site similar sample locations have seen and average soil benzene concentration decrease of 92.7%.

As with the soil boring data, the groundwater monitoring data is also trending downward across the site. In all wells benzene and toluene values are trending towards closure limits within the targeted eighteen-month time-line. Variations in the rate at which the targeted compounds are mineralized within the groundwater is directly related to the sorbed mass of contamination in and immediately upgradient from the individual wells.

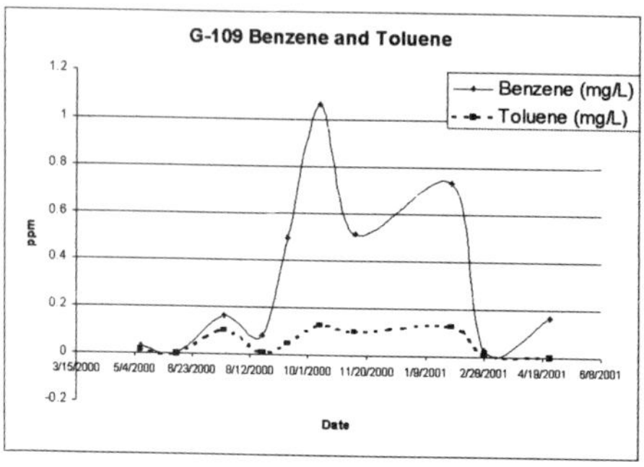

Figure 2. G-109 Benzene and Toluene (down gradient monitoring well)

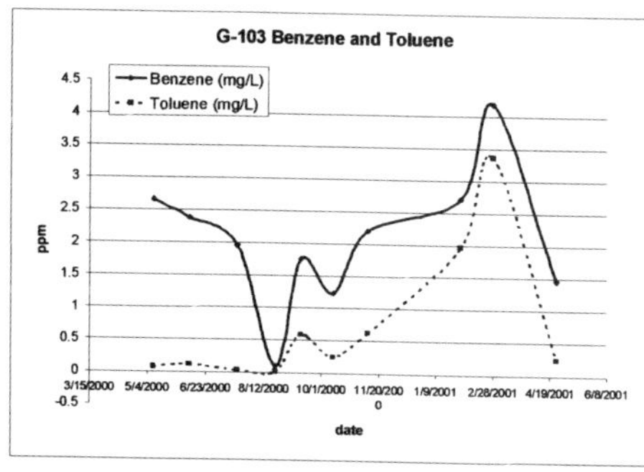

Figure 3. G-103 Benzene and Toluene (source area monitoring well)

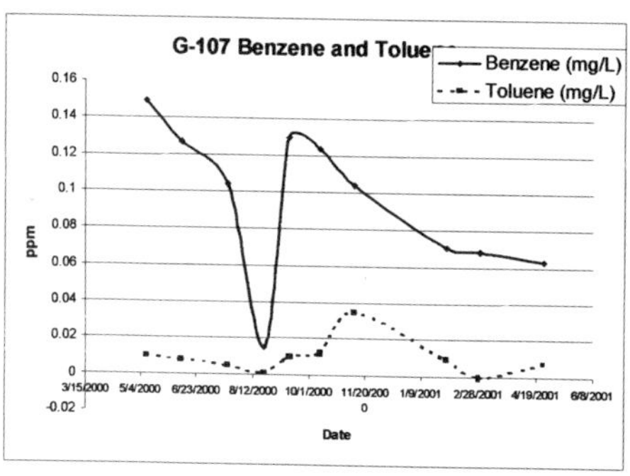

Figure 4. G-107 Benzene and Toluene (cross gradient monitoring well)

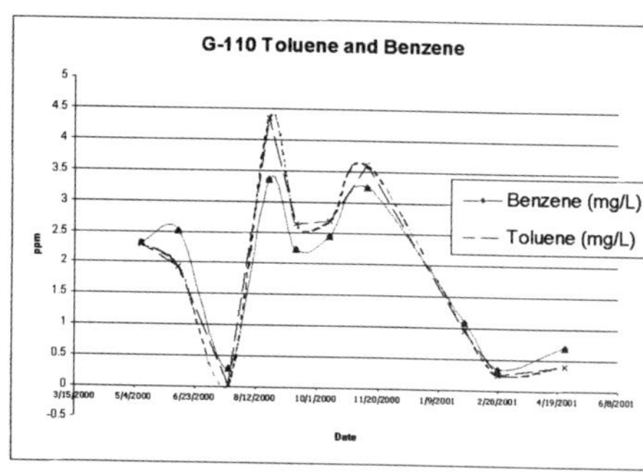

Figure 5. G-110 Benzene and Toluene (down gradient monitoring well)

In all wells a significant "rebound" is seen three to six months following the initial injections. The severity of this rebound, and the recovery period is seen to be related to the starting concentration, the soil type and the period of time the soils in the area of the monitoring well have been in contact with the petroleum hydrocarbon. A comparison of the cross gradient well, G-107 and the source area well, G-103, their rebound and recovery times and the overall rate of disappearence of the targeted compounds in the ground water may be seen in figure 3 and figure 4.

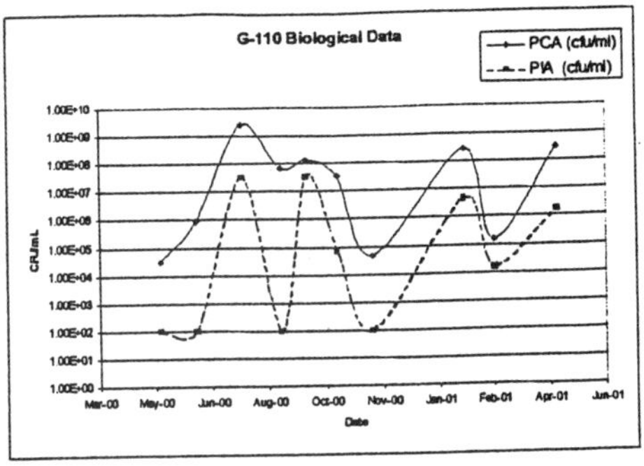

Figure 6. G-110 Biological Analyses

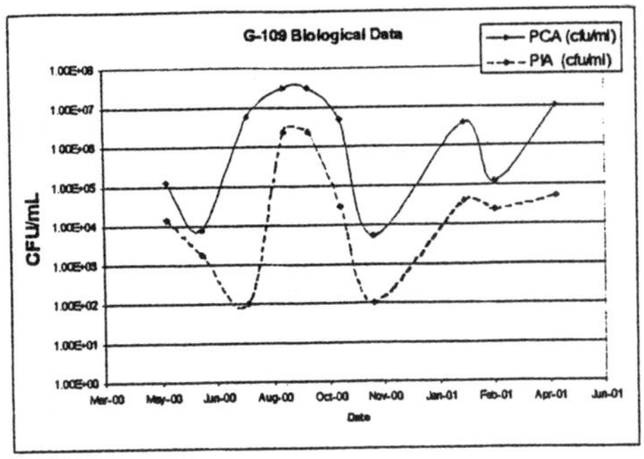

Figure 7. G-109 Biological Analyses

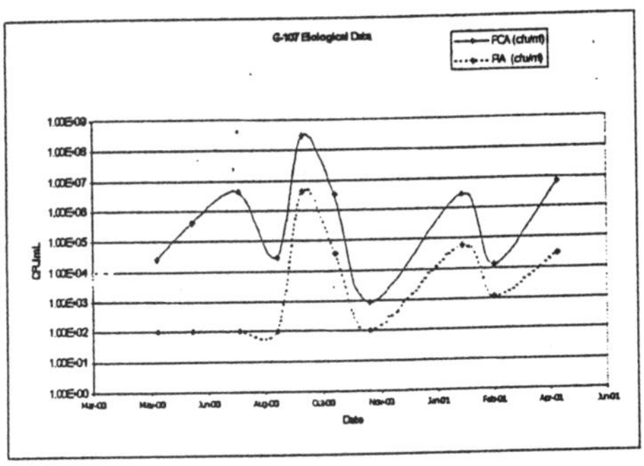

Figure 8. G-107 Biological Analyses

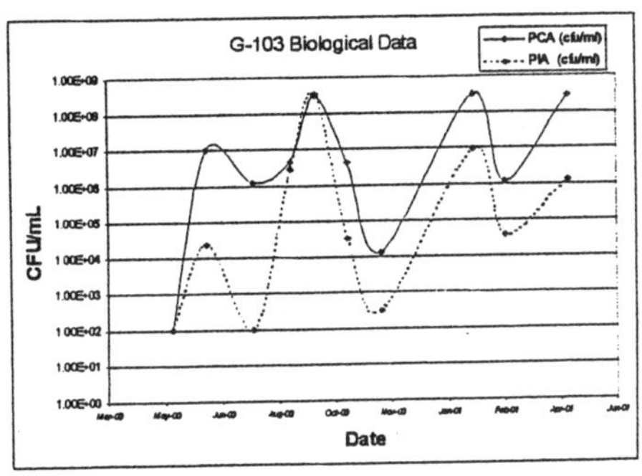

Figure 9. G-103 Biological Analyses

The continued monitoring of the total and selective biological counts throughout the project has allowed for correlations to be drawn between substrate concentration, nutrients and dissolved oxygen. Figure 9, the total and selective heterotrophic enumerations for G-103, a source area monitoring well, has maintained the highest biological counts through the project due to the maintained levels of nutrients, nutrients and growth substrate. In those wells where the substrate is becoming the rate limiting factor, such as G-107 (Figure 8), bacterial counts are several orders of magnitude less than source area or down gradient wells (G-110, Figure 6). Further, the application of vegetative cultures at the beginning of the project has allowed for better population management and a greater proportion of the total heterotrophs to remain as *Pseudomonas*. This population management is best seen as the project progresses in the source area and down gradient wells.

In order to obtain more rapid bacteria counts at a lower analytical cost, plate counts of *Pseudomonas* were used throughout the project in lieu of substrate specific enumerations. To confirm that the reported PIA count represents BTEX degrading cultures, the March 2001 monitoring well samples were further evaluated for substrate specificity. Toluene was chosen as a sole-source growth substrate for this evaluation. The results from this analysis verify the assumption that the PIA and/MacConkey's agar may be utilized to assess the selective culture counts.

Table 3. Total Viable Count (CFU/mL) vs. Pseudomonas Count (CFU/mL)on MAC vs. Toluene Count (CFU/mL)

Date of Analysis	Well #	Total Viable Count	Pseudomonas Count	Toluene Count
3/5/01	107	1.28 X 10^4	1.15 X 10^3	1.00 X 10^3
3/5/01	103	1.13 X 10^6	3.66 X 10^4	7.60 X 10^5
3/5/01	109	1.15 X 10^5	2.34 X 10^4	1.42 X 10^4
3/5/01	110	1.60 X 10^5	1.56 X 10^4	1.00 X 10^3

Throughout the project the limitations of essential nutrients have been seen to affect the biological robustness of the site's soils and groundwater. Increased total and selective counts have been observed following each injection of additional nutrients and peroxide. In that the analyses for biological enumeration were performed some several weeks following an injection, the increase in the biological counts must be attributed primarily to the additional nitrogen and ortho-phosphate made available rather than the increase in dissolved oxygen resulting from the hydrogen peroxide component of the bioslurry.

Throughout the project, dissolved oxygen within all the wells have been generally maintained above 0.7 ppm, with readings as high as 10 ppm noted. This sustained level of oxygen must be attributed to the effective introduction, distribution and disassociation of the ORC™ across the site. At no time has any monitoring well had a measure dissolved oxygen level below 0.65 ppm. As a tertiary confirmation that biological Mineralization of the BTEX compounds is occurring, site monitoring of pH has shown a general rise across all the monitoring wells. After ten months of the in-situ treatment program, pHs generally range from 7.0 to 7.7.

CONCLUSIONS

The notion that biologically based remediation technologies present scale-up issues and frequently do not offer reproducible results may be an outgrowth of the field practioner focusing on only a portion of the biological system. This field project, and others like it demonstrate that employing a "systems" approach to in-situ biological remediation necessitates that all the necessary components critical to in-situ environment need to be addressed. Further, the success of this project demonstrates the efficacy of both the proprietary delivery systems and products employed and capabilities of Oxygen Release Compound to supply sufficient terminal electron acceptor levels throughout the term of an in-situ project. If any of the components below are omitted or poorly managed the less likely the project is to succeed:

1) Uniform subsurface delivery to the effected areas,
2) Maintenance of bioavailable essential nutrients,

3) Maintained levels of terminal electron acceptors,
4) Effective population management and monitoring and
5) Effective interpretation of field data.

Of the five critical elements, four may be addressed by equipment and materials the fifth requires experience and sufficient field analytical data to base field activities. When applying biological solutions to remediate impacted soils and groundwater the interaction of all the in-situ elements must be evaluated. The equalibria between the aqueous phase and the soils cannot be ignored; the role the established biofilm plays in addressing the entrained BTEX compounds on the soils and in the groundwater requires thoughtful consideration. The bioavailablility of essential nutrients must be evaluated continuously and the overall health of the consortia monitored.

When applied as a system, bioremediation of BTEX contamination eliminates the residual source of the contamination, specifically the BTEX compounds entrained within the vadose zone soils, generating reproducible results for the practioner, his client and the property owner.

AOP/BIOREMEDIATION: A TECHNOLOGY BREAKTHROUGH WITH AQUIFER HEART-LUNG MACHINES

William B. Kerfoot (K-V Associates, Inc., Mashpee, Massachusetts, U.S.A.)

ABSTRACT: The rate of bioremediation is frequently controlled by the classic law of the minimum – whatever is present in lowest amount of any critical ratio for cellular metabolism becomes the critical element, whether carbon source, nutrient, or gas product. The rate of export of carbon dioxide and replenishment of oxygen must match the supply of the nutrients to indigenous bacteria. Simple injection and diffusion lead to time-phased depletions of essential elements. KVA has developed a gas/liquid delivery system that produces thin-layer coated microbubbles which introduce ozone, and later, oxygen-enriched air coated by a food/nutrient mixture into the aquifer.* The one, two punch of ozone/hydroperoxide sparging and introduction of nutribubbles creates a situation of initial reduction of indigenous bacterial populations, degradation of organics to more readily treatable forms, and enhancement of oxygen conditions to allow regrowth of bacterial colonies which quickly expand through the formation.

INTRODUCTION

Many bench-scale bioremediation tests are not satisfactorily transferred to field conditions because of the simple failure to replicate adequate nutrient availability and gaseous transfer. If an aquifer equivalent of a heart-lung machine were possible, which could transfuse liquids and permit gaseous exchange on a continuing basis instead of slug injection, the efficiency of field bioremediation could be greatly enhanced. Bench-scale test results are presented along with "field" testing set-up. Results show promise of an encouraging new form of treatment using KVA's gas/liquid delivery system for enhanced bioactivity with nutribubbles (thin-layer coated microbubbles).

TECHNOLOGY DESCRIPTION

A suitable area for remediation using the ozone/nutribubble technique is one in which the hydraulic conductivity of the geologic formation is between 10^{-1} and 10^{-6} cm/sec. The remedial area is first treated with ozone/air microbubbles coated with hydroperoxide, if necessary, for specific bond cleavage or increase in oxidative potential. Ozone reacts in an aqueous and gaseous form to degrade aromatic ring compounds (BTEX) and certain ethers (MTBE). It also breaks apart long-chain aliphatic compounds. Following removal of BTEX/MTBE compounds, oxygen-enriched air bubbles are coated with a nutrient mixture and

* U.S. Patent #5,855,775 and #6,083,407; other U.S. and foreign patents pending.

injected into the aquifer. The fine bubbles serve to assist in pumping the nutrients through the capillary structure of the formation being treated. The bubbles assist in removing CO_2 as well as supplying oxygen for respiration. Introduction of the ozone/nutribubbles is accomplished by means of specially-prepared Spargepoints®, which receive the simultaneous supply of gas and liquid. The pulsed injection of gas through the Spargepoints® eliminates the common problem of plugging of injection wells.

The introduction of ozone has an action similar to steaming of soils in reducing indigenous bacterial populations within 10 ft. (cm) of the injection well, while additionally degrading organics to more readily attackable forms. Ozone also naturally decomposes to oxygen, theoretically supplying up to ten times the dissolvable oxygen content found with direct air injection (CRC, 1972). Oxygen-enriched air injection with nutrients purges the remaining ozone residual, allowing indigenous bacterial species capable of organic residue metabolism to quickly expand through the formation.

For difficult organic compounds, the capacity to combine limited sterilization and pre-digestion to promote biodegradation with optimal species presents an intriguing challenge. The use of partial oxidation through an Advanced Oxidation Process (AOP) to initiate breakdown of difficult complex organics as a pre-step to biodegradation has been explored previously (Carberry, 1994; Miller *et.al.*, 1988). An AOP/BIO (Advanced Oxidation Process/Bioremediation) oxidative strategy would be particularly advantageous if the initial oxidation steps of the sequential toxic chemical biological mineralization process is initially step-rate limited, and could be carried out chemically rather than biologically (Carberry, 1994). Previously the use of the aqueous hydroxyl radical has been explored by stoichiometric means (Bishop *et. al.*, 1968; Carberry and Benzing, 1991). The development of Criegee-like ozone reactions with microencapsulated gas bubbles with the addition of thin-film aqueous ozone/hydroxyl radicals offers a set of powerful and efficient oxidants (Kerfoot, 2000) which may also be employed as pre-digesters. Carberry (1994) demonstrated that a PCP (polychlorinated phenol) aqueous system, partially oxidized by Fenton's Reagent, increased maximum substrate uptake rate to a level four times greater than that of an nonpretreated system. The time to maximum growth rate was also enhanced from two to four days in a nonpretreated control to one to two days using preoxidation. Suggested specific target organic compounds consisted of PCPs, PCBs (polychlorinated biphenols), and toluene. Of these only the last is appreciably water-soluble. KVA, from recent experience, would add MTBE and BTEX (benzene, toluene, ethyl benzene, and xylenes) compounds as well as certain polyaromatic hydrocarbons (PAHs), particularly naphthalene.

Microbiological degradation is a powerful means of degrading complex organic compounds. The rate of reaction depends upon the type of microbe, the substrate compound targeted as food, and the rate of gaseous exchange. Bubbling oxygen-enriched air through an aqueous solution of substrate compound, supplied with the nutrients nitrogen (N), phosphorus (P), and potassium (K), can encourage rapid metabolism and reproduction of microbes, which consume the organic compounds.

There are specific ratios of carbon (C), nitrogen (N), phosphorus (P), and potassium (K) to promote an efficient reaction. Often these reflect the molecular ratio of the elements in the growth of microbes and are used to expand the population numbers. During microbial degradation of the carbon compounds, oxygen is utilized and carbon dioxide (CO_2) is produced. If there is insufficient oxygen, the reaction will be self-limiting, abruptly slowing and eventually stopping.

The basic microbial process of biodegradation (aerobic) can be portrayed as a conversion of oxygen (O_2) to CO_2 and water plus more bacteria:

$$\text{Organic contaminant} \xrightarrow[O_2]{\text{bacteria}} CO_2 + H_2O + \text{more bacteria}$$

Whereas the reaction may proceed rapidly in vat or surface vessels, maintaining an efficient and continuous degradation is much more tricky in porous soils. Often evidence of natural biodegradation is shown by excess CO_2 in the overlying unsaturated soil zone (vadose zone) and low oxygen content in the saturated (aqueous) zone. The depletion of natural electron acceptors [O_2, NO_3, SO_4, Fe (III)], the depletion of natural electron donors (organic acids, *e.g.*, acetate, lactate, H_2), the buildup of anaerobic metabolism gases such as CO_2, and the depletion of mineral nutrients (NH_3, NO_3, PO_4, K) regulate the rate of biodegradation.

The ability to mix gases, electron donors, or nutrients with organic contaminates is severely limited by the porosity and hydraulic conductivity of saturated soils. Porous soils tend to encourage movement of liquids as slugs, not easily mixing with existing groundwater. The rate of natural movement is slow and determined by existing groundwater gradients. Velocities of natural flows commonly run .1 to 2 ft/day. The natural flow across a 100 ft wide contaminant zone may take 50 to 1000 days. The capability to remove product waste products is similarly hindered. The rate of biodegradation in natural formations is very slow compared to above-ground.

The need exists for an efficient means to provide reaction promoters and simultaneously to remove unnecessary products. Carbon sources which provide energy (electron donors) are available in liquid form. Nutrients also can be mixed with carbon sources to assure ready availability. If both can be present as a coating to oxygen-enriched air, bacteria can be provided with a very mobile lunch. In addition, the gaseous products such as CO_2 can be transported away as the gas rises.

The Law of the Minimum, which states that bacterial growth will stop when the nutrient that was present in the lowest concentration (relative to the requirement) is exhausted, becomes a problem since the rest of the mixture is useless. If that substance is replenished, growth will stop when the next substance is exhausted. By providing a means of readily mixing the constituents and having the capacity to modify the electron accelerator and nutrient ratios, using metabolic products as a guide, the rate of metabolism can be maximized. The use of a series of injection/extraction wells, or a trench, for reinjection with pumping of treated water has been proposed by Vogel (1998).

There are several new aspects to the KVA process: 1) gas/liquid oxidation (AOP) to predigest and sterilize around injection locations; 2) simultaneous injection of nutrients and food source with gas in proportion to optimal ratio for assimilation; 3) injection of nutrients to provide a coating (thin layer) on gas (oxygen-enriched air) being injected into porous soil capillaries; 4) introduction in a pulsed manner with microbubbles of 5 to 100 micron diameter sized to pores of soil to avoid fracturing of soil and to enhance transport through capillary-like soil voids; 5) use of specially-fabricated Spargepoint® for introduction of mixture into formation involving laminate microporous construction receiving simultaneously gas and liquid.

BENCH-SCALE TESTING

Bench-scale tests were conducted on soil samples collected from a site located in Paterson, New Jersey and augmented with PCBs (arochlor 1260). The purpose of the tests was to evaluate the response of volatile organic compounds and semi-volatile organic compounds contained in soil aliquots extracted from these soil samples to an aqueous environment containing various injected concentrations of sparged air with ozone or sparged air with ozone plus varying concentrations of injected hydrogen peroxide. During each test, treated and untreated aqueous samples were collected from the test cell at regular time intervals and field screened with portable gas chromatograph for ionizable compounds. Prior to, and following each test, soil and groundwater samples were collected from the test cell for confirmatory laboratory analyses.

Testing Procedure. Initial bench-scale testing included putting approximately 5 grams of soil sample in a 40 mL VOA vial with 30 ml of water and allowing headspace to develop, then screening the headspace in an HNu Systems Model 321 portable gas chromatograph. The purpose of this step was to extrapolate the volume of soil necessary to produce an instrument response in the GC that was noticeable for further testing.

Subsequent testing included stirring approximately 40 grams of soil sample in 500 mL of water contained in an Erlenmeyer flask and subjecting the entire solution to sparged air containing approximately 100 ppmv of ozone. This test was conducted under 0 psig with aqueous samples collected from the test cell at time intervals of 0, 2, 5, 10, 15, and 20 minutes and field screened with portable gas chromatograph for photoionizable volatile organic compounds.

Following the above preliminary bench-scale tests, further testing was conducted in a larger glass volume container to allow for split sampling of soil and groundwater samples to a certified laboratory for confirmatory analyses. This test cell included mounting a stopper on the 1,000 mL container and drilling holes in the stopper to allow for various tubing diameters to pass through the stopper to accommodate: an external pressure gauge T'd with a pressure relief valve; a sample guideport to permit below-water-level aqueous sampling; a gas line for a mini-Spargepoint®; and injection of treatment liquids such as a 3% solution of hydroperoxide.

The following four bench-scale tests were all conducted under 5 psig. Split groundwater samples were collected at the beginning and end of each test run and sent to a Massachusetts state-certified laboratory for volatile organic analysis. Soil samples were collected at the beginning of each test run and sent to a Massachusetts state-certified laboratory for volatile organic analysis. Soil samples were collected at the beginning of each test run and sent to a Massachusetts state-certified laboratory for volatile and semi-volatile organic analysis. A soil sample was collected at the end of each test run and sent to a Massachusetts state-certified laboratory for semi-volatile organic analyses.

Results. The soil samples appeared to contain weathered oil as well as other organic compounds. Review of the portable gas chromatograms indicate the nearly complete attenuation of ionizable compounds by the end of each test run. The last set, using a 300 ppmv ozone concentration and 2 mL/min hydroperoxide, gave the lowest baseline response of the four treatments. The introduction of hydroperoxide with ozone as a bubble coating appeared superior in treatment than a batch mixing with an added slug of equivalent volume.

A final additional test set was performed by increasing the oxidant concentrations to 300 ppmv ozone with 10 mL 10% hydroperoxide and 3000 ppmv with 30 mL 10% hydroperoxide as a bubble coating. The results of the test are presented in Table 1. After oxidation, the nutribubble introduction was started and continued until remaining concentrations drop below 10 µg/kg.

TABLE 1. Removal of PCBs and PAHs from contaminated sediment by ozone plus hydroperoxide, 30-minute bench-scale test.

Compound	Initial Concentration (µg/kg)	Ozone + Hydroperoxide 300 ppmv+ 10 mL 10%	3000 ppmv+ 30 mL 10%	% Removal
Arochlor 1260	23,300	3,900	5,000	79
Naphthalene	21,000	750	650	97
2-Methylnaphthalene	13,000	710	BRL	99
Acenaphthylene	15,000	BRL	BRL	99
Benzo(a)anthracene	18,000	5,800	2,400	86
Benzo(b)fluoranthene	17,000	5,800	3,100	82
Benzo(k)fluoranthene	17,000	5,400	2,700	84
Dibenzo(a,h)anthracene	17,000	4,900	2,300	86
Indo(1,2,3-c,d)pyrene	18,000	5,100	2,300	87
Phenanthracene	17,000	4,200	2,000	87
Pyrene	40,000	13,000	5,500	86

FIELD OPERATIONS –EASE OF IMPLEMENTATION

The process of administration of the liquid and gas would be *in-situ*, with use of laminar Spargepoints®. Figure 1 presents a schematic diagram of the KVA Criegee oxidizer with simultaneous liquid and gas supply. The individual

Spargepoints® are drilled into the contaminated aquifer region to be treated. The timer-controller delivers liquid and gas simultaneously. The number of Spargepoints® would depend upon the volume to be treated. Figure 2 shows an operating system being used to field treat about 1 cubic meter of saturated soil contained in a plastic trash bin. The system runs on house current (120 volts AC, 15 amp).

FIGURE 2. Field operating system showing laminar Spargepoint® with liquid and gas flowmeters.

CONCLUSIONS

The reactivity of ozone with benzene, phenanthrene, naphthalene, pyrene, and chrysene is well-documented in the scientific literature. Old diesel fuel or coal tar residues often contain these compounds. Not uncommonly, PCBs are encountered with these organics, presenting a difficult mix to remediate. The bench test of the coated microbubbles resulted in a removal varying from 79% to 99% for the mixture of PAHs and PCBs, with initial oxidant treatment. Switching to nutribubble injection was successful in bringing the total removal rate to less than 10 µg/kg of treated soil. The initial results offer an encouraging start for a new form of treatment.

REFERENCES

Bishop, D. F., et. al. 1968. "Hydrogen Peroxide Catalytic Oxidation of Refractory Organics in Municipal Wastewaters." *Ind.. Eng. Chem: Process Design and Development,* 7: 110.

Carberry, J. B. 1994. "Enhancement of Bioremediation by Partial Preoxidation." In D. L. Wise, and D. J. Trantiolo (Eds.), *Remediation of Hazardous Waste Contaminated Soils*, pp. 543-597. Marcel Dekker, Inc., New York, NY.

Carberry, J.B. and T. M. Benzing. 1991. "Peroxide Pre-oxidation of Recalcitrant Toxic Waste to Enhance Biodegration." *Water Sci. Tech. 23*: 367-376.

CRC, 1972. *Handbook of Chemistry and Physics*, 52nd edition. Chemical Rubber Company, Cleveland, OH.

Karpel vel Leitner, R. N. K. V., A. L. Papaihou, J. P. Grove, J. Payrot, and M. Dore. 1994. "Oxidation of Methyl tert-Butyl Ether (MTBE) and Ethyl tert-Butyl Ether (ETBE) by Ozone and Combined Ozone/Hydrogen Peroxide." *Ozone Science and Engineering, 16*:41-54.

Kerfoot, W. B. 2000. "Ozone Microsparging for Rapid MTBE Removal." In G. B. Wickramanayake, A. R. Gavaskar, and A. S. C. Chen (Eds), *Chemical Oxidation and Reactive Barriers: Remediation of Chlorinated and Recalcitrant Compounds*, pp. 187-194. Battelle Press, Columbus, OH.

Miller, R. M., Singer, G. M., Rosen, J.D., and Bartha, E. 1988. "Sequential Degradation of Chlorophenols by Photolytic and Microbial Treatment." *Env. Sci. Tech.,* pp. 1215-1219.

Vogel, S. 1998. "Natural Attenuation and Bioremediation of Soil and Groundwater Contamination." EBI Publication, LSP Course 1104, CTLEP-016, 44 pp., EBI, Cambridge, MA.

Verschueren, K. 1996. *Handbook of Environmental Data on Organic Chemicals,* 3rd Edition. John Wiley & Sons, Inc., New York, NY.

AIR SPARGING / SVE IN ITALY: SUSPENSE REPLACES EXPENSE

Michael D. Basel (Montgomery Watson, Des Moines, Iowa)
Simon Bluestone, Robert Dixon, and Giovanni Longoni
(Montgomery Watson, Milan, Italy)
Beth Lang (United Technologies Corporation, Hartford, Connecticut)

ABSTRACT: An air sparging / soil vapor extraction system was designed, constructed and implemented in Northern Italy. The system was installed inside and outside an operating facility in an environmentally sensitive area to remove trichloroethene and associated degradation products from the vadose zone and shallow groundwater. The system has operated for one year, has successfully removed a majority of the target compounds, and has lowered groundwater concentrations by as much as 75 percent at some locations. The system was installed using an abbreviated pilot test, with standard assumptions used to specify well spacing and configuration. This pilot test demonstrated that successful air sparging systems can be installed at lower cost without extensive pre-testing. At startup, the system operated at 98 percent of the design vapor extraction rate and 87 percent of the design air injection rate indicating that the design was accurate. The project also demonstrated that shared global knowledge and understanding of in-situ aerobic remediation techniques using United States resources can lead to successful achievement of unique remedial objectives set forth by regulatory frameworks in other countries, such as Italy.

INTRODUCTION

An air sparging / soil vapor extraction (AS/SVE) system was installed at an operating manufacturing facility in Northern Italy to address soil and groundwater constituents. The constituents to be remediated consisted of volatile organic compounds (VOCs), primarily trichloroethene (TCE) and its associated degradation products, *cis*-1,2-dichloroethene (DCE) and vinyl chloride. The site is located in an area of light industry within a region of Italy known for its agriculture. An evaluation of remedial alternatives indicated that natural attenuation would sufficiently remediate VOCs in groundwater. However, stakeholders decided that additional mass removal from areas of higher concentration would be beneficial. The remedial objective of the AS/SVE system was to supplement monitored natural attenuation (MNA) by reducing levels of VOCs as much as possible, operating the system to the limits of its mass removal capabilities. The AS/SVE system was the first phase of a multi-phase remedial program that was coordinated with the authorities; Phase 2 consisted of monitored natural attenuation.

The manufacturing facility occupies approximately 30,000 square meters (m^2), of which approximately 90 percent of the surface is paved or covered with buildings. A layout of the facility is shown on Figure 1. The vadose zone consists of fill overlying heterogeneous clays and alluvial silts. Shallow groundwater is encountered at approximately 4 to 6 meters below ground surface (bgs). The water bearing zone consists of 5 m of sand and gravel; the hydraulic conductivity has been approximated to be 10^{-2} to 10^{-3} cm/s with clay lenses

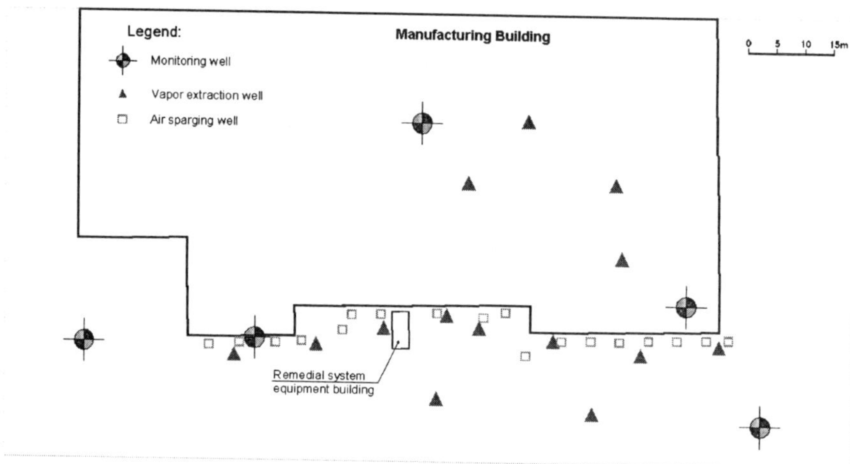

FIGURE 1. Schematic of the facility layout

causing local variations. The underlying aquitard consists of 9 m of clayey silt; groundwater constituents have not been detected in deeper groundwater.

Low levels of chlorinated VOCs have been detected in shallow soils around the facility, and concentrations decrease significantly with depth. VOCs have been detected in shallow groundwater at concentrations as high as 258 micrograms per liter (µg/L); VOCs primarily consist of TCE and associated degradation products. DCE and vinyl chloride have been detected at the highest concentrations, as high as 130 µg/L and 73 µg/L, respectively.

Potential soil and groundwater remedial alternatives were evaluated for the site. AS/SVE was selected as the recommended Phase I remedial alternative for the following reasons:

- Would address soil and groundwater in areas of highest concentration
- Contaminants would be either removed or destroyed
- High probability of success
- Provide significant remediation in short time frame (1 to 3 years)
- Relatively easy to implement

PILOT TEST

A pilot test of AS/SVE was conducted as the first step of implementing the technology full-scale. The pilot testing was conducted over three days in May 1999. Costs of the pilot test were minimized by conducting a series of brief tests to confirm the effectiveness of the technology and provide order-of-magnitude set points for operating parameters such as vapor extraction rates, air injection rates, and induced impacts on soil vapor and groundwater in the subsurface.

Extensive testing was not performed. As a trade-off, optimization of the system would be performed during start-up of full-scale if the pilot test indicated AS/SVE was effective. Extensive testing was also not performed because of the observation of the site engineers and other publicized experiences that air sparging systems can often be designed with minimal pilot testing using default design conditions (Gillespie, et al., 1999).

The pilot test consisted of one day of SVE pilot testing from various wells at different applied vacuums. The second day consisted of a continuous SVE test from one well at a constant applied vacuum. The third day consisted of air sparging at one location with and without SVE. The pilot testing indicated that AS/SVE could be effectively applied at the site with the following conclusions:

- SVE could extract vapor at approximately 65 cfm per well at relatively low applied vacuums (24 inches water). Air could also be successfully injected into the aquifer at 35 cfm per well.
- During SVE, induced vacuum was measured in monitoring points at least 10 m from extraction wells. During AS, increased pressures were observed in monitoring points 5 to 6 m from sparge points.
- Field screened concentrations in recovered vapor were as high as 430 ppm total VOCs.

AS/SVE SYSTEM

An AS/SVE system was installed with a design/build approach using a minimal design document and close coordination of design engineers and construction staff in the field during installation. Fourteen SVE wells and 18 AS wells were installed. A schematic of the AS/SVE system is shown in Figure 2.

The AS/SVE system was started up in November 1999. Since this time, it has been operated continuously using all 14 vapor extraction wells and 18 air sparge points. An 18-hp blower was installed for the SVE system with an expectation that it would extract about 420 cfm vapor from the extraction wells. Following optimization during startup, the SVE system stabilized at extraction of 414 cfm vapor at an applied vacuum of 21 inches of water. At these conditions, the SVE system was operating at 98 percent of the expected design rate. Flows have remained steady at these levels within 10 percent.

A 15-hp blower was installed for the AS system with an expectation that it would inject about 150 cfm air into the sparge points. Following optimization during startup, the AS system stabilized at an injection rate of approximately 132 cfm, which is 87 percent of the expected design rate. During operation, flow rates have decreased as the injection pressure has been lowered.

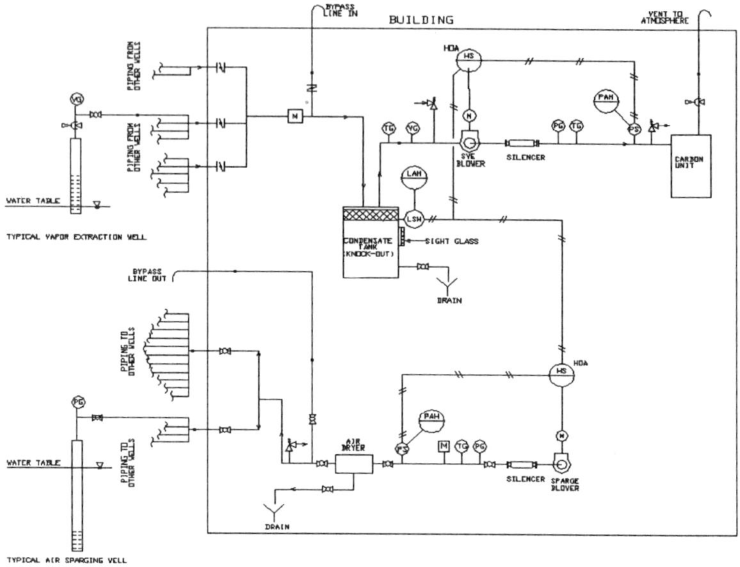

FIGURE 2. Schematic of AS/SVE system

RESULTS AND DISCUSSION

The effectiveness of the AS/SVE system was monitored by primarily tracking three parameters: 1) VOC concentrations in removed vapor, 2) total VOC mass removed, and 3) VOC concentrations in groundwater. Results of monitoring of these parameters for a period of 12 months are tabulated in Tables 1 and 2, and are described in the following paragraphs. Other parameters, such as dissolved oxygen in groundwater and groundwater elevations were also monitored during AS/SVE operations. These results are not discussed in this paper, but were generally consistent with those of other parameters.

Monitoring Event	**Pilot**	**Week**	**Week**	**Month**						
	Test	2	3	1	2	4	6	8	10	12
VOCs (μg/l)										
Tetrachloroethene	.2	.2	.3	.4	.8	.8	.3	.2	0.2	0.1
Trichloroethene	2.6	12	3.5	2.6	2.3	1.2	.5	.4	0.3	0.1
cis-1,2-DCE	18	8.6	4.7	4.6	4.3	3.2	1.5	1.1	0.7	0.2
Vinyl Chloride	2.2	.9	.5	.8	.9	ND	ND	ND	ND	ND
Total VOCs (μg/l)	24.2	22	9	8.5	7.6	5.2	2.3	1.7	1.3	0.4
Total VOCs (ppm)	6.27	4.81	2.06	2.02	1.67	1.29	0.52	0.37	0.27	0.10

TABLE 1. Vapor sampling results

	Upgradient			Cross-Grad.		Source Area			Down-grad.	
Well ID	1	3	13	14	15	E-1	E-2	E-3	O-1	O-2
BASELINE										
Tetrachloroethene	<1	<1	<1	3	<1	8	<1	<1	3	<1
Trichloroethene	<1	<1	<1	7	<1	4	2	2	2	<1
cis-1,2-Dichloroethene	<1	<1	<1	12	49	46	12	3	2	4
Vinyl Chloride	<1	<1	<1	<1	2	4	2	19	<1	2
Total Halogenated HCs	ND	ND	ND	22	51	62	16	24	7	6
AFTER 1 Year										
Tetrachloroethene	<1	<1	<1	1	<1	<1	<1	<1	2	<1
Trichloroethene	<1	<1	<1	4	1	1	<1	2	3	<1
cis-1,2-Dichloroethene	<1	<1	<1	2	35	4	1	37	1	2
Vinyl Chloride	<1	<1	<1	<1	3	<1	<1	2	<1	<1
Total Halogenated HCs	ND	ND	ND	7	39	5	1	41	6	2

TABLE 2. Groundwater sampling results (μg/L)

VOC Concentrations in Vapor. VOC concentrations in vapor were used to indicate the benefits of SVE. Vapor concentrations were detected at levels as high as 400 ppm during the pilot test. During long-term operations, VOC concentrations in vapor have declined to 0.1 ppm. The steady reduction in concentrations that has occurred from start-up (post pilot test) through the 12-month operating period is shown in Figure 3. This figure demonstrates that the SVE system is now recovering vapor with low concentrations of VOCs. The SVE system has reached the limits of its effectiveness under continuous operation.

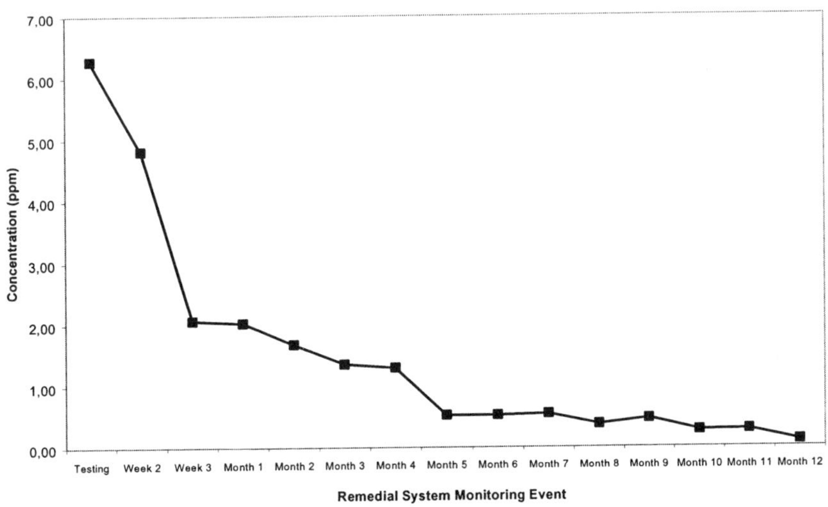

FIGURE 3. VOC concentration in removed vapor

VOC Mass Removed in Vapor. The primary purpose of the SVE system was mass removal, which was calculated based on vapor concentrations. As VOC concentrations have declined, the VOC mass removed has also decreased. As of December 2000, a total of 28.6 kg of VOCs have been removed in the extracted vapor since startup. This does not include the possibility of mass that has been degraded by the sparged air introduced into the groundwater. Figure 4 demonstrates that the mass removal rate has reached an asymptotic level, and it is anticipated that further SVE will not significantly enhance the mass removal rate.

VOC Concentrations in Groundwater. The apparent result of mass removal via AS/SVE has been to decrease VOC concentrations in groundwater. VOC concentrations have also decreased in groundwater samples collected from wells within the area remediated by the AS/SVE system. VOC concentrations as high as 258 μg/L were historically detected in monitoring wells. During the last sampling event in December 2000, total VOCs were detected at concentrations of 41 μg/L or less. Figure 5 indicates the VOC concentrations that have been monitored at key monitoring wells since July 1996. This figure demonstrates that concentrations have dropped an average of 82 percent. Monitoring will be continued after shutdown of AS/SVE to confirm that rebound does not occur.

CONCLUSIONS

In conclusion, the AS/SVE system has successfully operated over a 12-month period and has effectively removed mass from the subsurface and lowered VOC concentrations in vapor and groundwater. Costs of designing and testing the AS/SVE system were minimized by following a design/build approach using minimal pilot testing, with no adverse effects on the effectiveness of the system. Actual operating conditions were nearly exactly the conditions that were anticipated. Successful implementation of the AS/SVE system was attributed to close coordination of the entire project team including client and consultant participants from the United States and Italy. This project demonstrates that remediation projects in other countries such as Italy can be implemented effectively and practically combining local experience and regulatory knowledge with experience from the U.S.

REFERENCES

Gillespie, R., A. Leeson, C. M. Vogel, M. T. Wiley, A. Drucker, P. C. Johnson, R. L. Johnson, and R. E. Hinchee. 1999. Evaluation of the Multisite Air Sparging Demonstration." Presented at the 1999 In Situ and On-Site Bioremediation – Fifth International Symposium. April.

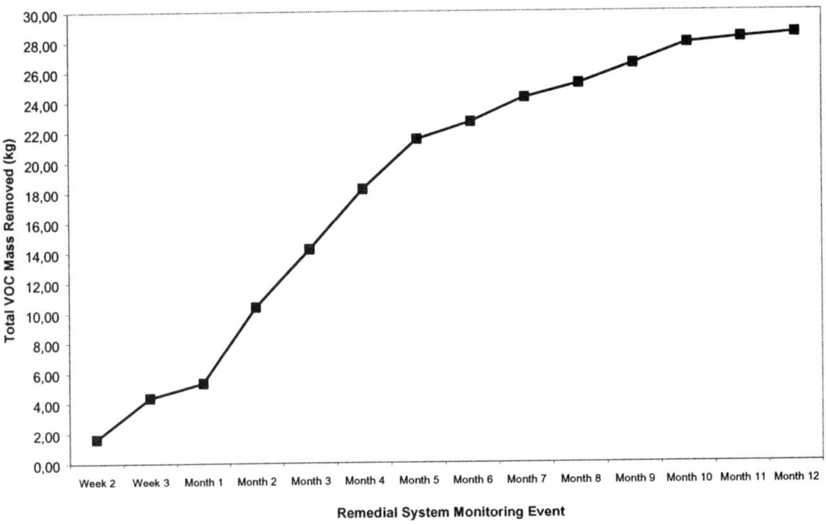

FIGURE 4. Demonstration of asymptotic SVE mass removal rate

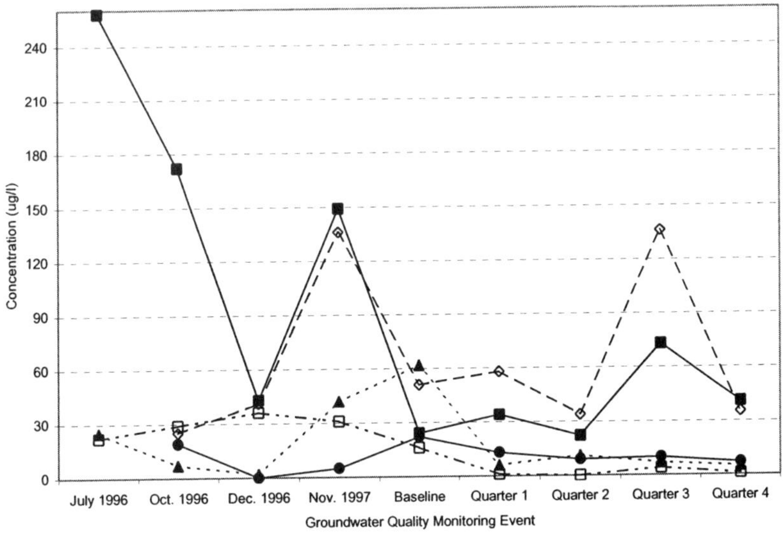

FIGURE 5. VOC concentrations in groundwater at key monitoring wells

AIR SPARGE/BIOREMEDIATION DEMONSTRATION IN FRACTURED OIL SHALE

James R. Covell and Mark H. Thomas
(EG&G Technical Services, Inc., Morgantown, WV)

ABSTRACT: A combined air sparge and bioremediation technology is being conducted to eliminate volatile organic contamination from a fractured oil shale aquifer at depths ranging from 65 to 95 ft (19.8 to 29.0 m). The remediation is being conducted at a site in southwest Wyoming for the National Energy Technology Laboratory (NETL) of the U.S. Department of Energy in Morgantown, West Virginia. Site characterization indicated that a non-dissolved contaminant source existed within the fractures and micro-fractures of the oil shale. This source material is introducing benzene, toluene, ethylbenzene, and xylene (BTEX) contamination into the groundwater. Pre-demonstration concentrations in 12 monitoring wells ranged from 31 to 3,960 µg/L and averaged 1,063 µg/L. The air sparge operations as of October 24, 2000 have reduced BTEX contamination to an average of 67 µg/L. This represents an approximate BTEX reduction of 94 percent.

INTRODUCTION

A combined air sparge and bioremediation demonstration is being conducted at an old oil shale retort test site in southwestern Wyoming. The demonstration is being conducted for the National Energy Technology Laboratory (NETL) of the U.S. Department of Energy. The test site is located approximately 7 miles west of Rock Springs, Wyoming, and is approximately 200 acres (0.8 km^2) in area. The test site was designated as the Rock Springs Oil Shale Retort Test Site.

The test site was used during the sixties and seventies to develop and evaluate "true in situ retorting" technology to extract oil from oil shale. Oil shale is a fine-grained layered shale that contains significant quantities of a solid organic material called kerogen (Hendrickson 1975). The term "retort" refers to the thermal process to convert kerogen in oil shale to oil and gas products. This technology requires the permeability of the native oil shale to be enhanced through methods such as hydraulic fracturing. After the permeability is enhanced, air is injected into the oil shale formation and the formation is ignited. A combustion front is formed within the formation and is supported by residual carbon and oil remaining after the extraction of oil from the oil shale. Hot gases produced from the combustion front move ahead of the front to heat and retort the oil shale. At temperatures greater than 800 ^{0}F (427 ^{0}C), kerogen is converted to non-condensable gases (e.g. carbon monoxide), condensable gases (shale oil and steam), and fixed carbon (residual solid carbon). As the hot gases pass out into the formation, the gases cool and the oil condenses. The oil is pumped from surrounding production wells.

Within the test site, there are 12 individual test sites. Most individual test sites underwent permeability enhancement; however, only five of the test sites had undergone thermal retorting. Subsequent to oil shale retort testing, organic contamination was detected in the groundwater at and adjacent to the test areas that had experienced thermal retorting. The contaminants included volatile and semi-volatile organics.

NETL requested EG&G to select and demonstrate a remediation technology to reduce organic contamination levels at Site 4 of the Rock Springs Oil Shale Retort Test Site. During 1968 and 1969, numerous permeability enhancement tests were conducted at the site. Of the various techniques applied to enhance formation permeability, hydraulic fracturing followed by explosive fracturing proved the most effective. Following the enhancement testing, retorting was performed. Approximately 8,000 gal (30,283 L) of oil were produced at the surface during retorting; however, an estimated 22,000 gal (83,279 L) remained in the formation. A monitoring well was installed at the site during 1997 (Harza, 1997). Analyses of groundwater samples showed benzene, toluene, ethyl-benzene, and xylene (BTEX) concentrations greater than 2,000 µg/L. BETX contamination became the target of the remediation demonstration. No remediation goal has been established for Site 4 by the Wyoming regulatory agencies.

SITE EVALUATION

Six characterization wells (IW-1 through IW-6 in Figure 1) were drilled during June 1998. The wells were drilled to 95 ft (29.0 m) and were cased to 65 ft (19.8 m). No screens were installed in the wells across the open well bore. These wells defined the original demonstration area. The area also included a previously drilled monitoring well (R-4). The demonstration area was expanded with 5 additional wells (IW-7 through IW-11) during June 1999 and with 7 additional wells (IW-12 through IW-18) during September 2000.

The estimated BTEX concentrations at Site 4 also are shown in Figure 1. The wells in the original demonstration area were sampled during June 1998 (prior to any air injection tests) and represent true pre-demonstration concentrations. Because the expansion wells (IW-7 through IW-18) were drilled after demonstration operations had started, the BTEX concentrations are estimates based on the highest observed concentrations. The actual pre-demonstration concentrations for these wells probably were higher than shown in Figure 1.

DEMONSTRATION DESCRIPTION

Based on the site evaluation at Site 4 and similar experience in a fractured coal aquifer (Covell and Thomas, 1997), an aggressive air sparge and bioremediation approach was selected to eliminate BTEX contamination. Based on the site evaluations, it was postulated that a non-dissolved contaminant source resided in the fractures and micro-fractures within the oil shale. Pumping tests showed that the source material was immobile within the formation. Because the contaminant source was transported to the deposition sites in a gaseous phase,

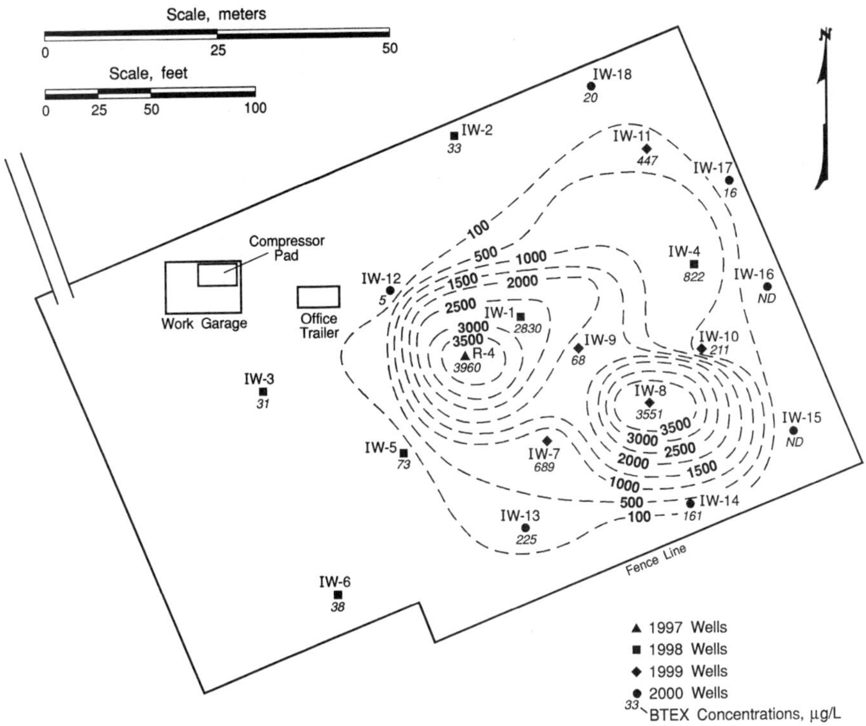

FIGURE 1. Rock Springs Site 4 well pattern and pre-demonstration BTEX concentrations.

injected air would potentially find the areas of source deposition. This provided direct source transfer to the air for venting to the surface.

Four 15-hp (11,185 J/s) compressors generated the air for the sparge operation and provided a combined flow rate of approximately 200 ft^3/m (94.4 L/s) at 30 psig (206.8 kP). Four wells were selected for air injection in the original demonstration area (IW-1, IW-3, IW-4, and IW-5). IW-2, IW-6, and R-4 were used as vent wells. Three injection wells (IW-7, IW-8, and IW-10) and two vent wells (IW-9 and IW-11) were added in the 1999 expansion.

DEMONSTRATION OPERATIONS

A series of short duration field tests (4 hours to 3 days) were conducted prior to the start of demonstration operations. The field tests occurred during July 1998, August 1998, December 1998, and January 1999. The field test included an extended period of contamination rebound monitoring after completion of one of the air injection tests. The field tests were conducted to determine operating parameters and to verify design assumptions.

Demonstration operations started on February 24, 1999 and are expected to continue through 2001. Air injection periods were alternated with contaminant rebound monitoring periods. Table 1 summarizes the periods of air injection operation and rebound monitoring. Air injection wells and flow rates were varied during operations. Four periods of rebound monitoring were conducted through November 2000. The rebound periods ranged from 33 to 67 days. The rebound period conducted during the field testing phase lasted 104 days.

TABLE 1. Air Injection and Rebound Monitoring Periods.

Air Injection Periods	Rebound Monitoring Periods
August 18 – 19, 1998	August 19 – December 1, 1998
February 24 – July 5, 1999	July 6 – August 9, 1999
August 10 – November 8, 1999	November 9, 1999 – January 11, 2000
January 12 – March 23, 2000	March 24 – May 11, 2000
May 12 – August 18, 2000	August 19 – October 24, 2000

DEMONSTRATION RESULTS AND DISCUSSION

Subsurface airflow patterns were dominated by natural and induced fractures. Most vent wells produced gases; however, the gas production rate varied significantly between vent wells. Wells near the center of the site (R-4 and IW-9) produced the most vent gases; however, several peripheral wells (IW-11 and IW-13) also produced significant vent gas flow. Hydraulic and explosive fracturing were concentrated near the site center. Air movement was also observed at another retort site. This site is located approximately 500 ft (152.4 m) north of Site 4.

Analytical results showed air sparge technology effectively removed dissolved BTEX and non-dissolved contaminant sources. Figure 2 shows BTEX concentrations from the October 24, 2000, groundwater samples. These samples represent 67 days of contaminant rebound. Concentrations for wells IW-12 through IW-18 are the same as shown for pre-demonstration concentrations in Figure 1. These wells were drilled during the last rebound period and were only sampled prior to the resumption of air sparge operations. For wells R-4 and IW-1 through IW-11, BTEX concentrations from the October 24 samples ranged from "not detected" (ND) to 222 μg/L and averaged 67 μg/L. Pre-demonstration estimates ranged from 31 to 3,960 μg/L and averaged 1,063 μg/L for these same wells. Based on average concentrations and the October 24 samples, BTEX reductions greater than 90 percent were achieved.

Both the original and expansion wells have shown substantial BTEX reductions compared to pre-demonstration concentrations. Figure 3 shows the average BTEX concentrations from three of the original demonstration wells (IW-1, IW-4, and R-4). These wells recorded the highest pre-demonstration concentrations of the original demonstration area. Based on the last groundwater sampling event (October 24, 2000), BTEX concentrations have been reduced from an average of 2,416 to 97 μg/L (96 percent reduction). Figure 4 shows the average BTEX concentration from wells IW-7 and IW-8. These wells recorded the highest BTEX concentrations of the 1999 expansion wells. Substantial BTEX

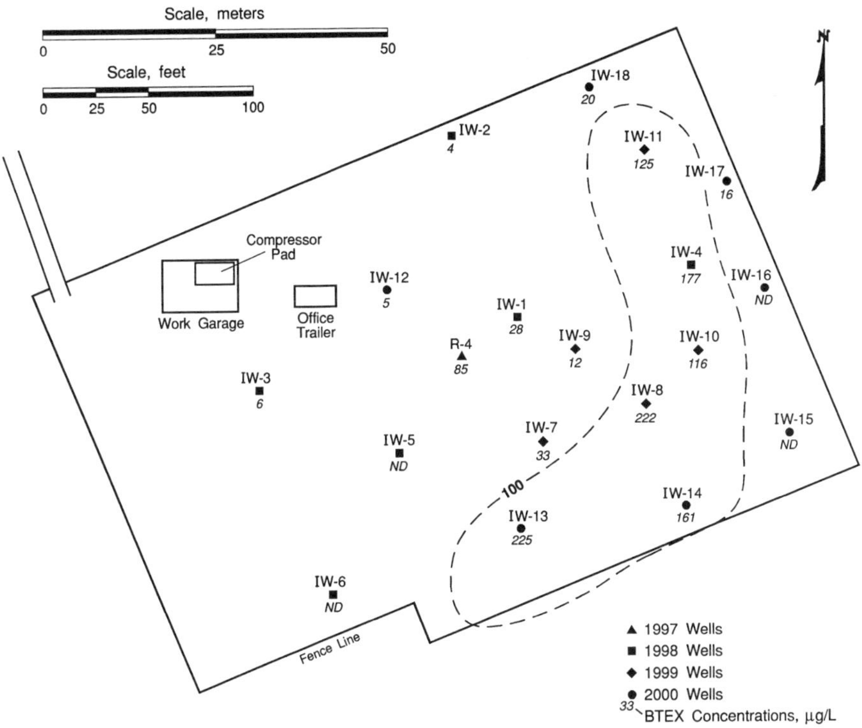

FIGURE 2. Rock Springs Site 4 BTEX concentrations from October 24, 2000, samples.

reductions were observed. Based on the October 24 samples, the BTEX average for these two wells was reduced from 1957 to 127 µg/L (93 percent reduction).

EG&G conducted a series of rebound monitoring periods to evaluate impacts on the non-dissolved source material. Contaminant rebound rate and extent should be directly related to the amount of source material present in a subsurface system. By comparing sequential rebound periods, the relative amount of contaminant source material can be inferred.

Figure 5 shows the sequential BTEX rebound for average concentrations of wells IW-1, IW-4, and R-4. These wells are part of the original demonstration area. The first rebound period was conducted during the field testing phase of the demonstration. The other rebound periods were conducted after demonstration operations had been initiated. Except for the fifth rebound period, contaminant rebound rates and extent decreased with successive rebound periods. Although the fifth rebound period showed a higher rate and extent than the fourth period, the BTEX average concentration appeared to be stabilizing.

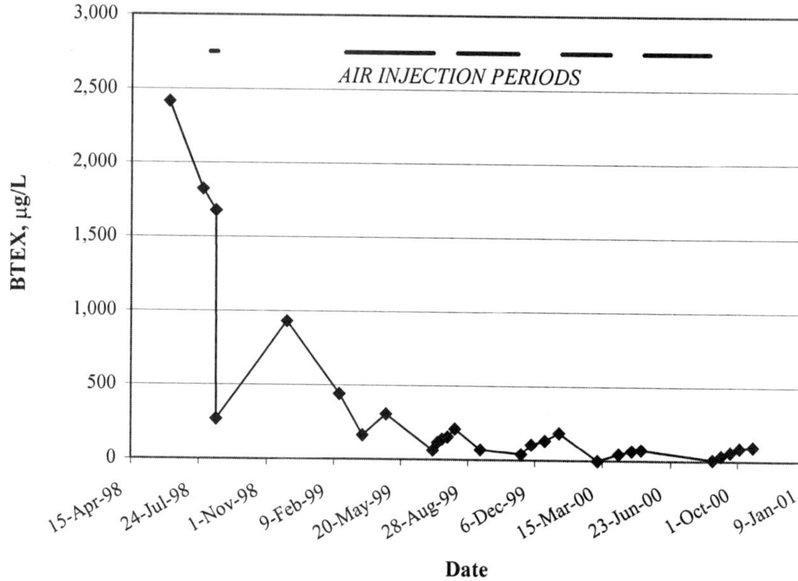

FIGURE 3. Average BTEX concentrations from wells IW-1, IW-4, and R-4.

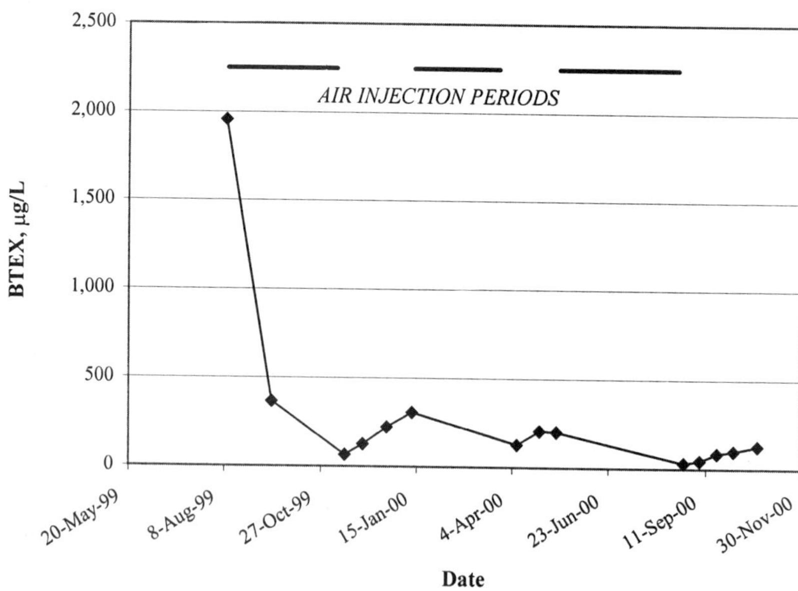

FIGURE 4. Average BTEX concentrations from wells IW-7 and IW-8.

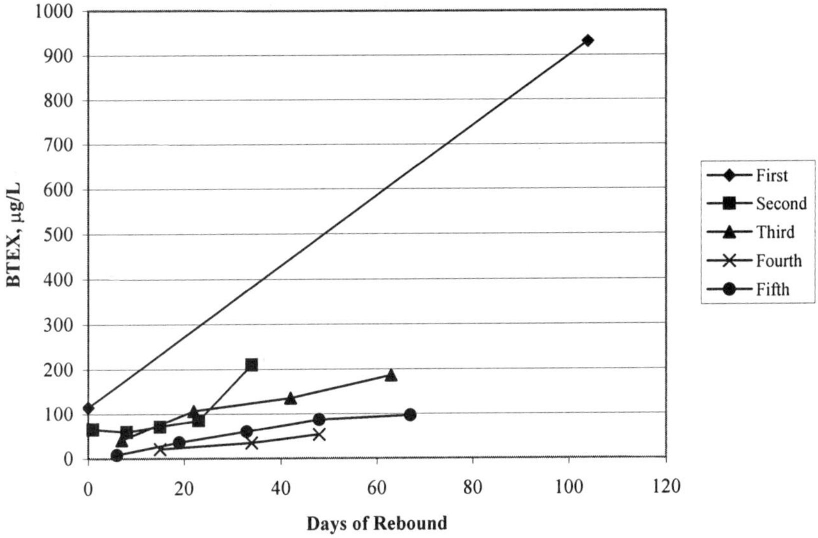

FIGURE 5. Average BTEX rebound concentrations from wells IW-1, IW-4, and R-4.

Figure 6 shows the average BTEX rebound concentrations for the 1999 expansion wells. Successive rebound periods show a decrease in rebound extent. Comparison of the rebound periods indicates significant and continuous degradation of contaminant source.

Significant increases in heterotrophic bacteria were observed during demonstration operations. Pre-demonstration concentrations averaged about 1,500 colony-forming units per milliliter (CFU/mL) for the wells in the original demonstration area. The bacteria increased to averages greater than 1,000,000 CFU/mL after air injection periods. These concentrations were maintained after significant time intervals. For example, heterotrophic bacteria concentrations averaged 1,400,000 CFU/mL 48 days after air injection had been suspended.

CONCLUSIONS

The aggressive air sparge and bioremediation approach used at Site 4 effectively reduced the groundwater contamination. Based on vent well production and site wide reductions in contamination, subsurface airflow patterns were sufficient to impact the entire site. BTEX rebound monitoring inferred significant decreases in non-dissolved BTEX sources. Because of the effectiveness of the approach, complete remediation is possible under the demonstration phase.

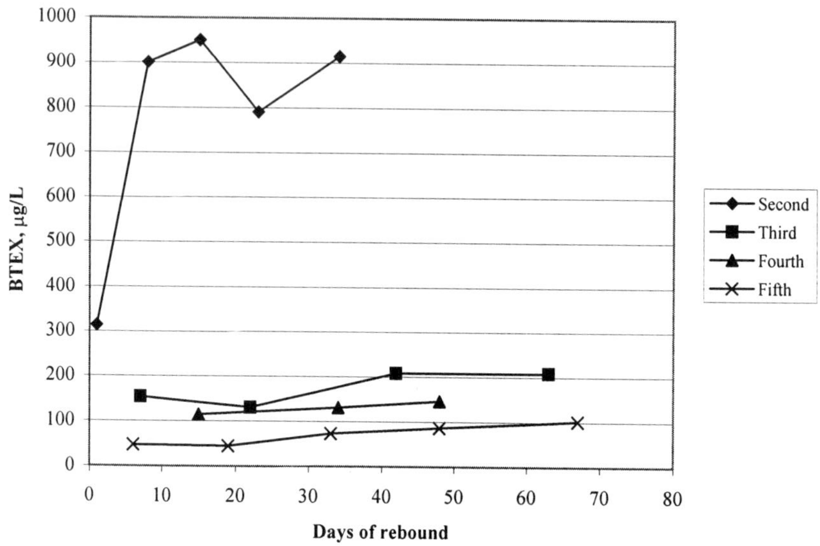

FIGURE 6. Average BTEX rebound concentrations from wells IW-7 through IW-11.

REFERENCES

Covell, J. C. and M. H. Thomas. 1997. "Air Sparge and Bioremediation Demonstration at an Underground Coal Gasification Site." *Fourth International In Situ and On-Site Bioremediation Symposium: Volume 1*, pp. 173-178. Battelle Press, Columbus, OH.

Harza Environmental Services, Inc. 1997. *Well Installation Report, Rock Springs In-Situ Oil Shale Retort Site Sweetwater County, Wyoming.* Draft report prepared for U.S. Department of Energy.

Hendrickson, D.A. 1975. *Synthetic Fuels Handbook.* Cameron Engineers Inc., p. 3.

ENHANCED *In Situ* OXYGEN SUPPLY FOR LNAPL BIOREMEDIATION

Jeonghyub Ha and ***Eric A. Seagren***
(University of Maryland, College Park, Maryland, U.S.A)

ABSTRACT: The objective of this work is to evaluate the hypothesis that a good technique for supplying oxygen to the saturated zone in the presence of light nonaqueous phase liquid (LNAPL) pool contamination at the water table is to pass air through the unsaturated zone above the pool. This hypothesis was evaluated in experimental studies performed using a bench-scale, sand-tank reactor. Steady-state abiotic experiments in the sand-tank reactor with air flowing through the reactor headspace demonstrated that oxygen supply through the water table interface into the saturated zone was enhanced when an LNAPL (dodecane) pool was present at the water table. These experimental results confirmed the hypothesis that an LNAPL pool can serve as a high concentration oxygen source to the oxygen-limited area beneath the pool and, as a result, enhance the *in situ* biodegradation rate.

INTRODUCTION

A useful and promising technology for treating subsurface contamination by nonaqueous phase liquids (NAPLs) is bioremediation. For example, the hydrocarbons in petroleum product light NAPLs (LNAPLs) are generally biodegradable, making them good candidates for bioremediation (Rittmann et al., 1994). As a result, engineered *in situ* bioremediation has been used successfully for their cleanup since the early 1970s (Lee et al., 1988). In applying the technique of engineered *in situ* bioremediation, limiting nutrients and other amendments are added to the subsurface via various means to enhance the activity of the native contaminant-degrading microorganisms. In the case of petroleum hydrocarbons, biodegradation is generally most rapid under aerobic conditions. Therefore, the main amendment that is often added is oxygen, and much of the work on engineered *in situ* bioremediation of petroleum hydrocarbons has focused on how to supply sufficient oxygen to the native microorganisms. Using water as the oxygen carrier has been challenging, the difficulty being the limited amount of oxygen (8 to 10 mg/L) that can be delivered to water from air. Alternatives such as using liquid oxygen, hydrogen peroxide, or oxygen release compounds to supply oxygen to the water have been used, but each of these techniques has its shortcomings (Chapelle, 1999). All of these techniques for adding oxygen to the aqueous phase are also plagued by the limited mixing in groundwater and the difficulty of delivering the oxygen to the desired location in the inherently heterogeneous and complex subsurface environment. Air can also be used as the carrier to supply oxygen to the saturated zone of the subsurface via the technique of air sparging--the direct injection of air under pressure into the saturated zone. However, with air sparging it can be difficult to predict where the injected air will travel, especially in heterogeneous environments and there can be resulting problems with controlling contaminant migration (Brown, 1994).

There is, however, one property of the hydrocarbon NAPL contaminants themselves that may provide an alternative approach to supplying oxygen to the contaminated saturated zone: oxygen solubilities in hydrocarbons have been shown to be much higher than in pure water. For example, the solubility of oxygen in *n*-hexadecane was measured to be 8 times higher than that in pure water (Ju and Ho, 1989). In fact, this property of hydrocarbons has been used to enhance the oxygen supply in laboratory bioreactors, by actually adding hydrocarbons to the culture medium (e.g., Ho et al., 1990; Mimura et al., 1969). Thus, in principle, if oxygen can be supplied to the source zone of LNAPL contamination in the subsurface, more oxygen can dissolve into the LNAPL than can into the water itself. As a result, the LNAPL can become a high-concentration source of oxygen right at the interface of the contaminant source and groundwater. In this way both of the key substrates (i.e., the electron-donor and the electron acceptor) for the hydrocarbon-degrading bacteria are provided in the same location. The overall goal of this research was to perform a "proof of concept" test of the hypothesis that a good method for supplying oxygen to the water-saturated zone in the presence of LNAPL contamination at the groundwater table is to pass air through the unsaturated zone above an LNAPL pool.

EXPERIMENTAL MATERIALS AND METHODS

Sand-Tank Reactor. Oxygen-supply experiments were conducted in a 50-cm long, 10-cm wide, 20-cm deep, sand-tank reactor. A schematic of the basic design of the flow-cell reactor assembly is presented in Figure 1. The tank dimensions were selected such that it is a "two-dimensional" tank, i.e., any material entering the tank at the water table will dissolve and form a the plume that varies only in the longitudinal and vertical directions.

The sand tank walls and bottom were made from untempered glass, held together using silicone caulk, and supported by wood frame. The lid of the tank was made from Plexiglas, with air inlet and outlet connections to allow air flow through the sand tank headspace. The tank was filled with sand (Mystic White #, pool filter sand, U.S. Silica, d_{50} = 0.75 mm), using a method similar to that described by Voudrias and Yeh (1994) to reduce the possibility of sand stratification. Flow in and out of the sand passed through clear wells at each end of the sand tank and was distributed across the sand through a permeable screen made from 80-mesh stainless-steel wire cloth (mesh opening = 177μm), held in a stainless steel frame. Influent flow was maintained with a peristaltic pump (Cole-Parmer Instrument Co. 7553-30) and the water level in the sand tank was controlled by a constant-head tank made of plexiglas in the effluent line. The LNAPL pool was emplaced in the sand tank using a technique similar to that described by Voudrias and Yeh (1994). The general precedure was as follows: (1) a 5-cm deep, 10-cm wide, 40-cm long section of sand above the table was excavated; (2) 250 mL of the LNAPL was applied at the leveled air/water/sand interface; (3) small amounts of dry sand were carefully applied to the LNAPL

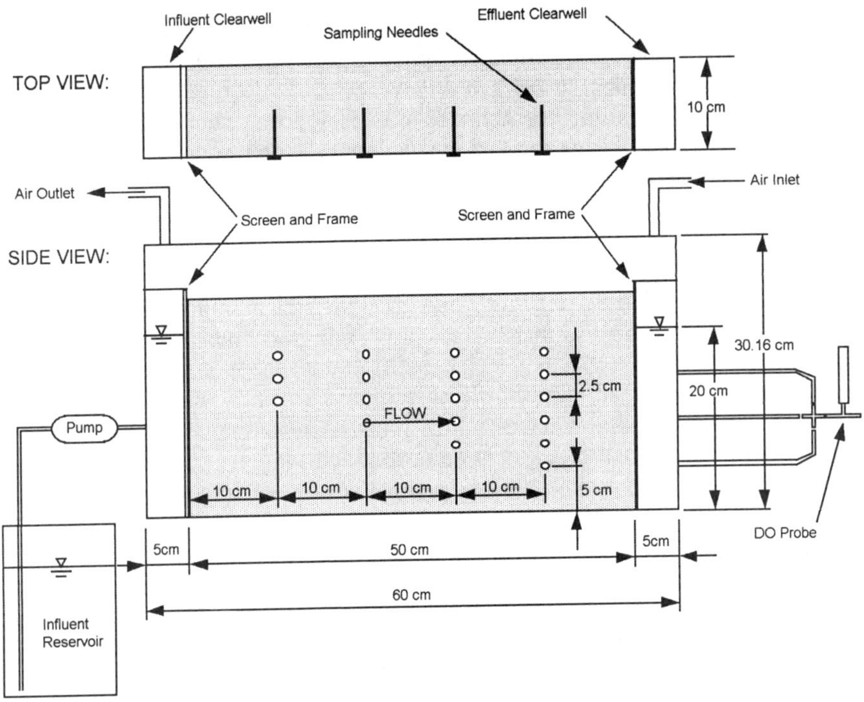

FIGURE 1. Schematic of the sand-tank reactor used in the oxygen-supply experiments.

phase; and (4) a layer of prewetted sand was carefully added to the top of the pool.

Aqueous Medium and Model LNAPL. The aqueous medium was deionized water, with $HgCl_2$ added at 0.4 mg/L as a general microbial inhibitor. All aqueous media were de-aerated by sparging with N_2 gas for one day and testing that the dissolved oxygen (DO) was nondetectable prior to using it as the influent feed source in the oxygen supply experiments described below. To conclusively evaluate the proposed hypothesis, a hydrocarbon LNAPL was selected that would maximize the impact of the LNAPL layer on the oxygen supply. Therefore, the LNAPL was selected based on key physical and chemical properties such as high oxygen solubility (Hesse et al., 1996) and high oxygen diffusion coefficient (Kowert and Dang, 1999). Based on these criteria, dodecane (Aldrich, Cat. No. D22,110-4, 99+%) was selected as the LNAPL. To allow the NAPL phase to be visually distinguished from the aqueous and sand phases in the sand tank, the dodecane was dyed red with the hydrophobic dye Oil Red O (Fisher Scientific, Cat. No. BP 112-10) at approximately 0.1g/L.

Oxygen-Supply Experiments. These experiments were performed at room temperature (average 21 °C) in the sand-tank reactor (Figure 1) to demonstrate the effect of the presence of an LNAPL pool on oxygen transfer under flowing conditions. Specifically, the experiments were designed to isolate the effect of the presence of the LNAPL pool on oxygen supply.

The general experimental approach was as follows. Humidified air was passed across the top of the unsaturated sand layer as the only source of oxygen to the system and de-aerated water was pumped horizontally through the sand. Care was taken to ensure that the air flow rate was not so high as to cause a pressure build-up in the sand-tank reactor headspace. Therefore, the water level in the effluent clear well was monitored, and a manometer device was attached to the sand-tank headspace. The selected air flow rate, Q_{air} = 11 ml/min was monitored daily using a soap-film flow meter. This flow rate was found to result in a pressure in the headspace approximately equal to atmospheric pressure. A water flow rate, Q_{water} = 3.2 ml/min, which corresponds to an average pore water velocity, v_x (=Q/nA, where n is the porosity and A is the cross-sectional tank area), of 0.5 m/d was used to represent an unconfined sand and gravel aquifer of the type that are commonly impacted by petroleum product contamination. With this velocity, it takes approximately 1 day to pass 1 pore volume of water through the sand tank. The flow rate of the water was monitored and adjusted at least three times per day during an experiment.

Three experiments were performed as summarized in Table 1, to demonstrate the transfer of oxygen to the de-aerated water flowing through the sand in the presence and absence of the LNAPL pool. During these sand-tank experiments influent and effluent samples were analyzed for DO. During the transient period, these measurements were made at least every other day, up to twice per day; during the quasi-steady state period, effluent DO was measured two or three times per day.

TABLE 1. Summary of experimental conditions for oxygen-supply studies Performed in the sand-tank reactor.

Experiment Number*	Feed Solution	Air Flow	LNAPL
1	Deaerated deionized water with $HgCl_2$	no	no
2	Deaerated deionized water with $HgCl_2$	yes	no
3	Deaerated deionized water with $HgCl_2$	yes	yes

*The experiments are numbered in the same order as they were performed.

Analytical Methods. Dissolved oxygen (DO) was measured using Standard Method 4500-06, Membrane Electrode Method (APHA et al., 1995). Specifically, DO was measured using a YSI Model 5730 probe and Model 57 DO meter. The probe and meter were calibrated using the air calibration method. Calibration was performed before the DO probe was placed in the small flow-cell reactor installed in the effluent line. Subsequently, the DO probe was removed and recalibrated every week. In addition, the membrane of the DO probe was changed every week.

RESULTS AND DISCUSSION

Abiotic Experimental Oxygen Dissolution Data. The oxygen-supply experiments were performed sequentially in the sand-tank reactor with no difference in the experimental setup, except for the addition of the LNAPL pool in the third experiment. The air flow rate, Q_{air}, the bulk aqueous flow rate, Q_x, and the oxygen concentration in the sand tank effluent were measured in each experiment. The effluent oxygen concentration was periodically measured until a quasi-steady-state DO concentration, C_{avg}, was reached. The effluent DO data for the abiotic experiments are presented in Figure 2. The experimentally determined Q_{air}, Q_x, and C_{avg} values for each experiment are presented in Table 2. The data used to calculate C_{avg} were those points collected after 72hr (~3 days or ~3 PV) in the experiments without a NAPL pool and after 96 hr (~4 day or ~4 PV) with the NAPL pool. Also summarized in Table 2 are the calculated values for the average pore-water velocity, v_x, and the average oxygen dissolution flux, J_{avg} (averaged

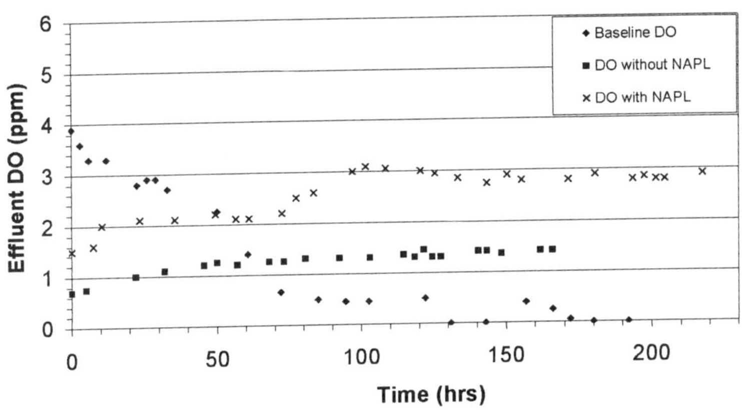

FIGURE 2. Effluent DO data for baseline experiment, abiotic experiment with and without NAPL

TABLE 2. Summary of measured and calculated values determined in abiotic O_2 supply experiments.

Expt.	Temp °C (Avg ± Std dev)	Q_{air} (Avg) (ml/min)	Q_x (Avg) (min/min)	v_x (m/d)	C_{avg} (Avg ± Std dev) (mg/l)	J_{avg} ($gO_2/m^2/d$)
1	21.1 ± 0.6	0	3.2	0.5	0.46 ± 0.03[1]	0.042
2	21.0 ± 0.8	11	3.2	0.5	1.4 ± 0.05[2]	0.13
3	21.4 ± 0.5	11	3.2	0.5	2.8 ± 0.06[3]	0.33

[1] n = 12 data points used to calculated C_{avg}
[2] n = 13 data points used to calculated C_{avg}
[3] n = 12 data points used to calculated C_{avg}

over the projected area of the unsaturated zone/saturated zone interface). v_x and J_{avg} were calculated from the experimental data as follows:

$$v_x = \frac{Q_x}{nA} \tag{1}$$

$$J_{avg} = \frac{Q_x C_{avg}}{L_x L_y} \tag{2}$$

where L_x and L_y are the length and width, respectively, of the projected area of the unsaturated zone and saturated zone interface. Eq.(2) can be used to calculate the steady-state experimental flux of oxygen into the saturated zone because the only major source of oxygen to the sand tank is through the water table interface, as shown in the base line Experiment 1 (Fig.2). Therefore, at steady state, the mass per unit time rate O_2 enters through the water table must equal the mass per unit time rate O_2 leaves in effluent of the reactor. Taking the mass per unit time (=Q_x x C_{avg}) of O_2 out the end of the reactor and dividing by the projected interfacial area at the water table then gives the steady state flux of oxygen through the water table ($M/L^2/T$).

These data clearly prove the research hypothesis: the presence of the LNAPL pool enhances the supply of oxygen to the saturated zone. This is demonstrated by the average effluent oxygen concentration with the LNAPL pool (Experiment 3), which is approximately 2 times the value without the LNAPL pool (Experiment 2). Also, the average oxygen dissolution flux from the water table interface with the LNAPL is approximately 2 times the value without the NAPL. These observations of an enhanced oxygen supply in the presence of the LNAPL pool can be explained by the fact that the oxygen solubility in dodecane in contact with air is greater than that for water in contact with air. Therefore, more oxygen partitions from air into the overlying dodecane layer than can partition from air into water directly. This results in the dodecane being a high concentration source of oxygen, which can then partition into the water. Because the mixing of this experimental set-up is not turbulent, other potential mechanisms of oxygen transfer enhancement (e.g., bubble size, boundary layer mixing and increased permeability of oxygen in the boundary layer) are not a factor (McMillan and Wang, 1987; Mimura et al., 1969).

CONCLUSIONS

These results confirm the hypothesis that a petroleum-hydrocarbon LNAPL can act as a high concentration oxygen source. This is an important observation with implications concerning engineered bioremediation approaches. Specifically, this indicates that a good technique for supplying oxygen to enhance biodegradation of solute in the saturated zone below a LNAPL pool may be to pass air through the unsaturated zone above the pool.

REFERENCES

APHA, AWWA, and WEF. 1995. *Standard Methods for the Examination of Water and Wastewater*. 19th edition. American Public Health Association, American Water Works Association, and Water Environment Federation, Washington, D.C.

Brown, R. 1994. "Treatment of Petroleum Hydrocarbons in Ground Water by Air Sparging." In R.D. Norris et al. (Eds.), *Handbook of Bioremediation*, pp.61-85. Lewis Publishers, Boca Raton, FL.

Chapelle, F. H. 1999. "Bioremediation of Petroleum Hydrocarbon-Contaminated Ground Water: The Perspectives of History and Hydrology." *Ground Water*. *37*(1): 122-132.

Hesse, P.J., R. Battino, P. Scharlin, and E. Wilhelm. 1996. "Solubility of Gases in Liquids. 20. Solubility of He, Ne, Ar, Kr, N_2, O_2, CH_4, CF_4, and SF_6 in n-Alkanes n-C_iH_{2I+2} ($6 \leq i \leq 16$) at 298.15 K." *J. Chem. Eng. Data*. *41*(2): 195-201.

Ho, C. S., L-K Ju, and R.F. Baddour. 1990. "Enhancing Penicillin Fermentations by Increased Oxygen Solubility through the Addition of n-Hexadecane." *Biotechnology and Bioengineering*. *36*: 1110-1118.

Ju, L-K., and C.S. Ho. 1989. "Oxygen Diffusion Coefficient and Solubility in n-Hexadecane." *Biotechnology and Bioengineering*. *34*: 1221-1224.

Kowert, B. A., and N.C. Dang. 1999. "Diffusion of Dioxygen in n-Alkanes." *The Journal of Physical Chemistry A*. *103*(7): 779-781.

Lee, M.D., J.M. Thomas, R.C. Borden, P.B. Bedient, C.H. Ward, and J.T. Wilson. 1988. "Biorestoration of Aquifers Contaminated with Organic Compounds." *CRC Crit. Rev. Environ. Control*. *18*: 29-89.

McMillan, J. D., and D.I.C. Wang. 1987. "Enhanced Oxygen Transfer Using Oil in Water Dispersions." *Annals New York Academy of Sciences*. 506: 569582.

Mimura, A., T. Kawano, and R. Kodaira. 1969. "Biochemical Engineering Analysis of Hydrocarbon Fermentation." *J. Ferment. Technol.* *47*(3): 229-236.

Rittmann, B.E., E. Seagren, B.A. Wrenn, A.J. Valocchi, C. Ray, and L. Raskin. 1994. *In Situ Bioremediation*, 2nd ed. Noyes Publications, Park Ridge, NJ.

Voudrias, E.A., and M.-F. Yeh. 1994. "Dissolution of a Toluene Pool Under Constant and Variable Hydraulic Gradients with Implications for Aquifer Remediation." *Ground Water*. *32*: 305-311.

PHOSter™ TECHNOLOGY TO REMEDIATE PETROLEUM CONTAMINATION USING PAY-FOR-PERFORMANCE CONTRACTING: BACKGROUND AND CASE STUDIES

James L. Brickell, Ph.D., P.E. (Earth Tech, Charleston, South Carolina)
Nick Ugolini, EIC, NAVFAC, (Southern Division, Charleston, South Carolina)
Lamar E. Priester, III, (Priester & Associates, Columbia, South Carolina)

ABSTRACT: The patented PHOSter™ Enhanced Bioremediation Technology has been applied to petroleum contaminated sites. With this innovative system, gaseous oxygen, nitrogen, and phosphorus are injected through wells, which stimulates naturally occurring microorganisms to rapidly remediate soil and groundwater. Two case studies are presented (1) *Post-remediation monitoring*: This project is conducted for the State of South Carolina and involves petroleum-contaminated soil and groundwater at the former Rocket Service Station, Abbeville, SC. After fifteen months of treatment, concentrations of all contaminants in all wells were reported at or below Site Specific Target Levels. (2) *Active remediation:* This project is conducted for the Naval Facilities Engineering Command, Southern Division, North Charleston, SC, and is a petroleum contaminated site at an operating gas station located on the Naval Air Station Jacksonville, FL. A retrofit was provided to the existing dewatering/soil vapor extraction system, which had proved to be ineffectual. An innovative "Pay-for-Performance" contract approach was used for both projects. Payment is made upon reaching target levels of treatment as specified in contract documents. Combining the two innovative aspects of these projects, technology and contracting, allows for site remediation at lower cost, in less time, and with lower risk than conventional remediation technologies.

INTRODUCTION

This paper addresses two aspects of a remediation project where a different approach has been taken to reduce cleanup times which subsequently reduces costs, and allows for a fixed cost estimate for budgeting purposes. Petroleum contaminated sites at the former Rocket Service Station and at the Naval Exchange gasoline station at Naval Air Station Jacksonville are used as examples to illustrate these aspects.

The first aspect is the implementation of an innovative technology, the PHOSter™ enhanced bioremediation process, which was developed by Department of Energy contractors at the Savannah River Site. This patented technology has proved to be faster and less costly than traditional remediation technologies.

The second aspect is a Pay-for-Performance method of contracting. Under this type of contract, a contractor will design, build, operate and maintain a remediation system for a fixed cost, and only get paid upon reaching contractually established treatment milestones.

PHOSter^{TM} TECHNOLOGY

Background. Bacteria that degrade organic material are naturally present in the subsurface environment. These heterotrophs metabolize organic compounds, such as benzene, as an energy source and more efficiently break down organic compounds in an aerobic environment than in an anaerobic environment. Therefore, oxygen is required to improve efficiency. In addition, other macronutrients such as nitrogen and phosphorous are required for rapid growth of microorganisms.

Technology Development. The PHOSter^{TM} technology was developed by a team from the Westinghouse Savannah River Technology Center, Oak Ridge National Laboratory, and the Evoca Corporation at the Department of Energy's Savannah River Site. Originally developed for use on groundwater contaminated with trichloroethylene, the PHOSter^{TM} system provides controlled addition of phosphorous, a critical nutrient, into sites contaminated with organic compounds which stimulates the growth of natural microorganisms.

The PHOSter^{TM} Process. Injecting air into the subsurface supplies the macronutrients oxygen and nitrogen to microorganisms. However, the lack of the macronutrient, phosphorous, can limit the rate of decontamination. The use of the PHOSter^{TM} process allows for the controlled application of all three macronutrients. It works with any system (e.g., bioventing, biosparging) that delivers nutrients in a gas phase to a microbial population in a contaminated site. Liquid triethylphosphate volatilizes in contact with the air stream and produces a gaseous form of phosphate. The phosphate in this gaseous form is highly available to the target microorganisms. The presence of organic groups attached to the phosphate controls dissolution of the nutrient into the groundwater which produces a time-release action. This delayed reaction reduces the risk of overstimulation (Looney et al., 1996).

CASE STUDY: ROCKET SERVICE STATION

Site Description. The former Rocket Service Station, located in Abbeville, SC is no longer in service. Contamination assessments were conducted by various contractors in 1996, 1997 and 1998 for the South Carolina Department of Health and Environmental Control (SCDHEC). These investigations revealed that three underground storage tanks (USTs) had been removed from the site. Petroleum from these USTs had impacted the groundwater at the site. In a competitively bid process, a contract was awarded in September 1998 to reduce contaminant concentrations to below risk-based Site Specific Target Levels (SSTLs) as defined in the bid documents for a total cost of $136,160.

PHOSter^{TM} Remediation System. The groundwater remediation system consists of a PHOSter^{TM} nutrient delivery system to deliver air, nitrous oxide, and vapor phase triethylphosphate (TEP) to the subsurface environment through injection wells. The average rate of vapor injection is 1.0 cubic foot per minute per injection well, at an average pressure of 8.0 pounds per square inch.

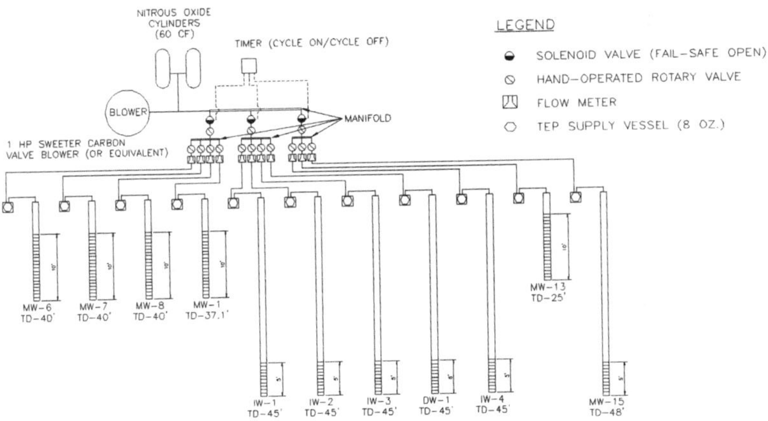

FIGURE 1. Remediation System, Former Rocket Service Station (Earth Tech, 1998).

To ensure effective areal coverage of the contaminated zone, four injection wells were constructed and seven pre-existing monitoring wells were converted to injection wells. All-above-ground remediation system components were housed in a pre-manufactured, skid-mounted shed located on-site. Equipment housed includes a blower, nitrous oxide tanks and a manifold system. The remediation system was started in January, 1999. The area treated was divided into three zones, and vapor injection was rotated to each zone eight hours per day. The system was monitored through pressure and flow indicators to ensure sufficient volume of the proper nutrient mix is provided to each injection point.

The chemicals of concern (COCs) for this site are benzene, toluene, ethylbenzene, xylenes, methyl tert-butyl ether (MTBE), naphthalene, and ethylene dibromide (EDB). Figure 2 presents the COC concentrations vs. time for all on site monitoring wells that have established SSTLs. Concentration levels decreased markedly in the first quarter of treatment as shown in the change in concentrations between the December 1998 and April 1999 sampling events. By March 1, 2000 (15 months of active remediation), concentrations in all monitoring wells were below SSTLs.

The site then entered the post-remediation monitoring phase. A slight rebound in benzene and MTBE concentrations above SSTLs in monitor wells MW-10 and MW-15 was seen in the first post-remediation groundwater sampling event requiring a portion of the remediation system to be restarted. After six months of additional treatment, all concentrations were below SSTLs and the treatment system was shut down in May, 2001 which restarts the post-remediation monitoring phase. Since the contract is pay-for-performance at a fixed cost, the additional treatment was provided at no additional cost to the client.

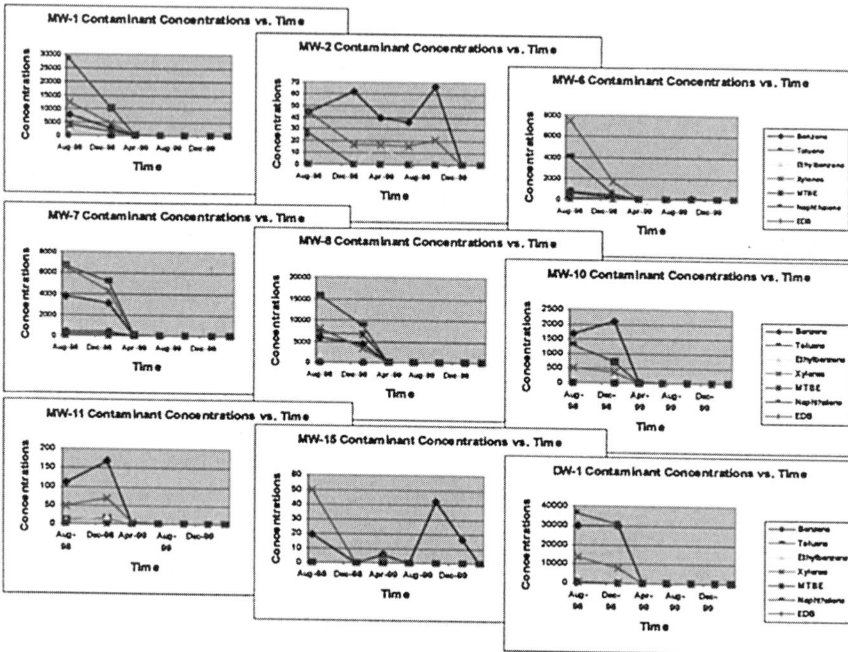

FIGURE 2. Concentration levels vs. time, former Rocket Service Station (concentrations in μg/L)(Earth Tech, July, 2000).

CASE STUDY: NAS JACKSONVILLE SERVICE STATION

Site Description. NAS Jacksonville is located approximately 8 miles south of Jacksonville, Florida. The Naval Exchange (NEX) Service Station site was constructed in 1975 and seven 10,000 gallon USTs were installed. In 1982 a leak in a fuel delivery line to the pump island released an estimated 2000 gallons of gasoline. In 1992, a sump gasket for one tank was found to be defective and repaired after an estimated 500 gallons of product were released and not recovered. In 1993, leaking piping were repaired after an estimated less than 40 gallons of product were released

In 1996 a Remedial Action Plan (RAP) was approved by the Florida Department of Environmental Protection (FDEP) which included pump and treat, dewatering and soil vapor extraction. This system was started in November 1998, however, equipment problems resulted in partial and intermittent system operation, which resulted in total shutdown in June 2000.

Remediation System Modification for PHOSter™ Application. In July, 2000, an amendment to the RAP was submitted to the FDEP requesting to discontinue the use of the existing remediation system, installation of two nested biosparging wells and the following conversions of the pre-existing remediation system: ten well points to injection wells; two monitoring wells to injection wells; and a groundwater recovery trench to an injection trench. A blower injects air, nitrous oxide and volatilized triethylphosphate into the subsurface. The capability to provide an additional carbon source (propane) to aid in building up the subsurface microbial population was included to be used if required. An enclosed trailer housing all the equipment and piping was assembled off-site such that electrical and piping connection could be rapidly made once the trailer was staged on-site (Figure 3). Construction was completed and the modified system was started on September 29, 2000.

FIGURE 3. Interior of Treatment Trailer (Earth Tech, October, 2000).

PHOSter™ System Performance. Concentration levels required for this project (target levels) are the maximum permissible levels set forth in the Florida Administrative Code. Benzene and MTBE are the only COCs that exceed target levels of 1 μg/L (benzene) and 35 μg/L (MTBE).

Figure 4 shows the concentration levels of benzene for monitoring well JAX 880-31, located in the source area of contamination.

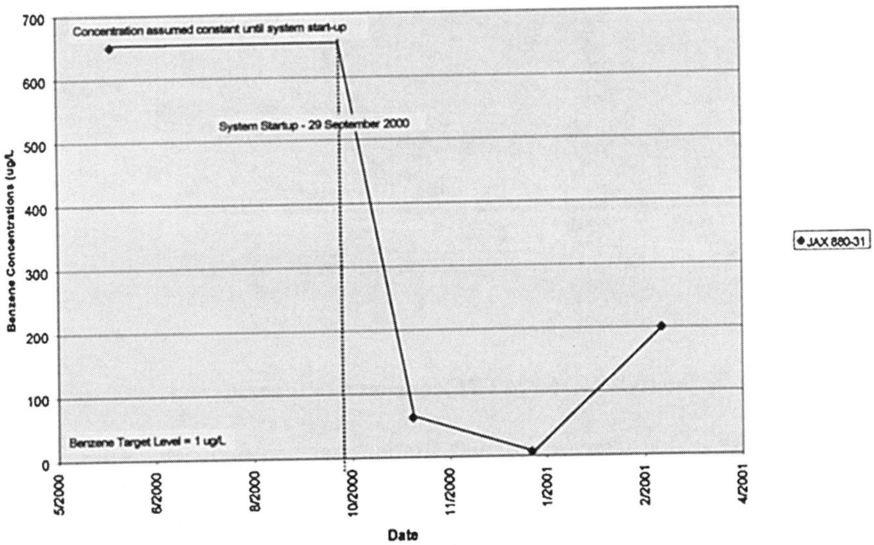

FIGURE 4. Benzene concentrations versus time, source area monitoring well (Brickell et al., 2001).

In June 2000, prior to start-up of the modified system, the benzene concentration was 650 μg/L. The benzene concentration was reduced an order of magnitude to 62 μg/L after 6 weeks, and another order of magnitude to 6.2 μg/L by 17 weeks (January 2001) of treatment. However in March 2001, the concentration level increased to 200 μg/L. It is suspected that this most recent increase in concentration may be due to the draught conditions recently experienced at the site. This situation will be further evaluated upon reception of subsequent sampling and analyses data.

The remaining wells in the sampled network are peripheral to or in the outermost fringe of the contaminated area. Benzene concentrations in four of the seven peripheral wells were reported as increasing above the target level of 1 μg/L, to as high as 12 μg/L. It is suspected that mounding of the groundwater table in the vicinity of the source area has allowed contaminated groundwater to flow toward the peripheral wells.

As seen in Figure 5, the concentrations of MTBE in all monitoring wells except in peripheral monitoring wells, JAX 880-8 and JAX 880-9, either showed reductions in concentration levels or remained below the MTBE target level of 35 μg/L. The increase is suspected to be due to mounding of the groundwater table, similar to what occurred with the benzene.

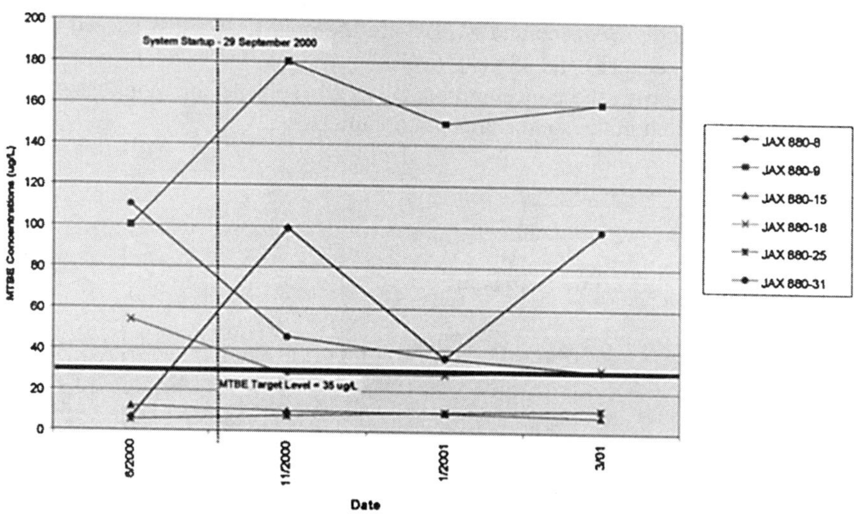

FIGURE 5. MTBE concentration versus time.

Because of the increase in concentrations seen in peripheral wells. The remediation system was modified in May 2001 to include an additional injection well. As with the previous case study, because the contract was based on pay-for-performance at a fixed cost, the additional treatment was provided at no additional cost to the client.

PAY-FOR-PERFORMANCE CONTRACTING

Contracting for remediation projects traditionally is done on a time and material basis, or a fixed cost for tasks as a portion of the entire project. Typically, when a design firm completes a remediation system design, there is an estimate of the length in time that is required for the system to operate in order to reach the remediation goals. However, if the system is not as effective as anticipated, and modifications to the system or length of time of treatment is required, it is the client who will bear the burden of funding these changes. An alternative way to contract for remediation services is a Pay-for-Performance concept.

To illustrate this contracting concept, for the NEX Service station at NAS Jacksonville, a proposal was submitted to Southern Division, Naval Facilities Engineering Command, to design, build, operate and maintain a remediation system and payment for services is based on pre-agreed upon levels of reductions of COCs which were written into the Statement of Work. Extracted from the Statement of Work, the following illustrates the conditions set forth in this contract:

A. Payment of 50% of the contract value is made upon completion of installation and start-up of the remediation system.

B. Payment of 30% of the contract value is made based on achieving interim total concentration reductions at the site as verified in specified monitoring wells for goals of 25, 50 and 75 percent reductions in Total Concentrations Above Target Levels (TCATL).
   (1) Payment of 10% of the contract value is made upon a 25 percent reduction in TCATL. This is achieved when there is a one-quarter reduction of the initial benzene and MTBE concentrations above target levels.
   (2) Payment of 10% of the contract value is made upon a 50 percent reduction in TCATL. This is achieved when there is a one-half reduction of the initial benzene and MTBE concentrations above target levels.
   (3) Payment of 10% of the contract value is made upon a 75 percent reduction in TCATL. This is achieved when there is a three-quarter reduction of the initial benzene and MTBE concentrations above target levels.

C. The final 20% of the contract value is paid when all benzene and MTBE concentrations are below target levels, and all above ground equipment installed by the contractor is removed from the site.

This method of contracting transfers risk from the customer to the contracting firm. Therefore the contractor has to balance the risk versus reward. This produces an incentive to consider innovative technologies that are low-cost, effective and predictable. When compared with traditional remediation systems such as pump and treat, air sparge and/or soil vapor extraction, the PHOSter™ system has much lower capital, installation and operation & maintenance (O&M)

costs. These lower costs when combined with confidence in the process allows for a fixed cost that is reasonable to the customer, even with contingency costs included in the price. A benefit to the customer is that the cost is fixed and known, which simplifies the customer's budgeting process.

For the NEX Service Station, the contracted fixed cost to install the retrofit and provide O&M on the system until the remediation is complete is $90,011. Because of the rapid decrease (three-quarter reduction of the initial benzene and MTBE concentrations above target levels) in concentrations seen at the site, the contractor was awarded 80% of the contract value after the first three months of treatment. The final 20% of the contract value will be paid when all benzene and MTBE concentrations are below target levels, and all above ground equipment installed by the contractor is removed from the site.

SUMMARY

The remediation of two petroleum sites was contracted for a fixed-cost, using a pay-for-performance concept. Utilizing the PHOSter™ technology allows for a high degree of confidence in the length of clean-up time and the successful remediation of the site. By combining known cleanup times based on the selected technology with innovative contracting, a guaranteed price to the customer could be offered for site cleanup to a no-further-action status.

REFERENCES

Brickell, J. L., N. Ugolini, and L. E. Priester. 2001. "Application of PHOSter™ Technology Using a 'Pay-For-Performance' Contract to Retrofit an Existing Remediation System at Naval Air Station, Jacksonville." In *A New Era for Federal Environmental Leadership, Management and Technology, 27th Environmental Symposium & Exhibition Proceedings*, pp. 629-644. National Defense Industrial Association, 2111 Wilson Boulevard, Suite 400, Arlington, VA 22201-3061.

Looney, B. B., T. Hazen, K. Lombard, T. Phelps, S. Pfiffner. 1996. *Westinghouse Savannah River Technology Center 1996 R&D 100 Entry, PHOSTer™*. Submitted for the 1996 R&D 100 Award.

Earth Tech, Inc. *Corrective Action Plan/Engineering Report, Former Rocket Service Station, Abbeville, SC. Site No. 00046.* Prepared for South Carolina Department of Health and Environmental Control, Columbia, SC, October, 1998.

Earth Tech, Inc. *Corrective Action Monitoring Report, Former Rocket Service Station, Abbeville, SC. Site No. 00046.* Prepared for South Carolina Department of Health and Environmental Control, Columbia, SC, July, 2000.

Earth Tech, Inc. *As-Built Report/Baseline Sampling, NEX Service Station-Building 880, Naval Air Station, Jacksonville, Jacksonville, FL.* Prepared for Southern Division, NAVFAC, Charleston, SC, October, 2000.

IN-SITU BIOREMEDIATION OF A GAS WELL SITE WITH PERCARBONATE-Q30

Scott Noland (NESCO, Denver, Colorado)
Bob Gec (Degussa Corporation, Burlington, Ontario)
Christoph Brossmer (Degussa Corporation, Allendale, New Jersey)

ABSTRACT: A field trial has been in progress since May 2000 to demonstrate the efficacy of a new oxygen release compound, Percarbonate-Q30, manufactured by Degussa Corporation. The test site is an active natural gas well property where historical releases of condensate resulted in high levels of soil and groundwater contamination. Contaminants of concern include benzene and total petroleum hydrocarbons (TPH). Special procedures were developed for installation of the material and approximately 63,000 pounds have been installed across the site during two treatment events. Based on results from continued groundwater monitoring and two soil-sampling events, Percarbonate-Q30 performance has exceeded expectations, and it is anticipated that site closure will be possible in the near future.

BACKGROUND

Site History. Natural gas, at the well head, contains a variety of constituents other than light alkanes, most notably water and condensate. Condensate is a light petroleum liquid consisting primarily of low molecular weight alkanes, however the liquid typically includes a significant aromatic fraction along with a variety of heavy saturated alkanes. A natural gas well was drilled at the subject site in September of 1992 in what was then a rural area. A records review in early 1997 indicated that the produced water vault was possibly leaking and repairs were performed in April of that year. In the following year, modifications were performed on all ancillary process equipment at the site, including excavation of the produced water vault and replacement of the tank battery and associated piping. At that time, hydrocarbon constituents were detected in soil adjacent to the vault where concentrations of TPH up to 6,300 mg/kg were noted.

Within the last three years, a community development has been constructed surrounding the site. In the spring of 1998, geotechnical soil borings were advanced by a homebuilder approximately 300 feet east of the tank battery. Two of these borings encountered hydrocarbon contamination in the soil and groundwater. In the summer of 1998, the well owner commissioned a site investigation to delineate the lateral and vertical extent of subsurface contamination in soil and groundwater. Based on subsequent findings, limited remedial measures were implemented later that year to prevent further migration of contamination.

Site Stratigraphy. In general, low permeability clay soils underlie the entire site from the near surface to bedrock. The bedrock elevation map prepared for the site clearly indicates the presence of a paleochannel, running west to east starting near the original location of the tank battery and water vault. To the west, bedrock is encountered at approximately 12 feet bgs. It slopes downward to the east and is encountered at roughly 30 feet bgs at the far east-end of the site.

The groundwater gradient, as determined from measurements taken at each of the monitoring wells, confirms that groundwater flow is from west to east. Shallow groundwater at the site appears to be present only in discontinuous perched zones, even though static water levels in monitor wells indicate that it should be encountered at approximately 16 feet bgs. Soil samples taken from depths of 20 to 30 feet bgs in the vicinity of wells to the east are moist, but do not show evidence of saturation. Saturated soils are not normally encountered until just above bedrock.

Distribution of Contamination. It appeared that the paleochannel was the primary factor controlling migration of contamination from the tank battery area. As a result, the impacted area measured approximately 150 feet in width, but was over 750 feet long. Principal contaminants of concern included benzene, toluene, ethylbenzene, and xylenes (BTEX) as well as TPH. Maximum soil concentrations of over 8,000 mg/kg TPH existed near the former vault location where benzene in the groundwater was measured at 20 mg/L. High concentrations persisted throughout the plume along the paleochannel to the east and levels of TPH exceeding 2,000 mg/kg were not uncommon. Figure 1 graphically depicts the areal extent of contamination. Key regulatory cleanup standards applicable to the site are benzene at 5 µg/L and TPH at 1,000 mg/kg.

TECHNICAL APPROACH

Selection of Remedial Technology. A groundwater interceptor trench was constructed to collect and remove impacted groundwater, and a solar powered air sparging system was installed in early1999. Only limited success was realized by these systems, and advancing surface development activity demanded a more aggressive approach. At this point, the site was surrounded by a public pool, a golf course, and residential construction.

In addition to being cost prohibitive, excavation of impacted soils within a residential community raised a host of health and safety concerns. As a consequence, various technologies amenable to *in-situ* cleanup were investigated. Accelerated bioremediation using slow release compounds was a strong candidate because it was a passive technology, which once installed, required minimal oversight. Cost appeared to be an insurmountable difficulty however, as the high concentrations required in excess of 170,000 pounds of the REGENESIS oxygen release compound (ORC®). Consequently, alternative products were investigated that might offer a more cost-effective solution. At a projected cost of approximately half that of conventional materials, a product that had been successfully used for *ex-situ* bioremediation of hydrocarbon contaminated soil was identified.

Oxygen Release Material – Sodium Percarbonate (SPC-30). Percarbonate-Q30 is a granular peroxygen product with a number of advantages in aerobic degradation protocols compared to the "classical" use of hydrogen peroxide or other oxygen release materials. Oxygen supplied to soil is often the single most important factor limiting bioremediation of petroleum containing soils. While microbial populations, nutrients and moisture are typically present, most sites are deficient in oxygen, which results in a site where bioremediation proceeds anaerobically. Addition of an oxygen source significantly increases the remediation rate, typically by one or two orders of magnitude.

Degussa Corporation's Percarbonate-Q30 is manufactured in a proprietary fluid bed granulation process. It slowly releases hydrogen peroxide at a rate that depends on the soil moisture content. Hydrogen peroxide is further decomposed into water and oxygen in the soil through the action of inorganic catalysts or in direct contact with the microbes through enzymatic decomposition by catalases or peroxidases. These enzymes are present in every living cell. The overall stoichiometry is shown in the following reaction:

$$Na_2CO_3 \bullet 1.5H_2O_2 \rightarrow Na_2CO_3 + 1.5\ H_2O_2 \rightarrow 1.5\ H_2O + \tfrac{3}{4}\ O_2$$

A summary of the physical and chemical properties of Percarbonate-Q30 is provided below.

Appearance:	White, odorless free flowing solid
Chemical Formula:	$Na_2CO_3 \bullet 1.5\ H_2O_2$
Molecular Weight:	157.0 gm/mole
Active Oxygen:	13.0% minimum
pH :	10.6 (1% aqueous solution)
Solubility:	(In water at 20°C) - 140gms/L
Bulk Density:	0.9 – 1.2 kg/L
Particle Size Distribution:	> 1.4mm 5% , < 0.15m 2%

FIELDWORK

Pilot Installation. Conventional procedure for subsurface installation of oxygen release material involves suspension of the material in water, as a 10 to 50-weight percent slurry, followed by injection using medium to high-pressure pumps. One reason that ORC® has been successful is that the rate of oxygen release is very slow, occurring over a period of several months. In addition, the material is virtually insoluble in water, so it is easily handled and installed.

As indicated above, Q30 is very soluble in water, so conventional procedures would not be useful for installation of this material. As a result, pilot work was performed at the site to develop special techniques for subsurface injection of this material and to determine optimal equipment specifications. A procedure was developed that allowed successful placement of the material using high pressure pumps to create and propagate horizontal fractures, or "fracs," in the formation.

Phase 1 Installation. The procedure developed during the pilot work was scaled up, and an injection plan was prepared for full-scale installation of Q30 at the site. Figure 1 shows the Phase 1 borehole layout across the site. Approximately 30,000

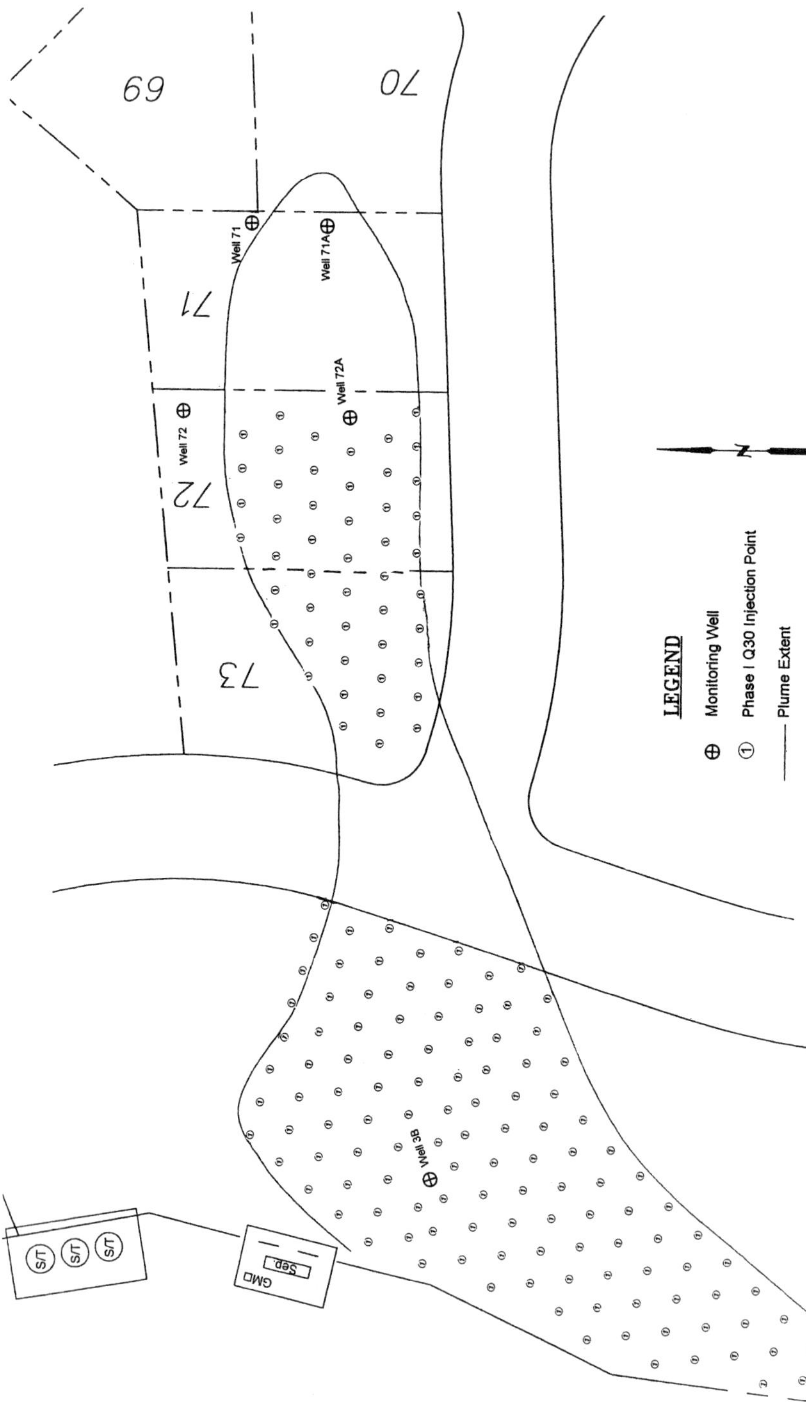

FIGURE 1. Phase I injection points.
(Rough Scale: 1 inch equals 60 feet/Injections: 15-foot centers)

pounds of material were injected throughout the affected formation over a three-week period in May 2000. The amount of material installed was insufficient to address contaminant loading in the soil completely, and a second installation was anticipated.

Since oxygen release is a strong function of soil moisture, and site soils within the affected formation were typically low in moisture content, an aqueous solution of cultured bacteria and nutrients was subsequently injected to assist this release and jumpstart degradation of contaminants. The solution was interspersed between prior injection points to avoid co-location of Q30 and bacteria.

Soil Sampling. To evaluate progress, a sampling plan was devised to find residual "hot spots" above a specified volume at a 95 percent level of confidence. Continuous soil cores were taken at each borehole from the surface to depths of 20 to 24 feet bgs, depending on site location. Cores were logged and field PID measurements were performed over the entire core to identify the depth of apparent maximum contamination and to locate depths where contamination began and ended. Samples were taken from each core at 2-foot intervals and from the area of maximum contamination, temporarily stored in coolers, and then submitted to the laboratory for total volatile petroleum hydrocarbon (TVPH) analysis each day.

Figure 2 shows borehole locations, and the denoted values (ppb) indicate the maximum residual TVPH level. The sampling was performed in late July 2000, approximately 60 days after the Phase 1 work.

Phase 2 Installation. After the post-Phase 1 soil sampling was completed, a very detailed picture of contamination as a function of position and depth emerged. As a result, a retreatment plan was prepared targeting specific depths for placement of additional material and specifying loading for Q30 at each location and depth. During Phase 2 work, an additional 33,000 pounds of Q30 was installed, and as before, bacteria and nutrient injections followed approximately two weeks later. This work was performed in August and September of 2000.

RESULTS AND DISCUSSION

Initial Installation. As discussed above, significant amounts of contamination persisted after completion of Phase 1 work, and it was difficult to determine precisely what portion of the initial contaminant had been mineralized, since historical data did not lend itself to this calculation. Cross-sections of the site were prepared based on laboratory and field PID data to understand better what effects had occurred. (See Figure 3.) The upper section of Figure 3 describes the depths where contamination began and ended prior to treatment, moving across the site from west to east. In general, the thickness of the impacted zone averaged about 14 feet and ranged from 9 to 20 feet.

The lower section of Figure 3 depicts a comparable plot of contamination after completion of the Phase 1 work. Significant differences are evident that strongly suggest a reduction in contamination occurred as a result of the initial installation of Q30. The plume now appears to be discontinuous, with clean areas

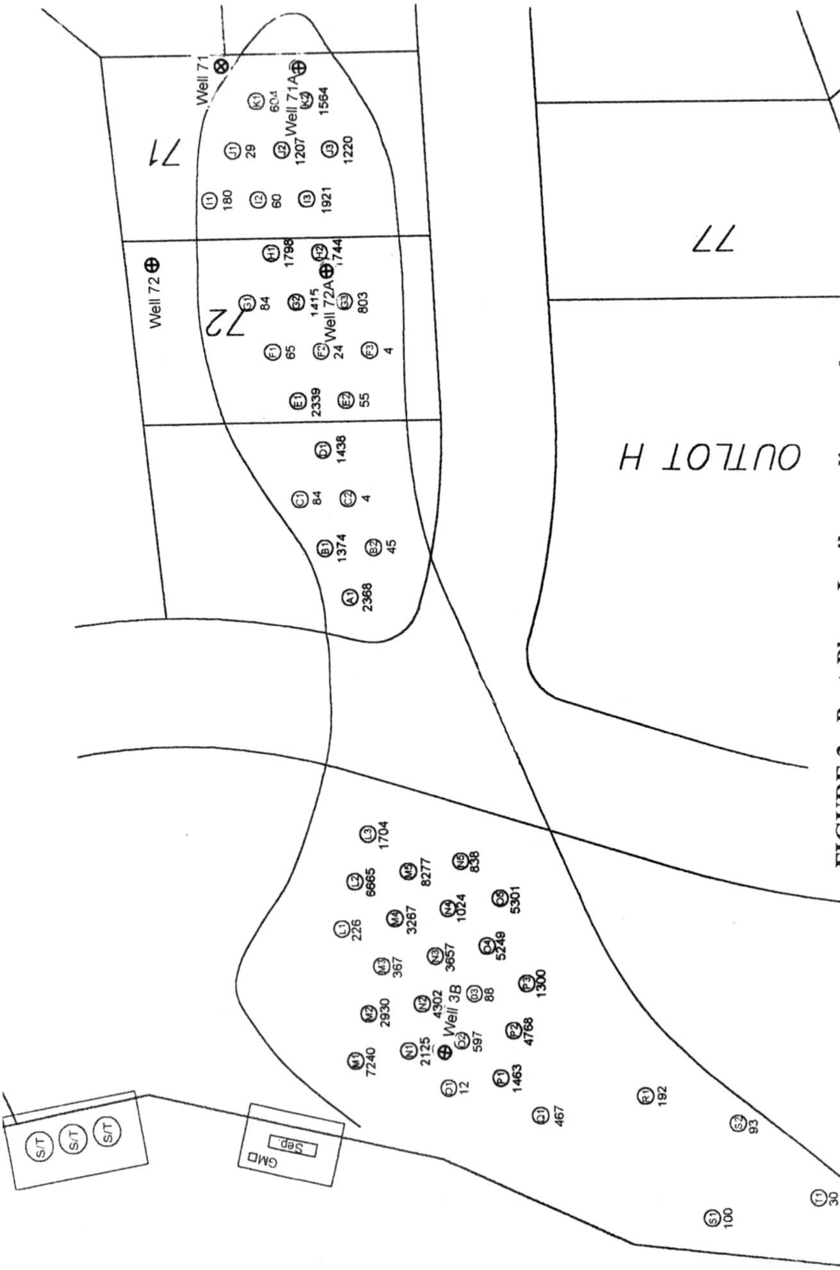

FIGURE 2. Post-Phase I soil sampling results.
(Rough Scale: 1 inch equals 60 feet/Sample results are in ppb.)

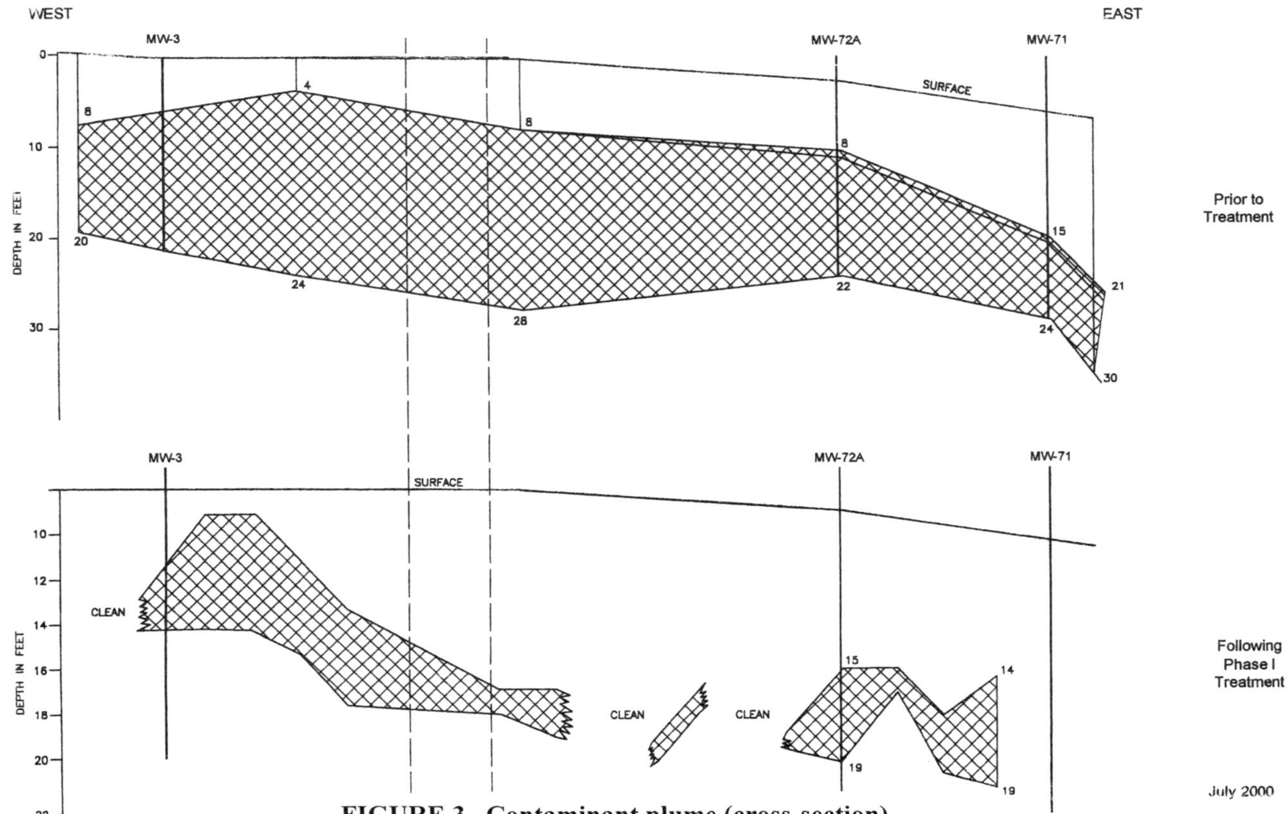

FIGURE 3. Contaminant plume (cross-section). (Rough Horizontal Scale: 1 inch equals 60 feet)

in-between zones of residual contamination. Also, the contaminated formation is dramatically thinner, and the overall appearance of the plume cross-section is very irregular.

Soil cores were carefully inspected in the field for evidence of Q30, because no one knew how long the material might persist in the soil to provide usable amounts of oxygen. Intact beads of Q30 were found in many of the cores and visual confirmation of the "frac" geometry and placement of material was obtained. Comparison of this field intelligence with the three-dimensional plot of residual contamination indicated that significant reductions in hydrocarbon coincided with placement of Q30 in nearly every case.

Phase 2 Installation. An effort was made to complete cleanup of the site as a result of this second treatment of impacted soil and groundwater. Q30 loading was calculated to address residuals, and detailed three-dimensional plots of contamination in the subsurface allowed precise placement of material in close contact with contamination. As previously mentioned, the fieldwork was performed in August and September of 2000, just after a scheduled quarterly sampling event of all site point-of-compliance (POC) monitor wells. Compiled data associated with two of the key wells is provided in Figures 5 and 6.

Figure 4 shows data from well MW-3B, which is located near the former produced water vault and is historically the most contaminated well on site. BTEX was temporarily eliminated after Phase 1 work, although some rebound was observed in August 2000 for benzene only. Benzene once again was reduced after the Phase 2 installation and no significant rebound was detected at the December 2000 sampling event. When this well was sampled in February 2001, no detectable BTEX was present.

Figure 5 shows data from well MW-71A, which is located toward the far east-end of the contaminant plume. A steady decline in BTEX is evident since commencement of fieldwork in May of 2000, although residual benzene in the well remains above the 5 μg/l regulatory standard as of the February 2001 sampling.

Figure 6 is of particular interest, since it shows dissolved oxygen (DO) levels at all five POC monitor wells over the last year. Since monitor wells MW-71 and MW-72 have never shown any evidence of contamination, it may be assumed that they represent the natural ambient condition. The DO in MW-72A remains depressed, indicating that the bacterial population in this area continues to consume available oxygen while degrading the hydrocarbon load. The most striking result is that groundwater at wells MW-3B and MW-71A remains super-saturated with DO a full three months after the last Q30 application.

Verification Soil Sampling. Concurrent with the December 2000 quarterly groundwater sampling event, verification soil sampling was performed targeting residual "hot spots" delineated during the post-Phase 1 sampling work. Figure 7 shows borehole locations, and the denoted values indicate the residual TPH concentrations. Comparison of TPH values with the applicable regulatory guideline (1000 mg/kg TPH) shows that only four locations remain with hydrocarbon levels exceeding the standard.

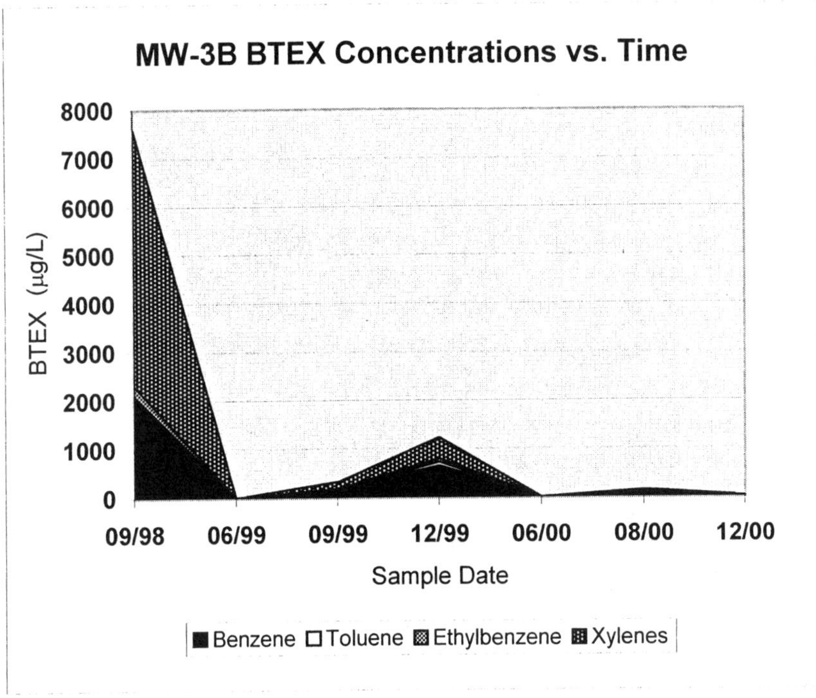

Date	Benzene μg/L	Toluene μg/L	Ethylbenzene μg/L	Xylenes μg/L
09/98	2100	0	210	5300
06/99	0	0	0	0
09/99	160	2.2	13	149.3
12/99	650	77	69	440
06/00	0	0	0	0
08/00	140	0	0	0
12/00	1.6	0	0	8.1

FIGURE 4. Monitoring results – MW-3B. Monitoring well 3B (MW-3B) is located in the western outlots, near the original location of the produced water vaults, the suspected source of the release. The total BTEX concentration at MW-3B has been reduced from a high of 7,610 μg/L to 9.7 μg/L. The groundwater at this well currently meets all standards.

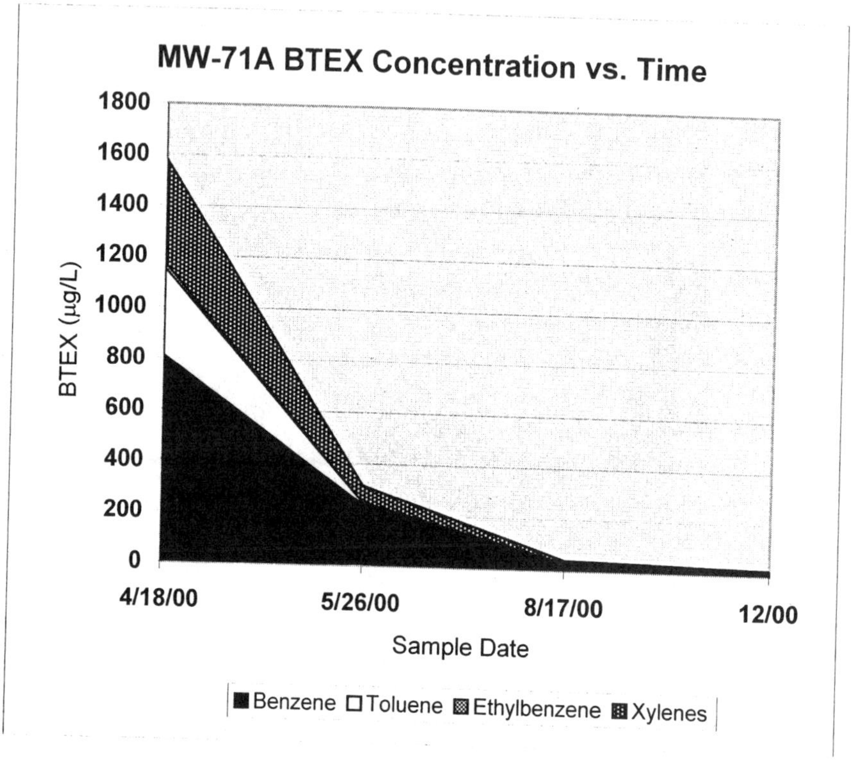

Date	Benzene µg/L	Toluene µg/L	Ethylbenzene µg/L	Xylenes µg/L
4/18/00	810	340	19	410
5/26/00	240	7.9	0	65
8/17/00	30	0	0	10
12/00	14	0	0	6.5

FIGURE 5. Monitoring results – MW-71A. Monitoring well MW-71A is located near the leading edge of the plume. The total BTEX concentration at this well has continuously decreased from a maximum of 1,579 µg/L to the current level of 20.5 µg/L, a 98.7% decrease. For the most recent sample, the 14 µg/L benzene concentration slightly exceeds the groundwater standard of 5 µg/L. The toluene, ethylbenzene, and xylene concentrations observed at this well have never exceeded the groundwater standards.

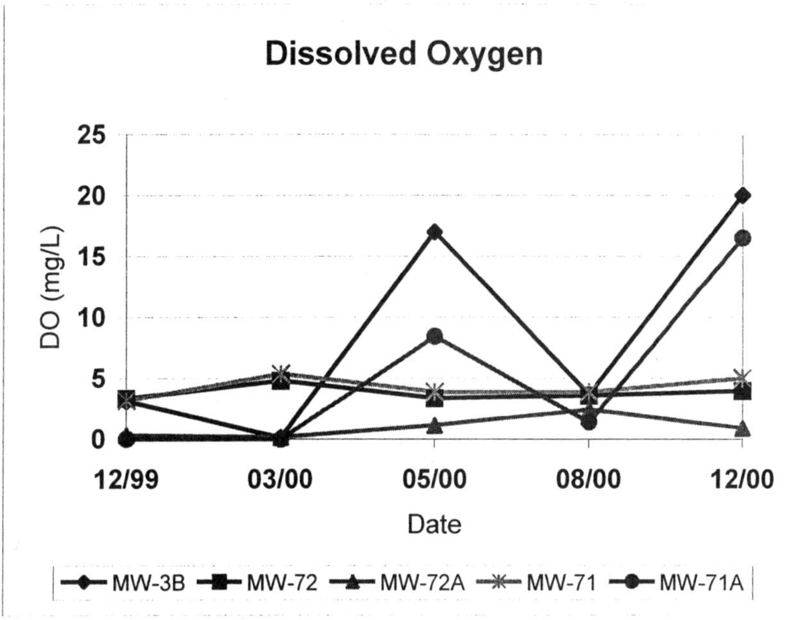

Date	MW-3B	MW-72	MW-72A	MW-71	MW-71A
12/99	3.1	3.34	0.31	3.16	0
03/00	0.11	4.8	0.16	5.35	0
05/00	17	3.36	1.13	3.87	8.4
08/00	3.64	3.57	2.44	3.87	1.4
12/00	20	3.95	0.90	5.00	16.5

FIGURE 6. Monitor well dissolved oxygen data. This chart presents the dissolved oxygen concentrations measured at the monitoring wells. Since wells 71 and 72 have not been impacted by the hydrocarbon release, it can be assumed that they represent the natural ambient conditions. The groundwater DO for wells 71 and 72 typically ranges between 3 and 5 mg/L. Saturated water typically has a DO near 9 mg/L. As shown, the DO measurements at MW-3B and MW-71A have ranged from levels below ambient conditions to super-saturated levels greater than 16 mg/L. As expected, the periods of super-saturated DO levels correspond with little or no dissolved BTEX concentrations. The high levels of DO measured at MW-71A in December 2000, three months after the last Q30 application, indicate that the residual hydrocarbons detected in this well should continue to degrade. The measured DO at MW-72A has consistently been below the natural background levels, indicating that the bacterial population in this area continues to consume the available oxygen while degrading the hydrocarbon load.

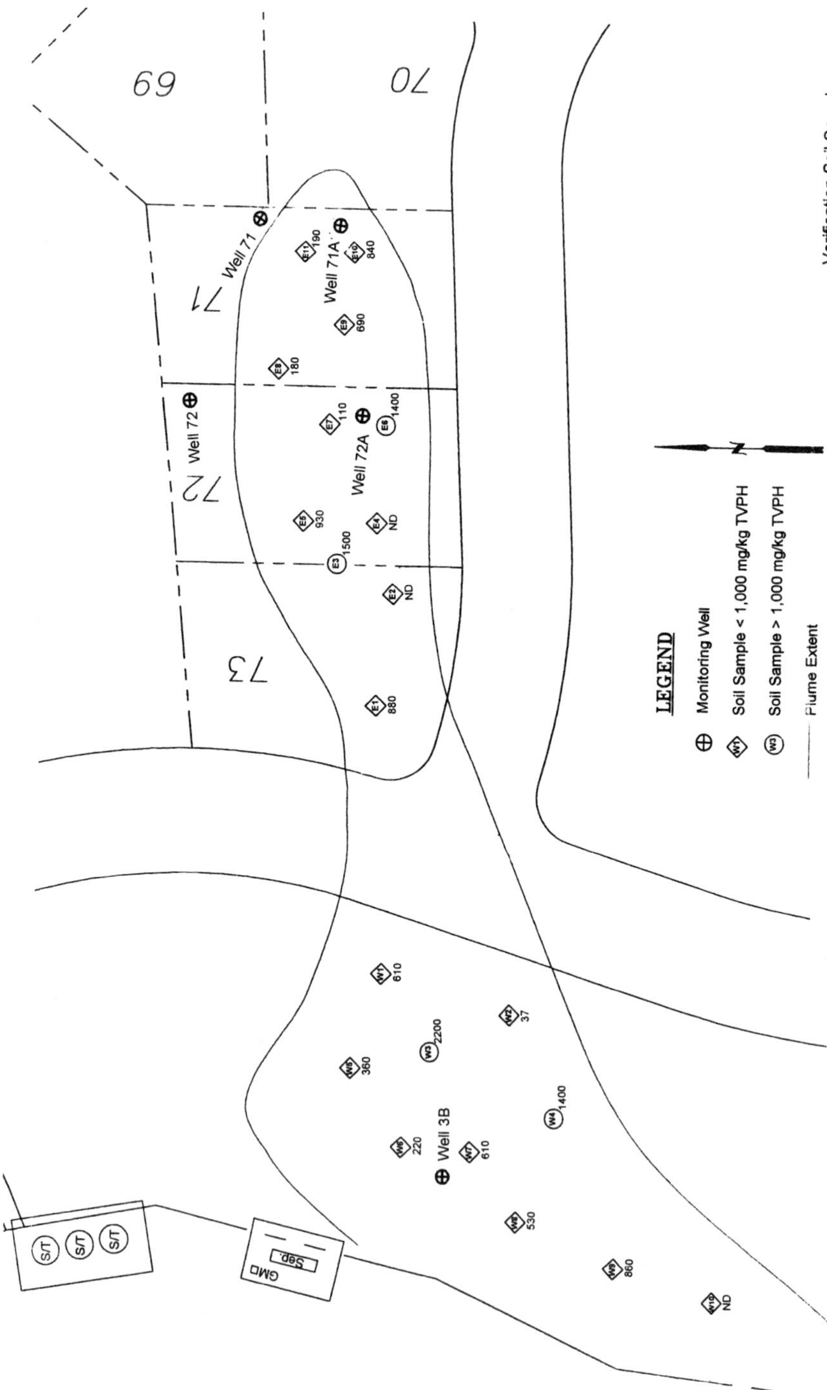

FIGURE 7. Verification sampling results. (Rough Scale – 1 inch equals 60 feet/Sample results are in ppb.)

Based on results from continued groundwater monitoring and the verification soil sampling, a final treatment plan has been prepared and fieldwork will be completed in April of this year. Given the results obtained over the last nine months, residual contaminant levels are expected to fall below both the groundwater and soil standard shortly after treatment and closure of the site should follow. If this schedule holds up, approximately 75,000 pounds of Q30 will have been required to achieve cleanup.

REFERENCES

Degussa Corporation. 2000. Off-site biormediation of TPH contaminated soil. A case study with Percarbonate Q30. Bio-10.doc rev.1/12/00

Degussa Corporation. 1999. Bioremediation of contaminated soil with Sodium Percarbonate Q30. Bio-8.doc 8/20/99

Vesper, S. J., Lawrence, C. M., Sam, H., and Wendy, J. D-H. 1994. Solid oxygen source for bioremediation in subsurface soils. Journal of Hazardous Materials, 36 (1994) 265-274. Elsevier Science B.V., Amsterdam.

APPLICATION OF OXYGEN RELEASE COMPOUND: A SIX-YEAR REVIEW

Stephen Koenigsberg (Regenesis, San Clemente, CA)
Craig Sandefur (Regenesis, San Clemente, CA)

ABSTRACT: Oxygen Release Compound (ORC®) is proprietary formulation of intercalated magnesium peroxide, marketed by Regenesis Bioremediation Products, that releases oxygen slowly and facilitates the aerobic degradation of a range of environmental contaminants including petroleum hydrocarbons, certain chlorinated hydrocarbons, ether oxygenates and nitroaromatics. Since 1994 ORC has been used on over 6,500 sites worldwide and has been the subject of an extensive body of literature. This technology, which can be further characterized as one that employs a "time release electron acceptor", has now been clearly established as a sensible strategy for engineering accelerated bioattenuation on sites where design, capital and management intensive options are either undesirable or contraindicated. ORC can be configured as a permeable reactive barrier, applied as a broader plume treatment and emplaced post-excavation as part of the backfill. Some guidelines for using ORC have also emerged. It is contraindicated at sites where the Biological Oxygen Demand (BOD)/ Chemical Oxygen Demand (COD) load, seasonal or otherwise, is excessive or poorly understood, i.e, the technology is best applied to dissolved phase plumes and moderate levels of residual Non Aqueous Phase Liquid (NAPL) once the majority of the source is removed by more mechanically intensive means. With regard to the range of compounds that can be addressed, ORC was first used for the remediation of benzene, toluene, ethylbenzene and xylene (BTEX) and total petroleum hydrocarbons (TPH). Groundwater contamination and other applications have since been made, with variable results, on an array of other aerobically degradable compounds such as vinyl chloride (VC), pentachlorophenol (PCP), polyaromatic hydrocarbons (PAHs) and methyl tert-butyl ether (MTBE). With respect to MTBE, as early as 1996 consultants using ORC noticed that MTBE concentrations decreased at a higher than expected rate. Working on this foundation, in concert with published evidence that ethers are aerobically biodegradable, additional field experiments demonstrated that oxygen can indeed enhance the remediation of MTBE; a concept that has since been verified in other quarters.

INTRODUCTION

Shaping a New Paradigm in Aquifer Restoration. It is clear as we enter the new millennium that we can manipulate the atom and the electron at will, so a call to return to simple solutions for complex problems appears to be counterintuitive and misplaced. To rally around more moderate strategies becomes a challenge. When our aquifers became unhealthy, restoration efforts began with available and palpable mechanical solutions. The approach was direct but heavy-handed—just pump out the water and clean it up; but this strategy was somewhat misguided or, at a minimum, over-applied. Eventually most mechanical systems were seen as a long-term, uneconomical, and incomplete response to the realities of environmental restoration. Then, in an almost reactionary turn of events, someone said, "The emperor has no clothes."

Contaminated groundwater plumes were suddenly seen as self-limiting, and regulators began to loosen the reins as the lexicon and mathematics of "exposure pathways" and "acceptable risk" took hold. Eventually the healing forces of nature were marshaled into service to complete the "revolution". So, now we have monitored natural attenuation, which counsels that, under the right conditions, we may be able to just watch the problem go away. In reality, it does nothing to improve the environment. It only serves to attempt to understand, quantify, and accept the risk that is already present. Unfortunately, this paradigm has the aura of a "do-nothing" technology and, while it has merits, it is constantly on the defense against these charges.

We now have a full spectrum of remediation options which bracket problem. On the one hand, we know that we can overdo it and, on the other hand, we are counseled that we can, after all, afford to eliminate the heroic efforts. The open question is—can we find a middle ground? Can that position be a strategy in which we incorporate the useful elements of monitored natural attenuation, recognizing that nature has a solution, and combine it with some level of proactive response to avoid both the stigma that nothing is being done and the continuing risk posed by the untreated contamination? Can we propose a level of action that is catalytic in nature so that a little bit of effort can rapidly accelerate the rate of natural processes and have a major impact on the problem? In other words, can a moderate strategy, between the extremes, be a serious contributing factor to solving our groundwater problems? Is this plausible in a world where we have grown accustomed to controlling atoms and electrons at will?

The paradox is that, in spite of our technological sophistication, we are still humbled by the fact that aquifer remediation is a macroscopic, multi-variable problem. There are exceptions; but, in general, groundwater contamination problems are hard to understand, characterize, monitor, and solve. Brute force is not usually the answer—at least not at an affordable price, and we seem to be on the right track by giving nature a chance to work.

The "New Paradigm" being advanced here is that we should simply offer nature a helping hand. In many circumstances it is the sensible and proactive thing to do to close sites more quickly, efficiently, and economically. Consequently, one of the main tools in this effort is the use of time-release electron acceptors and electron donors. The subject of this paper is ORC and a discussion of the time-release electron donor HRC® (Hydrogen Release Compound) for anaerobic systems appears elsewhere in this Proceeding.

ORC - A Time-Release Electron Acceptor. ORC is a source of oxygen, and oxygen is an electron acceptor - meaning that oxygen sits at the bottom of the mitochondrial electron transport chain and catches or accepts the "spent" electrons from contaminant molecules that are being degraded by microbes for energy and raw materials. A variety of peroxygen compounds can serve as electron acceptors, however, ORC is a unique formulation of magnesium peroxide that releases oxygen slowly when hydrated.

ORC is insoluble and releases its oxygen while being converted to ordinary magnesium hydroxide, which is also insoluble, in accordance with the following reaction:

$$MgO_2 + H_2O \rightarrow \frac{1}{2} O_2 + Mg(OH)_2$$

Both magnesium peroxide and magnesium hydroxide are environmentally benign and actually safe enough to ingest. In fact, they are both recognized as medicinal antacids - the more common one being magnesium hydroxide, also know as Milk of Magnesia.

ORC is not simply magnesium peroxide, but rather a patented formulation of magnesium peroxide intercalated with a food-grade phosphate ion. This gives ORC the time-release properties that are critical in a passive, low-cost oxygen application system. The term intercalation defines the fact that the phosphates permeate the magnesium peroxide crystal, partially inhibiting the transmission of water into the structure. Controlling the rate of hydration of the crystal effects a "timed" oxygen release. The concentration of phosphate, and the manner in which it is reacted with and permeates the magnesium peroxide crystal, creates a product that is well matched to the needs of subsurface remediation. Finally, intercalation facilitates the release of all the oxygen by keeping a "hydroxide rind" from sealing up individual particles during the hydration reaction. Intercalation also increases shelf life stability; ORC can be stored for several years without risking significant product degradation.

Remediation experts, looking for a long-lasting oxygen source without incurring the cost of having to install continuous oxygen injection systems, or having to return to the field for multiple re-injections of other oxygen releasing chemicals that do not have slow-release capabilities, have employed ORC. ORC has been actively used for about 5 years and to date has been applied at over 6,500 sites in 50 states and 11 countries.

The ORC technology melds the intercalation concept with electron acceptor chemistry. We can therefore characterize the status of this technology as the development of a time-release electron acceptor for engineered, accelerated bioattenuation. This technology is a useful strategy insofar as the aquifer microorganisms involved in bioremediation are not present in very high numbers, in comparison to other environments, and they are often challenged with only a very small mass of contaminant fluxing through the system. Consequently, only a small amount of electron acceptor is necessary to drive bioremediation – but it must be present on a constant basis.

Since ORC is an insoluble powder, it can be packaged in socks composed of a specially designed filter fabric. These "filter socks" are contacted with contaminated groundwater via an array of wells or trenches. ORC can also be mixed directly with water to form a slurry for permanent injection applications in the saturated zone. Finally, ORC can be used as a slurry or dispersed in powdered form for the in-situ or ex-situ treatment of soil. All of these methods help accelerate the natural bioremediation of aerobically degradable compounds.

ORC-mediated oxygenated zones have been shown to generally last from four months to over a year as a function of contaminant flux. The average lifetime for an ORC application in the field is in excess of six months. Usually only one application is necessary depending on rebound and regulatory demands.

The ORC can be installed as an "oxygen barrier" to create a permeable reactive barrier across a plume, thereby cleaning the groundwater as it passes through. A partial treatment, at lower cost, can significantly reduce contaminant mass and bring the control point of the plume back toward the source and reduce risk.

An array of treatment points, in which ORC is injected or backfilled into borings, has also been shown to be an effective dissolved phase plume (source) treatment or a method of establishing a barrier. The points can be implemented with low-cost, small-bore, push-point or standard auguring technologies to achieve full remediation or risk reduction objectives. One option that has emerged is a reusable injection point that can generate even more cost-effective treatments. ORC powder or slurry is applied several ways into biopiles or into the floor of tank pit excavations. In the latter case, the objective is to address residual hydrocarbons and the recharge of new contaminated water into the clean fill material. Full engineering application details can be found at www.regenesis.com.

In the evolution of ORC technology, there has been a transition from the use of filter socks in wells for plume cut-off to the use of the injectable slurry for direct source treatment and for the barriers. The slurry injection method has since proven to be lower in cost and faster acting. On a parallel evolutionary track, the first applications of ORC were for the treatment of benzene, toluene, ethylbenzene, and xylenes (BTEX) and other light petroleum hydrocarbon fractions. Use has now expanded to the treatment of heavier fractions such as heating oil and some of the polyaromatic hydrocarbons (PAHs). More recently, ORC has been used to bioremediate the highly mobile and problematic gasoline oxygenate methyl-*tert*-butyl-ether (MTBE) and has been applied to sites impacted with chloroaromatics, nitroaromatics and some of the lower-order chlorinated hydrocarbons that can be treated aerobically—most notably dichloroethylene (DCE) and vinyl chloride (VC).

RESULTS

Overview. There are about 100 publications in the peer-reviewed and conference literature on ORC; further details can be found at www.regenesis.com. Just focusing on the presentations made at Battelle In-Situ and On Site Bioremediation Conferences between 1995 and 2000 we have the following representative citations grouped by contaminant and application as follows:

- BTEX barriers (Chapman et al., 1997; Johnson and Odencrantz, 1997; Kao and Borden, 1994) and source treatments (Bohan and Schlett, 1997; Mysona and Hughes, 1999),
- BTEX/TPH source treatments (Duffy et al., 1999),
- BTEX/MTBE source treatments (Boyle et al., 1999; Buzea and DeStefanis, 1999; Defibaugh and Fischman, 1999),
- Heating oil source treatments (Oudijk et al., 1999),
- Nitroaromatic and chloroaromatic source treatments (Hicks et al., 1999), and
- VC barriers (Tacy, 1999).

At the Battelle In-Situ and On Site Bioremediation 2001 Conference there are additional relevant presentations grouped by contaminant and application as follows:

- BTEX/TPH barrier (Muniz et al., 2001) and source treatment (Fischer et al., 2001),
- BTEX/MTBE source treatment (Hicks et al., 2001),
- JP-4/JP-8 source treatment (Reed et al., 2001),
- Heating oil source treatment (Bennett and Ochs, 2001),
- PCP source treatment (MacEwen et al., 2001), and
- DCE/VC barrier (Larson and Voegeli, 2001).

SPECIFIC CASE STUDIES

A representative sample of unpublished cases is presented below in synopsis form, to convey some of the general performance characteristics of ORC across a range of contaminants and site characteristics.

TPH, Benzene and Naphthalene Source Injection Remediation in Oregon. Leaks from underground storage tanks and associated infrastructure at a car rental facility in Oregon resulted in gasoline-contaminated groundwater. The major objective was to accelerate the natural attenuation of the TPH contamination as well as the lesser concentrations of benzene and naphthalene components at the site. An interim program of pump and treat combined with SVE was implemented, and was carried forward to a practical limit, when ORC was then employed to polish the site to closure. The aquifer material consists of a silty sand interbedded with clay. Groundwater is 5 feet below ground surface. A total of 1900 pounds of ORC was injected as a slurry into the site in a rectangular array involving 32 push point applications for a total cost of less than $30,000. Two injections were made a year apart. The impact of the ORC on TPH in sentinel well MW-4 was to drive it to ND, where, after a 40% rebound event and the re-application, a final TPH reduction of 97% was achieved. At this point the residual levels of benzene were reduced from 500 ppb to 75 ppb and naphthalene was reduced from 220 ppb to ND. The site was then closed.

Permeable Barrier for Plume Control in Canada. BTEX contamination was controlled effectively in an ORC barrier in Canada. A series of eight, 4-inch diameter wells were installed at the front of the plume, each containing 10 feet of ORC socks. Groundwater was shallow and flowed in an unconfined silty, fine-grained sand aquifer at approximately 3 feet per day. Prior to the installation of the ORC barrier system, fourteen bundle-style piezometers were installed downgradient at five discrete depths. Four months following the installation of the ORC barrier, measurements of Eh and DO at monitoring points at the front, within and downgradient of the ORC treatment zone indicated that DO was being released into the groundwater. BTEX compounds went from an initial 934 ppb to 3.5 ppb and were maintained at these levels, which were below the drinking water standards, for seven months (consistent with ORC's known longevity range). Maintaining the barrier with two socks change-outs per year would cost $7,500.

LIMITATIONS OF ORC TECHNOLOGY

With regard to the lessons that have emerged from 6500 applications of ORC, a high level of success has been achieved at most sites when the site conditions are adequately understood and a proper design can be implemented. At one point, earlier in the technology development and roll-out process, the first thousand or so sites were implemented directly with Regenesis' oversight and 600 of those projects were subsequently tracked. In this population there were reports of 20 full technical failures – a rate of 3%. In a full technical failure we are referencing the complete inability for an ORC application to impact contamination at a site.

Technical failures, when they arise, can be classified as "intrinsic" and "non-intrinsic". Intrinsic problems are related to conditions on the site that are independent of proper characterization of contaminant concentrations and locations. These are rare events and are essentially traced to very high levels of dissolved iron or natural organic matter in the system, i.e., unmanageable COD/BOD. Non-intrinsic problems are related to any one of a number of issues related to poor site characterization, such as unidentified and unremoved sources, poor understanding of groundwater flow dynamics and in rare cases a limiting biological and chemical background in the aquifer.

There is another class of problems that involve limited performance and we can classify those sites as having technical difficulties. This is invariably related to rebound and is often tied to a single application of ORC. Rebound is normally quite manageable with additional applications of ORC and, due to the low cost of an ORC treatment, even multiple applications if necessary can still be economically feasible with regard to the next available technology. The nature of rebound and its manageability is governed by the status of non-dissolved contamination, e.g., the nature of the source. Simple hydrophobically sorbed materials are highly manageable, whereas residual sorbed material (smear zone NAPL) is somewhat manageable as a function of total mass in the system, rates of desorbtion and availability and application rates of ORC. Free-product conditions are essentially beyond the scope of ORC to control.

REFERENCES

Bennett, D.C. and L.D. Ochs. 2001."An Integrated Approach for Cleanup of Home Heating Oil Spills". *6th International Symposium on In Situ and On-Site Bioremediation*. June 4-7, 2001, San Diego, California (in press).

Bohan, D.G. and W.S. Schlett. 1997. "Enhanced Natural Bioremediation Using a Time Release Oxygen Compound". *In Situ and On-Site Bioremediation, Volume 5*. Battelle Press, Columbus, OH, pp. 475-480.

Boyle, D., M. Goers, S. Mandava, P. Hicks, K. Lemons, D. Pendrell and M. Canfield. 1999. "Enhanced In-Situ Bioremediation of Groundwater at MacDill Air Force Base, FL". *In Situ Bioremediation of Petroleum Hydrocarbon and Other Organic Compounds*. Battelle Press, Columbus, OH, pp. 289-294.

Buzea, D.C. and E.J. DeStefanis. 1999. "Accelerated Bioremediation as an Alternative to Conventional Remedial Technologies". *In Situ Bioremediation of Petroleum Hydrocarbon and Other Organic Compounds*. Battelle Press, Columbus, OH, pp.427-432.

Chapman, S.W., B.T. Byerley, D.J. Smyth, R.D. Wilson and D.M. Mackay. 1997. "Semi-Passive Oxygen Release Barrier for Enhancement of Intrinsic Bioremediation". *In Situ and On-Site Bioremediation, Volume 4*. Battelle Press, Columbus, OH, pp. 209-214.

Defibaugh, S.T. and D.S. Fischman. 1999. "Biodegradation of MTBE Utilizing a Magnesium Peroxide Compound: A Case Study". *In Situ Bioremediation of Petroleum Hydrocarbon and Other Organic Compounds*. Battelle Press, Columbus, OH, pp. 1-6.

Duffy, B.E., G. Oudijk and J.H. Guy. 1999. "Enhanced Aerobic Bioremediation of Petroleum UST Releases in Puerto Rico". *In Situ Bioremediation of Petroleum Hydrocarbon and Other Organic Compounds*. Battelle Press, Columbus, OH, pp.319-324.

Fischer, N.M., T. Reed and D. Camrur. 2001. "BTEX/TPH Remediation Using an Oxygen-Release Compound (ORC®)". *6th International Symposium on In Situ and On-Site Bioremediation*. June 4-7, 2001, San Diego, California (in press).

Hicks, P., J. Kubal and S.S. Koenigsberg. 1999. "Bioremediation of Nitrochlorobenzene, Nitroaniline, Chloroaniline and Other Organics in Groundwater". *Bioremediation of Nitroaromatic and Haloaromatic Compounds*. Battelle Press, Columbus, OH, pp.223-227.

Hicks, P. M.R. Pahr, J.P. Messier and R. Gillespie. 2001. "Aerobic Bioremediation of MTBE and BTEX at a USCG Facility". *6th International Symposium on In Situ and On-Site Bioremediation.* June 4-7, 2001, San Diego, California (in press).

Johnson, J.G. and J.E. Odencrantz. 1997. "Management of a Hydrocarbon Plume Using a Permeable ORC® Barrier". *In Situ and On-Site Bioremediation, Volume 4.* Battelle Press, Columbus, OH, pp. 215-220.

Kao, C.M. and R.C. Borden. 1994. "Enhanced Aerobic Bioremediation of a Gasoline-Contaminated Aquifer by Oxygen-Releasing Barriers". *Hydrocarbon Bioremediation.* Lewis Publishing, Boca Raton, FL, pp. 262-266.

Larson, J.R. and V.J. Voegeli. 2001. "Aerobic and Anaerobic Bioremediation of 1,1-DCE and Vinyl Chloride in Groundwater". *6th International Symposium on In Situ and On-Site Bioremediation.* June 4-7, 2001, San Diego, California (in press).

MacEwen, S.J., Fadullon, F. and D. Hayes. 2001. "Evaluation of Aerobic and naerobic Degradation of Pentachlorophenol in Groundwater". *6th International Symposium on In Situ and On-Site Bioremediation.* June 4-7, 2001, San Diego, California (in press).

Muniz, H., S.S. Koenigsberg and K.A. Lapus. 2001. "Accelerating Aerobic Biodegradation of GRO/DRO: A Five-Year Study". *6th International Symposium on In Situ and On-Site Bioremediation.* June 4-7, 2001, San Diego, California (in press).

Mysona, E.S. and W.D. Hughes. 1999. "Remediation of BTEX in Groundwater with LNAPL Using Oxygen Releasing Materials (ORM)". *In Situ Bioremediation of Petroleum Hydrocarbons and Other Organic Compounds.* Battelle Press, Columbus, OH, pp. 283-288.

Oudijk, G., B.E. Duffy and L.D. Ochs. 1999. "Enhanced Aerobic Cleanup of Home Heating Oil". *In Situ Bioremediation of Petroleum Hydrocarbon and Other Organic Compounds.* Battelle Press, Columbus, OH, pp. 271-276.

Reed, T., M.H. Allen and J.G. Peabody. 2001. "JP-4/JP-8 Bioremediation Using ORC® Pumped into Permanent Injection Wells". *6th International Symposium on In Situ and On-Site Bioremediation.* June 4-7, 2001, San Diego, California (in press).

Tacy, Jr., P.D. 1999. "Vinyl Chloride Attenuation by Direct Injection of Peroxygen". *Engineered Approaches for In Situ Bioremediation of Chlorinated Solvent Contamination.* Battelle Press, Columbus, OH, pp. 15-19.

JP-4/JP-8 BIOREMEDIATION USING ORC PUMPED INTO SEMI-PERMANENT INJECTION WELLS

Tom Reed (URS Corporation, Denver, Colorado, USA)
Mark H. Allen, (URS Corporation, San Diego, California, USA)
Jack Peabody (Regenesis, Inc., San Clemente, California, USA)

ABSTRACT: At Nellis Air Force Base, Nevada, a site investigation revealed that two separate floating product plumes and a larger dissolved plume consisting of JP-4 and JP-8 jet fuel existed. Floating product thickness ranged from less than one-half inch to almost three feet. The length of the dissolved plume extended about 700 ft, but was well contained within the boundary of the base. A variety of contaminants of concern (COCs), primarily benzene, toluene, ethylbenzene, and xylenes (BTEX) existed in the dissolved phase as a function of the JP-4/JP-8 releases. Specific remedial actions consisted of operating a mobile pump system to remove floating product in the source areas and deployment of Oxygen Release Compound (ORC®) in the downgradient, lower concentration areas of the plume. Because ORC placement through conventional direct-push technologies was not possible due to the cemented nature of the underlying soils, the ORC was deployed through 32 injection points consisting of 1½-inch steel pipe set in 4-inch diameter drilled holes. The injection points were arranged in two rectangular arrays to act as barriers to the downgradient movement of contaminants. Preliminary quarterly groundwater monitoring results have revealed notably lower contaminant concentrations in those wells located directly downgradient of the ORC arrays.

INTRODUCTION

Nellis AFB is located in the northeastern part of Las Vegas Valley, within Clark County, Nevada. The Fighter Revetments are located approximately 1,500 feet southeast of the southern end of the runways on the southeastern portion of Nellis AFB. There are 25 fighter jet Revetments designed for loading ordnance and fuel onto aircraft. Fighter Revetments 1 through 12 were constructed in 1989 and Revetments 13 through 25 were constructed in 1994. An 8-inch fuel line trends along the east side of the Fighter Revetments and connects to fuel headers at each Revetment. The fuel system was filled in Revetments 1 through 12 in November 1993, but was not used until March 1997 (Haarklau, 1996). Nellis AFB switched from JP-4 to JP-8 jet fuel in October 1993. In contrast to JP-4, JP-8 does not contain compounds with less than 10 carbon atoms, such as BTEX (Wilson, 1997).

A fuel leak was first suspected at the Revetments in May 1995 based on pipeline testing and inventory discrepancies. Leak location testing was conducted in June and July 1995. A leak was subsequently found at Revetment 5 in August 1995. The area around the piping at Revetment 5 was excavated in September

1995 and a cracked pressure-relief line was found (Haarklau, 1996). Between 1,500 and 4,000 gallons of jet fuel were estimated to have been released.

In order to characterize the release the consultant conducted an assessment at Revetment 5. The assessment was completed in two phases: Phase I between December 1995 and March 1996, and Phase II between November 1996 and April 1997.

The Phase I scope of work included drilling and sampling one soil boring (B-1) and the installation of two monitor wells (MW-1 and MW-2). Wells were installed after free-phase floating product was encountered in boring B-1. Monitor well MW-1 was located adjacent to boring B-1 and well MW-2 was located about 250 feet southeast of well MW-1 in the presumed hydraulically down-gradient position. Floating product accumulated in monitor wells MW-1 and MW-2 to thickness of 0.29 and 3.01 feet, respectively. Analyses of product samples from both wells by the USACE's laboratory indicated that the product was JP-8, along with a possible fraction of JP-4.

A second, separate fuel release (other than at Revetment 5) was suspected because of the greater product thickness that accumulated in down-gradient monitor well MW-2. As a result, the Phase II scope of work was implemented. Eleven soil borings (B-2 through B-12) were drilled to delineate the extent of the floating product; installation of three additional groundwater monitor wells (MW-3 through MW-5); and the first-time sampling of wells MW-1 through MW-5 to delineate the down-gradient extent of dissolved jet fuel components in groundwater.

The Phase II data indicated that there were two separate free product plumes at the Revetments -- a larger product plume originating from Revetment 5 and a smaller plume located to the southwest. The groundwater sampling results indicated the dissolved plume extended about 750 feet downgradient (southeast) of Revetment 5, but stayed well within the Nellis AFB boundary.

After completion of the assessment, the second product release near Revetment 3 was confirmed in July 1997 by leak testing. The leak occurred at a lateral pipeline connection from the main line to Revetment 3.

Monitor wells MW-6 and MW-7 were installed in October 1997. Well MW-6 was intended to monitor product thickness within the larger plume. Well MW-7 was designed to better locate the down-gradient edge of the larger product plume.

Monitoring well MW-8 was installed 50 feet southeast of a vault in August 1998 to assess for jet fuel in soil and groundwater after a leak within the vault overflowed onto the surrounding soil. The boring was located adjacent to the vault. TPH-contaminated soil was found in the soil boring to a maximum depth of 40 to 50 feet below ground surface (bgs). A groundwater sample from well MW-8 was non-detect for all volatile organic compounds (VOCs). Based on those results, it was concluded that groundwater had not been impacted and that monitor well MW-8 should be sampled on an annual basis.

The natural slope of the topography in this part of the Las Vegas Valley is less than 0.0037 ft/ft to the south-southwest. Groundwater flow within the deeper aquifers is from the west and northwest onto the main Base, but then shifts toward the south while moving beneath the eastern side of the air force base. The depth to water at the Fighter Revetments ranges from about 54 to 60 feet bgs and the gradient is toward the south-southeast.

ORC APPLICATION

To facilitate the aerobic biodegradation of fuel contaminants in groundwater, ORC was pressure-injected into the saturated zone directly downgradient of MW-2 and MW-6 in an attempt to create two oxygen-enriched zones to enhance the reduction of contaminant concentrations. Because ORC placement through conventional direct-push technologies was not possible, due to the cemented nature of the underlying soils, the ORC was applied via semi-permanent injection points. This approach not only allowed for the effective deployment of ORC, but it also provided a delivery system for subsequent injections.

In total, 32 soil borings were drilled into the saturated zone. Sixteen injection points were evenly spaced in each of two areas measuring about 20' x 80' and situated southeast (downgradient) of wells MW-2 and MW-6 (Figure 1).

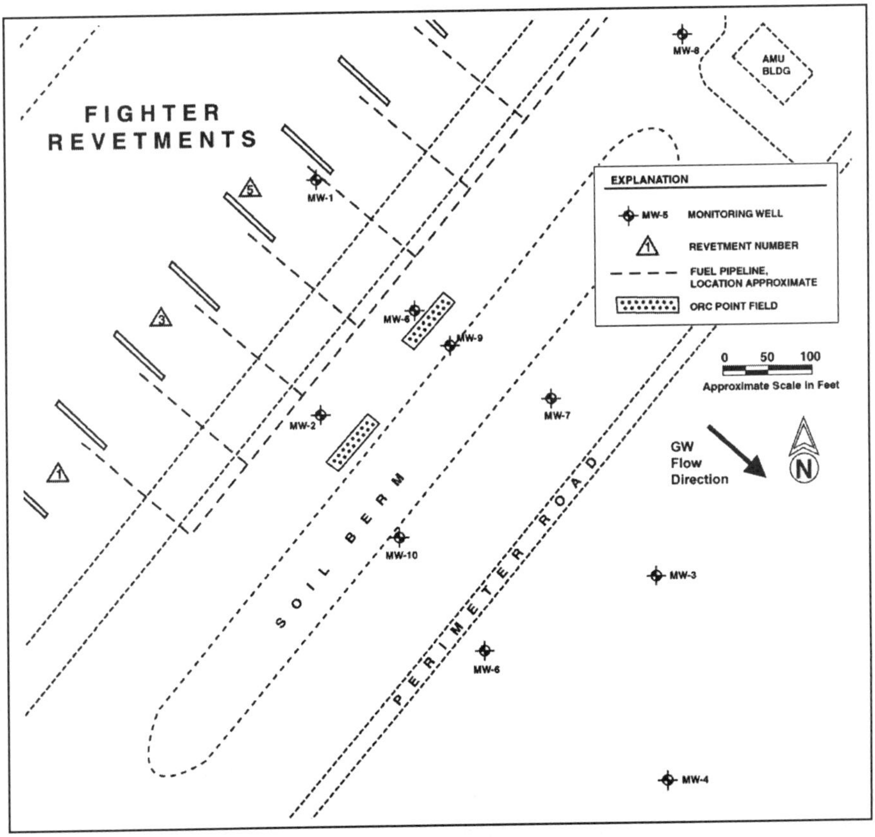

FIGURE 1. Nellis AFB site map.

Each rectangular-shaped area contains two lines of injection points, each spaced 10 ft apart, perpendicular to the direction of groundwater flow. The two ORC well fields are located in the areas of greatest dissolved fuel concentrations.

Prior to injection of the ORC, potable tap water was injected into the formation using a high-pressure pump to help fracture the soils and enhance dispersion pathways. This was accomplished with the use of a pump capable of producing pressures up to 1,500 pounds per square inch (psi). The pump was connected to the threaded end of each injection point and 20 gallons of water was forced into the formation. Only 300 psi was required to sufficiently open the formation as determined by pressure sensing instruments.

The ORC injection point borings were drilled using 4¼-inch diameter, solid stem augers, with a 5-inch diameter bit. Drill cuttings were screened in the field using a photo ionization detector (PID) to segregate contaminated soil cuttings for disposal. The ORC injection points were constructed of 1½-inch diameter, galvanized steel pipe with 5 feet of 0.060-inch slotted screen at the bottom. The tops of the screens were set 2 to 4 feet below the top of the water table (approximately 55 to 56 feet deep). Three centralizers were evenly placed along each pipe to ensure the pipe was centered in the borehole and that the cement grout would evenly seal the pipe in the borehole.

The ORC suspension was a mix of 5 gallons of water per 30 lb of ORC. The slurry was constantly agitated in a mixing hopper prior to injection to prevent the ORC from settling out of suspension or developing lumps.

The ORC slurry was injected under pressure into the formation using the high-pressure pump. After the slurry was placed, an additional 10 gallons of water and 2 lb of nitrogen- rich nutrient were pumped into the injection point to clear the ORC slurry from the steel pipe to allow for re-applications and to benefit the indigenous microbes and enhance biodegradation of the dissolved hydrocarbon contamination.

The initial ORC application took place from February to March 1999. A total of 960 lb of ORC (a rate of approximately 6 lb/vertical foot) was applied along with 64 lb of nitrogen-rich nutrient. A second application took place in September 1999. Similar to the first application, 960 lb of ORC was applied along with 64 lb of nutrient.

RESULTS

The monitoring program included sampling on a quarterly basis. In relation to the JP-4/JP-8, COCs at this site include BTEX, naphthalene, methyl-*tert*-butyl-ether (MTBE) and trimethylbenzene. By comparing the monitoring results of the wells upgradient of each barrier versus the wells downgradient of each barrier, we can see the effect of ORC at the site.

Comparing MW-6 and MW-9 (the upgradient and downgradient wells, respectively, to the northern ORC barrier application), we see a significant difference in the extent of degradation. In upgradient MW-6, total BTEX concentrations increased by 18% (from 1,422 μg/L at baseline to 1,671 μg/L at 13 months), MTBE decreased by 4% (from 230 μg/L to 220 μg/L), naphthalene increased by 55% (from 330 μg/L to 510 μg/L), and trimethyl benzene concentrations did not change (640 μg/L both at baseline and at month 13).

Comparatively, in downgradient MW-9, total BTEX concentrations decreased 75% (from 1,900 μg/L at baseline to 478 μg/L at month 13), MTBE decreased by 20% (from 110 μg/L to 88 μg/L), naphthalene decreased by 52% (from 330 μg/L to 160 μg/L), and trimethylbenzene decreased by 67% (from 730 μg/L to 238 μg/L).

Complete results for MW-6 and MW-9 are shown in Table 1. Likewise, the results for MW-2 and MW-10 (the upgradient and downgradient wells to the southern ORC barrier application) are displayed in Table 2.

TABLE 1. Northern ORC Barrier Results.

	UG WELL (MW-6)			DG WELL (MW-9)		
Compound	**Jan-99**	**Feb-00**	**% removal**	**Jan-99**	**Feb-00**	**% removal**
Benzene	**160**	**210**	-31%	**120**	**66**	45%
Ethylbenzene	260	300	-15%	220	39	82%
Toluene	62	41	34%	200	13	94%
Total Xylenes	940	1,120	-19%	1,360	360	74%
Total BTEX	1,422	1,671	-18%	1,900	478	75%
MTBE	230	**220**	4%	110	88	20%
Isopropylbenzene	42	48	-14%	40	16	60%
p-Isopropyltoluene	15	13	13%	15	5.7	62%
Naphthalene	330	510	-55%	330	160	52%
sec-Butylbenzene	13	16	-23%	14	5.3	62%
n-Propylbenzene	83	94	-13%	83	27	67%
1,2,4-Trimethylbenzene	520	510	2%	600	190	68%
1,3,5-Trimethylbenzene	120	130	-8%	130	48	63%

TABLE 2. Southern ORC Barrier Results.

	UG WELL (MW-2)			DG WELL (MW-10)		
Compound	**Jan-99**	**Feb-00**	**% removal**	**Jan-99**	**Feb-00**	**% removal**
Benzene	**35**	**31**	11%	**110**	**75**	32%
Ethylbenzene	58	38	34%	220	72	67%
Toluene	46	4.2	91%	4.9	1.1	78%
Total Xylenes	196	98	50%	1,080	480	56%
Total BTEX	335	171.2	49%	1,410	628.1	55%
MTBE	nd	nd	-	nd	nd	-
Isopropylbenzene	11	9.2	16%	36	21	42%
p-Isopropyltoluene	8	4.8	40%	15	6.7	55%
Naphthalene	74	66	11%	370	120	68%
sec-Butylbenzene	5	5.8	-16%	12	8.3	31%
n-Propylbenzene	31	27	13%	81	42	48%
1,2,4-Trimethylbenzene	230	140	39%	540	250	54%
1,3,5-Trimethylbenzene	67	38	43%	140	69	51%

CONCLUSION

Naturally occurring microbial populations, nutrients, and moisture are typically present in the saturated soils at most sites to facilitate aerobic biodegradation. However, oxygen is usually a limiting factor. Once the oxygen is depleted, biodegradation changes to anaerobic, but is about one or two orders of magnitude slower than aerobic biodegradation. By providing a constant supply of oxygen via ORC, accelerated aerobic biodegradation of the petroleum hydrocarbon contamination was achieved.

In addition to its ability to enhance aerobic degradation, ORC was also the preferred treatment technology because of its ability to be applied via semi-permanent injection wells. Due to the underlying cemented soils, direct-push injection at the site was not feasible. Once semi-permanent well probes were installed, ORC had the added benefit, if necessary, of multiple injection events.

REFERENCES

Dames & Moore. 1992. *Hydrogeologic Analysis, Nellis Air Force Base, Nevada.* Prepared for the U.S. Air Force.

Dames & Moore. 1997. *Final Fighter Revetment Release Assessment, Nellis Air Force Base, Nevada.* Prepared for the U.S. Air Force and U.S Army Corps of Engineers, Omaha District.

Haarklau, D.J. 1996. Personal Communication. U.S. Air Force, 99 Air Base Wing, Environmental Management Directorate, Compliance Division.

Matti, J.C., S.B. Castor, J.W. Bell and S.M. Rowland. 1993. Las Vegas NE Quadrangle Geologic Map. Nevada Bureau of Mines and Geology.

Plume, R.W. 1989. Ground-water Conditions in Las Vegas Valley, Clark County, Nevada: Part I, Hydrogeologic Framework. Water-Supply Paper 2320-A, USGS.

Wilson, T. 1997. Personal Communication. U.S. Army Corps of Engineers, Omaha District.

BIOREMEDIATION OF SOIL AND GROUNDWATER IMPACTS AT A FORMER PETROLEUM PRODUCTS STORAGE TERMINAL

Devin T. Shay (Groundwater & Environmental Services, Inc., Syracuse, NY)
Vincent Maresco (Groundwater & Environmental Services, Inc., Syracuse, NY)

ABSTRACT: Since 1993, a former petroleum products storage facility in Upstate New York has undergone several phases of site characterization and remediation. Investigation results suggested that a two-phase approach to site remediation should be employed. The primary remediation phase consisted of targeted shallow soil excavation in hydrocarbon source areas. A second phase of site remediation, targeting adsorbed phase impacts to soils below the water table, was designed based on a bench scale biofeasibility study. This study determined that hydrocarbon utilizing bacteria were present, and that stimulation of this community could be accomplished through the addition of nutrients and oxygen. An injection well network was installed to deliver alternating rounds of nutrient injection and air sparging to treat over 11,000 metric tons of impacted soil. Interim confirmatory soil samples collected after the first nutrient injection and 4 months of air sparging have shown a minimum total contaminant reduction of 74%, with the majority of samples falling between 93% and 100% total contaminant reduction. Reduction of dissolved phase impacts has also been documented over the course of the remediation.

INTRODUCTION

Between 1907 and 1998, the subject site was operated as petroleum products storage terminal on 13 acres of land in the city of Syracuse, New York. Hydrocarbon impacts were identified during site investigations, which included 162 direct-push soil borings, and the installation of 72 groundwater monitoring wells. Over 300 soil samples have been analyzed for petroleum indicator parameters (EPA Method 8021 and 8270 New York State Department of Environmental Conservation STARS Memo #1 listed compounds) (NYSDEC, 1992; U.S. EPA, 1986), and groundwater samples have been collected at all wells.

Geology at the site is typified by young sediments (approximately 12,000 years old) deposited in a glaciolacustrine setting. Three distinct units are present at the site: a brown silt and clay unit with varying amounts of fine sand [2 to 6 ft (0.6 to 1.8 m)]; a brown peat unit [6 to 11 ft (1.8 to 3.4 m)]; and a gray fine sand unit containing organic matter and mollusk shells [11 ft (3.4 m) and deeper]. Fill of varying composition overlies these sediments and extends from surface up to 6 ft (1.8 m) below grade. Depth to water at the site averages approximately 6 ft (1.8 m) below grade.

Initial Site Remediation. The first phase of site remediation focused on soils from 0 to 6 ft (0 to 1.8 m) below grade exhibiting impacts greater than 20 mg/kg total EPA Method 8021 NYSTARS compounds. Subsurface investigations delineated 8 areas where shallow soil excavation would remove likely hydrocarbon source material. Excavation was conducted from grade to a depth approximately 2 ft (0.6 m) below the average depth to water in the individual areas. A total of 15,321 metric tons of soil were excavated and disposed.

Following the soil excavation phase of work, 7 areas of the site where soil impacts below the water table would require remediation were identified (hereafter referred to as Areas A through G). Based on site conditions and the magnitude of soil impacts (Table 1) an in-situ bioremediation program was considered the most attractive remedial option. Initial ideas for remedial action included increasing the soil nutrient levels (Lord et al., 1995), in combination with air sparging (Brown, 1994; Brown, et al., 1994; Marley et al., 1992). The goal of this remediation program was to reduce the adsorbed phase impacts to soils located below the water table in these 7 areas.

TABLE 1. EPA Method 8021 and 8270 NYSTARS compounds and the typical concentrations of these compounds (in ug/kg) at Areas B and D.

EPA Method 8021 NYSTARS Compounds	Area B Typ	Area D Typ
MtBE	ND<100	ND<200
Benzene	590	4,000
Toluene	300	ND<500
Ethylbenzene	1,700	16,000
m,p Xylene	1,700	11,000
o Xylene	130	6,000
Isopropylbenzene	330	39,000
n-Propylbenzene	1,800	18,000
1,3,5-Trimethylbenzene	4,500	13,000
1,2,4-Trimethylbenzene	22,000	18,000
sec-Butylbenzene	1,300	18,000
p-Isopropyltoluene	630	ND<2,000
n-Butylbenzene	6,700	21,000
EPA Method 8270 NYSTARS Compounds		
Naphthalene	790	ND<170
Acenaphthene	ND<33	240
Fluorene	42	340
Phenanthrene	110	1,000
Anthracene	ND<33	320
Fluoroanthene	ND<33	780
Pyrene	49	1,500
Benzo (a) anthracene	ND<33	660
Chrysene	ND<33	780
Benzo (b) fluoroanthene	ND<33	830
Benzo (k) fluoroanthene	ND<33	270
Indeno (1,2,3-cd) pyrene	36	670
Dibenz (a,h) anthracene	ND<33	ND<170
Benzo (g,h,i) perylene	38	870
Benzo (a) pyrene	52	1,000

MATERIALS AND METHODS

Soil Characterization. A bench-scale study was undertaken to determine the feasibility of in-situ bioremediation at the site. Composite soil samples were collected from 3 different areas of the site where soil impacts below the water table were a concern. Samples were shipped to Biorem Technologies of Ontario, Canada for a soil characterization study that included Total Petroleum Hydrocarbon (TPH) analysis, pH, moisture, nitrate, ammonium, total nitrogen, total organic carbon, magnesium, phosphate, potassium, total microbial counts, and hydrocarbon utilizing bacteria counts.

Biodegradation Study. Following the soil characterization, a biodegradation study was performed on soil samples from the 2 areas of the site that contained the highest TPH concentrations. Samples were screened through a 0.25 inch (10.2 cm) screen and slurried in a 100:70 ratio of liquid to soil by weight in a reaction flask. Samples were prepared in duplicate with an organic nutrient amendment, an inorganic nutrient amendment, a control with no nutrient amendment, and an abiotic control prepared with mercuric chloride. The reaction flasks were shaken for 12 hours on an orbital shaker, after which the pH was adjusted to fall within a range between 6.5 and 8.0. Shaking continued for 16 days, and samples were removed on day 0, 4, 9, and 16 and analyzed for TPH. Results of the biodegradation study were used to determine which amendment, if any, was most effective at accelerating the degradation rate of TPH in the soil.

Nutrient Injection Network. Following confirmation of the biofeasibility study, an injection well network was designed to deliver both a nutrient and water mixture, and also atmospheric air during sparging. The well network was installed in 7 specific areas to be remediated. A total of 121 2-inch (5 cm) inside diameter (ID) schedule 40 PVC injection wells were installed. Screen length and total depth of screen were determined on an area-specific basis, but on average wells were installed to 14 ft (4.3 m) below grade and were screened from 10 to 14 ft (3.0 to 4.3 m) below grade with 10 slot screen.

Nutrient/Water Injection. Nutrient enriched water was injected during 2 separate events during November 1999 and August 2000. Nutrient injection was accomplished in 40,000 gallon (151,420 L) batches of nutrient and water mixture. Water was pumped from a nearby surface water body via a 6 inch (15.2 cm) diesel powered pump into two 21,000 gallon (79,500 L) holding tanks connected in tandem. Nutrients were dissolved in water in a plastic 55 gallon (208 L) drum and slowly bled into the water being pumped into the holding tanks. Following completion of the holding tank fill, the tanks were recirculated using the pump for approximately 30 minutes to ensure the nutrients were mixed and dissolved, and also to increase the dissolved oxygen of the delivery water. Nutrient-enriched water was then pumped from the tanks into a 6 inch (15.2 cm) ID HDPE piping system to deliver the batch to the area being injected. Manifolds were constructed at each remediation area for every 10 injection wells. Wells in each area were connected to their respective manifolds via 2 inch (5 cm) ID black polyethelyne pipe. Flow to each manifold, and each individual well, was controlled via 2-inch (5 cm) ball valves. Flow meters were used to determine total liquid delivery and flow rate to the manifold, and also at select individual wells. Flow rates to wells were balanced manually via the 2-inch ball valves. Injection pressure, which was maintained between 15 and 20 psi (1.1 to 1.4 kg/cm^2), was monitored using pressure gauges installed in front of the manifold. Injection pressure could be modified by recirculating the injection mixture back into the holding tanks.

Air Sparging. Following the nutrient/water injections, the injection well network was sparged with atmospheric air. The first injection event was followed by 4 months of sparging. Three blowers were employed for air delivery, including two 30-horsepower rotary lobe blowers and one 15-horsepower rotary lobe blower. Sparge pressure averaged 8 psi (0.6 kg/cm^2), and total airflow delivery per well averaged 8 cfm (0.2 m^3/min). Airflow was roughly balanced by manually adjusting the ball valves to each individual well. In order to inhibit the development of preferential pathways, airflow was shut down to individual manifolds on a rotating basis. Following the second nutrient injection event, sparging was conducted for 1 month. This month long sparge event was accomplished using one 30-horsepower rotary lobe blower and one 5-horsepower regenerative blower. Sparge pressure averaged 8 psi, and total airflow delivery per well averaged 3 cfm (0.08 m^3/min).

RESULTS AND DISCUSSION

Soil Characterization / Biodegradation Study. The TPH concentrations from soil samples collected as part of the soil characterization study were 124 mg/kg, 882mg/kg, and 2,426 mg/kg. Total microbial counts and hydrocarbon utilizing bacteria counts, determined using a Standard Plate Count method, are summarized in Table 2 below (reported as colony forming units per gram, or cfu/g). Values reported are averaged from 3 samples collected per area. Results indicated that microbial counts were normal for soil, and therefore not indicative of a toxic environment. Hydrocarbon utilizing bacteria (HUB) were present at an average of 19% of the total microbial population, confirming the site as a viable candidate for an in-situ bioremediation program.

TABLE 2. Total microbial counts, hydrocarbon utilizing bacteria counts, and TPH concentrations for reported in the soil characterization study.

	Total Microbial Count (cfu/g)	HUB Count (cfu/g)	TPH (mg/kg)
Area B	1.40E+06	3.30E+05	124
Area C	8.00E+06	1.10E+06	882
Area E	2.70E+06	5.70E+05	2,426

The biodegradation study, performed on samples collected from Areas C and E, determined that TPH degradation was stimulated through the addition of nutrients. Both the organic and the inorganic nutrient amended samples had greater reductions in TPH concentrations than the control and the abiotic control (Table 3). Results from the biodegradation study also indicated that the inorganic nutrient amendment was more efficient at reducing the TPH concentrations in the soil sample collected from Area E, whereas the organic nutrient amendment performed more efficiently in the soil sample collected from Area C. However, the beginning TPH concentrations at Area E and Area C were significantly different (2,426 mg/kg and 882 mg/kg, respectively). The 77% reduction in TPH concentrations at the Area E sample was indicative of a 3,300 mg/kg decline over the course of the degradation study, whereas the 55% reduction in the Area C sample was indicative of a 200 mg/kg decline over the course of the study. Based on this data, it was determined that the addition of inorganic nutrients would provide the best overall amendment to the site subsurface.

TABLE 3. Average TPH reduction measured over the course of the 16 day biodegradation study.

Treatment	Average TPH Reduction (%)	
	Area C	**Area E**
Inorganic Nutrient	41.3	77.2
Organic Nutrient	54.5	48.4
Control	33.7	32.7
Abiotic Control	11.9	6.9

Soil nutrient concentrations, pH, and moisture data identified several limiting factors which, if amended, could help with biodegradation of petroleum hydrocarbons. Soil pH, reported in Table 4, was greater than the optimal range of 6.5 to 8.0 in Area E, and also was on the high end of the range in Areas B and C. Phosphate levels were low in Areas B and E, and it was determined that they should be increased, based on the biodegradation study results, to stimulate the activity of the microbial community. It was also determined that nitrogen levels should be augmented to account for the high total organic carbon levels in Areas B and C, and also to augment the lower nitrogen levels recorded in samples collected at depths below 10 ft (3 m). For the three separate areas, it was determined that an average of 20 mg/kg of ammonium nitrate and 5 mg/kg of phosphate should be added to stimulate microbial activity. An acidic form of phosphate (P_2O_5) was selected for amendment in order to lower the soil pH to optimum levels.

TABLE 4. Selected results of the soil characterization study, reported as averages for Areas B, C, and E.

	pH	NH4-N (mg/kg)	NO3-N (mg/kg)	Total N (%)	TOC(%)	Phosphate (mg/kg)
Area B	7.8	11.60	2.17	0.06	2.07	2
Area C	7.7	50.79	2.17	0.39	6.34	7
Area E	8.5	11.42	1.80	0.005	0.33	1

Translation From Bench Scale to Field Scale. Results from the bench scale study were used to quantify the total kilograms of ammonium nitrate and phosphorous necessary to raise soil nutrient levels to the desired amount. Table 5 presents a summary of the estimated metric tons of soil being treated in each individual area, the kilograms of nutrients required to raise soil nutrient levels to the desired concentration, and the total gallons of water/nutrient mixture injected per area. A total of approximately 640,000 gallons (2,422,000 L) of nutrient/water mixture were injected per event.

TABLE 5. Summary of the mass of soil to be remediated per Area with the respective volume of injection water and mass of nutrients added.

Target Area	Soil Treated (metric tons)	Nutrient/Water Mixture (Liters)	Injection Wells/Area	Nitrogen (Kg)	Phosphate (Kg)
A	1,725	360,000	15	35.1	17.2
B	3,900	814,000	33	78.4	39.1
C	726	151,000	15	14.5	7.3
D	1,725	360,000	15	35.2	17.3
E	1,815	379,000	24	36.1	18.1
F	1,180	246,000	10	23.5	11.8
G	545	114,000	9	11.0	5.4

Remedial Progress - Groundwater. Groundwater samples were collected at a network consisting of 23 monitoring wells on a quarterly basis, including periods before, during, and after the active remediation phase. Samples were field tested for dissolved oxygen (DO) concentration at each sampling event. Prior to any remedial activity, DO at the site monitoring wells averaged 3.2 mg/L. During the first phase of air sparging, average DO increased to 7.4 mg/L. Three months after the completion of the first air sparging phase, average DO decreased to 1.7 mg/L. Groundwater samples were also sent to the laboratory to be analyzed for EPA Method 8021 and 8270 NYSTARS compounds. Laboratory results are reported in Table 6 for the February 1999 monitoring event (pre-remediation) and the December 2000 monitoring event (post-remediation) for 3 wells exhibiting the highest dissolved phase concentrations prior to the start of active remediation. The data collected illustrates a minimum of 1 order of magnitude decrease in total dissolved phase impacts between pre- and post-remediation ground water samples. The steady decrease in the dissolved phase impacts is likely due to the reduction in adsorbed phase impacts in soils located in remediated areas.

TABLE 6. Pre- and post-remediation comparison of typical total EPA Method 8021 NYSTARS compounds in groundwater at Areas B, C, and E.

	Area B	Area C	Area E
8021 Compounds Pre-Remediation (ug/L)	3,373	188.6	3,277
8021 Compounds Post-Remediation (ug/L)	69.9	69.0	130

Remedial Progress – Soil. Interim confirmatory soil samples collected following the first round of nutrient injection and air sparging showed significant remedial progress in eliminating adsorbed-phase soil impacts. Seven pre-remediation soil sample locations were chosen as repeat locations for remedial progress confirmation. These locations were selected based on location in a remediation area, and also to provide a range in the original concentrations of EPA Method 8021 NYSTARS compounds. The goal of the interim confirmatory soil sampling was to examine remedial progress in soils exhibiting the highest adsorbed phase concentrations at the site (>160 mg/kg), but also to examine if progress was made at locations where adsorbed phase concentrations were lower (<1 mg/kg). The original boring locations were surveyed by a licensed surveyor, and the same surveyor was employed to stake the location of the original boring for the resampling effort. The original boring location was duplicated, as was the original 2 ft (0.6 m) split spoon interval. Table 7 shows the concentration of EPA Method 8021 NYSTARS compounds in the 7 interim confirmatory soil samples as compared with the original soil sample collected before remediation. Total percent reduction in 8021 NYSTARS compounds ranged between 74% and 100%. The lowest percent reduction corresponded with the soil sampling location exhibiting the highest concentration of EPA Method 8021 STARS compounds prior to remediation.

TABLE 7. Comparison illustrating the total EPA Method 8021 NYSTARS compounds of pre- and post-remediation soil samples for Areas A, B, D, and G, with percent reduction.

	Total 8021 Compounds (ug/kg)	Percent Reduction
Area A Pre-Remediation	568	
Area A Post-Remediation	ND<8	100%
Area B Pre-Remediation	45,080	
Area B Post-Remediation	933.2	98%
Area D1 Pre-Remediation	162,060	
Area D1 Post-Remediation	9,127	94%
Area D2 Pre-Remediation	229,630	
Ared D2 Post-Remediation	60,520	74%
Area D3 Pre-Remediation	79,800	
Area D3 Post-Remediation	5,805	93%
Area D4 Pre-Remediation	175,000	
Area D4 Post-Remediation	1,753	99%
Area G Pre-Remediation	15,400	
Area G Post-Remediation	101.6	93%

CONCLUSIONS

In-situ bioremediation of petroleum hydrocarbon impacted soils was attempted through the addition of nutrients via injection and oxygen via air sparging. Over the course of the remediation project, a steady decline in dissolved phase concentrations of petroleum indicator parameters was noted at the areas of concern. Interim confirmatory soil samples collected following the first of two rounds of nutrient injection and air sparging indicated a minimum of 74% reduction in contaminant concentration. Six of the seven interim confirmatory soil samples exhibited greater than 93% reduction in contaminant concentration. A second nutrient injection and air sparging event was completed after the interim confirmatory soil sampling. A final confirmatory soil sampling plan has been submitted for regulatory review, and includes the collection of 38 confirmatory soil samples at repeat locations.

REFERENCES

Brown, R.A. 1994. "Treatment of Petroleum Hydrocarbons in Groundwater by Air Sparging." In Norris et al. (Eds.), *Handbook of Bioremediation*, pp. 61-85. Lewis Publishers, Boca Raton, FL.

Brown, R.A., R.J. Hicks, and P.M. Hicks. 1994. "Use of Air Sparging for In-Situ Bioremediation." In R.E. Hinchee (Ed.), *Air Sparging for Site Remediation*, pp. 38-55. Lewis Publishers, Boca Raton, FL.

Lord, D., L. Jiyu, M. Chapdelaine, J. Sansregret, and B. Cyr. 1995. "In Situ Air Sparging for Bioremediation of Groundwater and Soils." In R.E. Hinchee et al. (Eds.), *In Situ Aeration: Air Sparging, Bioventing, and Related Remediation Processes*, pp. 121-126. Battelle Press, Columbus, OH.

Marley, M.C., D.J. Hazebrouck, and M.T. Walsh. 1992. "The Application of In Situ Air Sparging as an Innovative Soils and Groundwater Remediation Technology." *Groundwater Monitoring Review 12*(2): 137-145.

New York State Department of Environmental Conservation. 1992. *STARS Memo #1: Petroleum Contaminated Soil Guidance Policy.*

U.S. EPA. 1986. *Test Methods for Evaluating Solid Waste – Physical/Chemical Methods.* SW-846.

LONG-TERM BIODEGRADATION OF UNDERGROUND AND AQUIFER CONTAMINATION AT TRECATE

Roland Mage, ***Augusto Porta*** (Battelle, Geneva, Switzerland)

ABSTRACT: Underground pollution in rice fields was created following an oil-well blowout due to oil penetrating through a network of existing water infiltration pits. Oil floating on the water table has been distributed in the ground (smear zone) by considerable seasonal water-table movements. These have greatly helped the progress of oxidative oil degradation associated with bioventing. Detailed analyses of oil composition changes over time confirm preferred biodegradation of lighter and linear petroleum hydrocarbons down to at least seven meters.

The TRECATE 24 oil well blowout has been the object of several remedial and monitoring activities the results of which have been presented and commented in various papers in the last seven years.

The topic selected for this presentation is related to the underground pollution, how the actions taken influenced its evolution in qualitative and quantitative terms, its impact on the free phase and dissolved plume in the aquifer.

The TRECATE oil well blowout case is interesting for several reasons:

- it is a major inland release of a large volume of crude oil
- it relates to fresh oil, since from the moment of the blowout data have been collected for characterizing pollution levels in different matrices (surface soil, underground soil, underground water, etc.)
- the data and information gathered cover an extended period of time (eight years now).

The geology consists of quaternary fluvio-glacial sediments represented by sandy gravel and silt lenses. Some large gravel exploitations are still in activity in the east of the site.

In order to grow rice, exogenous sediments (silt) have been brought into the fields (about 100 years ago) to reduce the infiltration capacity of the sandy gravel. The depth of these "artificial" soils is between 50 cm and 1 m.

In order to evacuate the rainwater during heavy rains, several infiltration pits have been dug.

It has to be mentioned that the larger portion of the underground contamination was caused by the presence of these infiltration pits in the rice fields. Through these pits, which correspond to spots of high permeability compared to the relatively impermeable bottoms of rice field, the oil went down and reached the aquifer in the days immediately after the blow out.

Figure 1 shows the distribution of these infiltration pits. The hydrocarbon concentrations in the underground have been monitored through successive sampling surveys carried out with Geoprobe® equipment in the years 1995, 1997, 1998, 1999 and 2000.

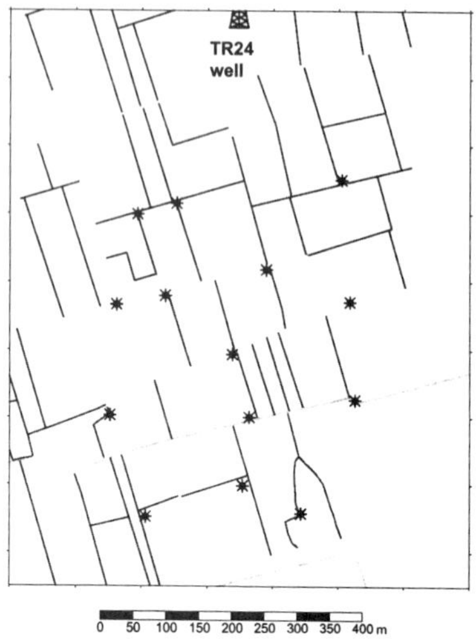

Fig. 1 Infiltration pit positions (✱) in the main zone

Also, since right from the beginning oil had reached the aquifer in some points, floating oil has been determined periodically and mapped in order to follow the evolution of the free-phase plume. The water table itself fluctuates considerably during a year, reaching the lowest point around March, before the flooding of rice fields, and the highest around October (Figure 2).

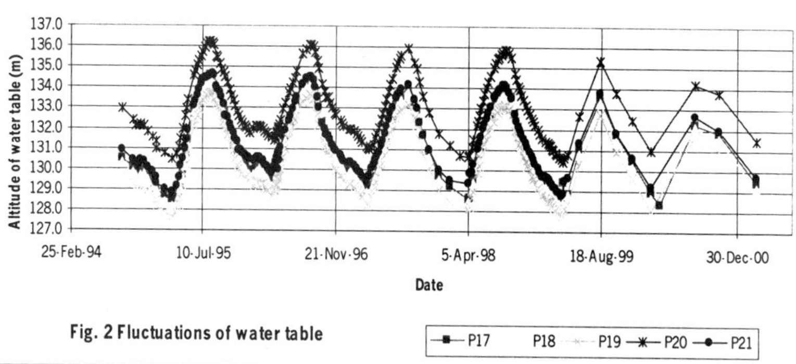

Fig. 2 Fluctuations of water table

The extension of the floating oil plume during the low water table period is presented in Figure 3 as a function of time. It appears from the map that oil from the vadose zone may have been "feeding" the floating-oil plume until 1998; in the following years the thickness and extent of the plume have started to decrease.

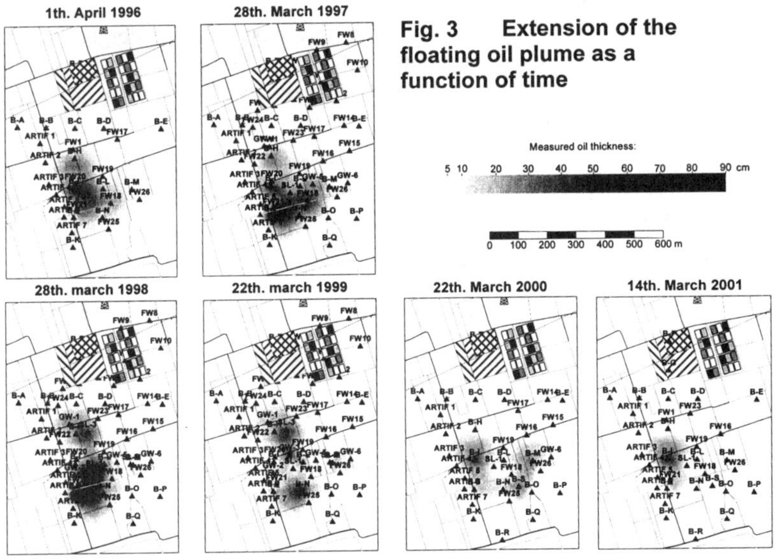

Fig. 3 Extension of the floating oil plume as a function of time

Figure 4 correlates the plume thickness and aquifer position in two of the piezometers: B-N (top) and B-H (bottom)

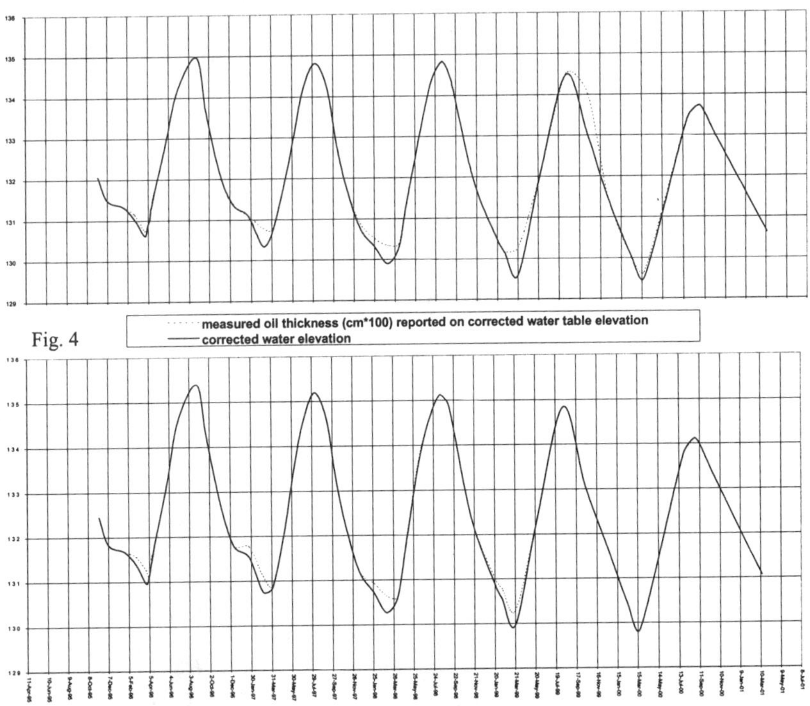

Fig. 4

When comparing the maps of free-oil thickness and underground soil contamination at depths corresponding to the bottom of the vadose zone, we find a good match between the presence of higher concentrations of total hydrocarbons and the highest thicknesses of free phase.

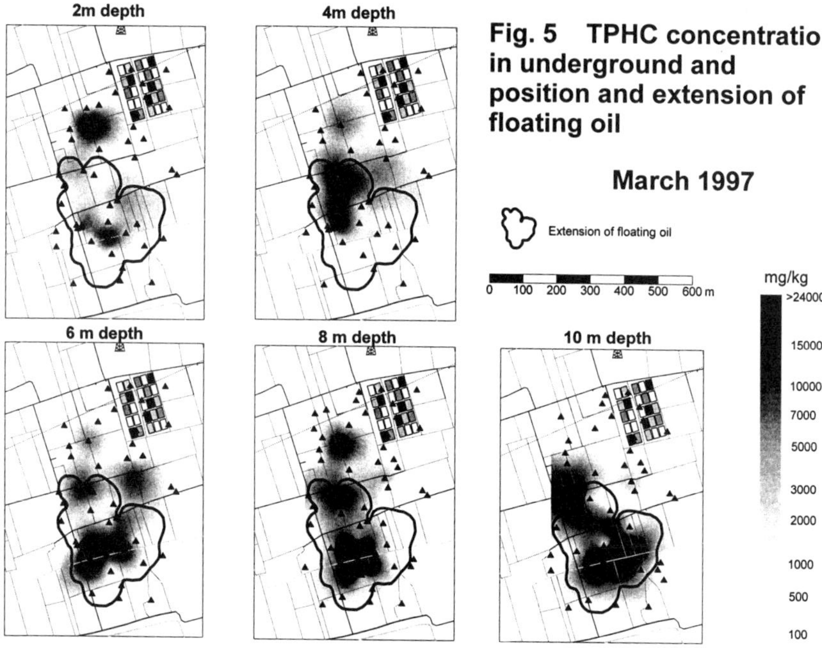

Fig. 5 TPHC concentration in underground and position and extension of floating oil

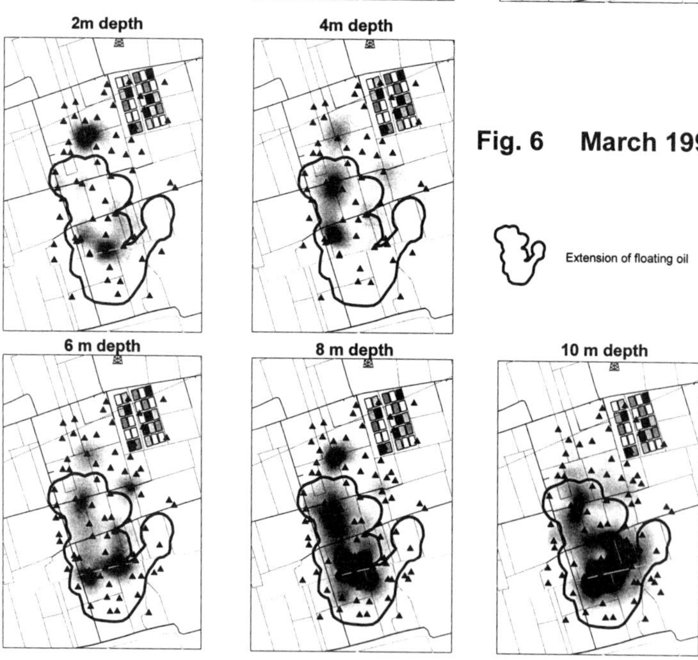

Fig. 6 March 1999

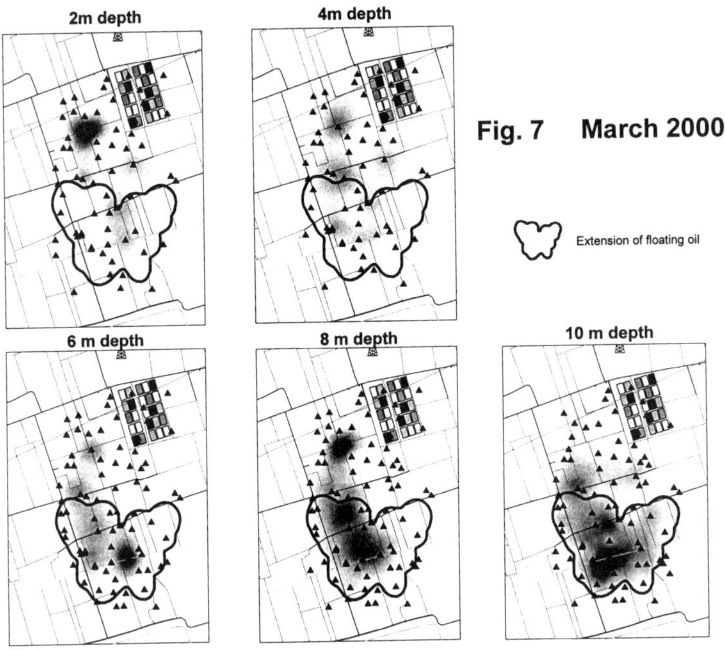

Fig. 7 March 2000

These maps (Figures 5 to 7) also show that, under the effect of bioventing, concentrations of total hydrocarbons have been declining considerably in the vadose zone particularly between the top surface and strata 6-7 meters deep. Based on the analytical results for soil samples collected with a Geoprobe® between April 95 and October 99, we have calculated and represented in Figure 8, as a function of volume percent decrease, the total hydrocarbons present in the underground down to a depth of seven meters. The figure shows different residual volumes within the envelopes corresponding to different concentrations. One can see that within the contour enveloping soils with more than 500 ppm of total hydrocarbons, about 70% have been degraded, while in the volume of soil affected by higher concentrations this percentage is even higher.

It is also of interest to analyze the results of our investigations in terms of the qualitative (not merely the quantitative) evolution of the oil in the ground.

When we compare the GC/MS and GC/FID chromatograms of the fresh oil after the spill and of the oil floating on the aquifer at the beginning of this year (2001), we see that:

- the original oil exhibited a nice distribution of hydrocarbons between C_9 and C_{40}. The linear hydrocarbons represented about 33% relative to the total including branched + unresolved compounds
- the weathered oil sampled in February 2001 also shows a distribution of hydrocarbons, which now ranges from C_9 to C_{35}. The percentage of linear versus branched + unresolved has dropped to 20%.

In addition, chromatograms show an appreciable decrease of the fraction between C_{19} and C_{25}, while the profile of the fraction C_9 - C_{18} remains more or less unchanged.

When one looks at the percentage ratio of the fraction C_8 - C_{12} versus total petroleum hydrocarbons, a detailed analysis of the results of soil samples of 1999 and 2000 reveals that the highest values of this ratio correspond to samples taken at depths beyond 9 meters, i.e. corresponding to the lower portion of the smear zone directly above the water table. These values are often around 10%, but in some cases they are as high as 25%. In the upper portion of the vadose zone the values are often lower than 5%; this indicates that biodegradation by bioventing has principally affected the lighter fractions of hydrocarbons, as one could have expected.

The influence of the bioventing system on petroleum hydrocarbon degradation in the underground, which was assessed on the basis of the evolution of the residual hydrocarbons according to analytical data obtained with Geoprobe® samples (illustrated in Figure 8), appears to be significant over the period between 1995 and 1998. Since the beginning of 1999, the degradation rates have decreased, and the effect of bioventing appears to become marginal.

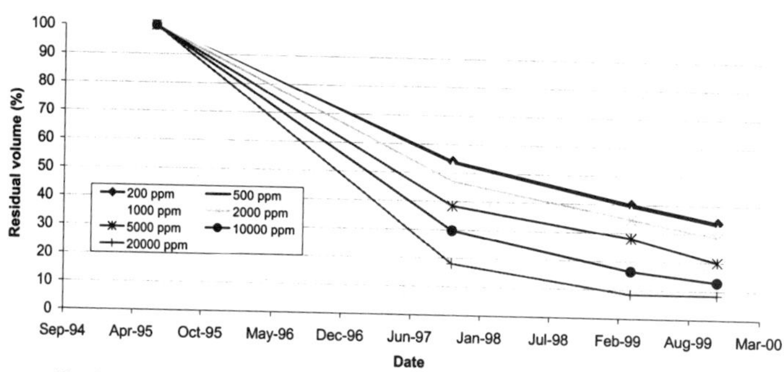

Fig. 8 Residual percentual hydrocarbons volume in the underground between the top soil and 7 meters depth

In conclusion, air injection bioventing as used in Trecate has contributed in treating the smear zone in which most of the oil mass is concentrated; due to the large water table fluctuations, even the deeper portions of the smear zone are aerated during the seasonal period of low water.

In addition, air injection probably is depressing to some degree the water table, resulting in increased smear zone treatment.

Bioventing has also been effective in adequately aerating the vadose zone, where hydrocarbon concentrations have dropped considerably over the first two years.

Modeling of dissolved-plume migration reveals a rather conservative evolution, while the health risk assessment study has demonstrated a largely safe and acceptable situation. Therefore, no additional active remedial activities are required on this site. Intrinsic (natural) remediation with medium-term monitoring and control of land and groundwater use will suffice to gradually achieve clean-up targets compatible with a resumption of agricultural activities.

TREATABILITY STUDY FOR TOTAL PETROLEUM HYDROCARBONS (TPH) CONTAMINATION AT GASOLINE-ALLEY

Altaf Wani (Applied Research Associates Inc., Vicksburg, MS 39180)
Lance Hansen, Jeffrey Talley, Scott Waisner, David Ringelberg (US Army Engineer Research & Development Center, Vicksburg, MS 39180)

ABSTRACT: This study discusses the biotreatability options for remediating total petroleum hydrocarbon (TPH) contaminated soils at Gasoline-Alley, Fort Drum, NY. The main objective of the study was to evaluate alternative bioremediation technologies for follow on full-scale remediation activities on the basis of intrinsic TPH degradation capability of native subsurface consortia. A two-phase treatability study comprised of microcosm evaluation using a single soil core, and a bench-scale column evaluation comparing natural attenuation (NA), bioventing (BV), and biosparging (BS), using three soil cores was performed. Microorganisms in soil samples from the smear zone (10.5 feet bgs) mineralized about 37% of ^{14}C-acetate to $^{14}CO_2$ in 12 days. The extant microbiota taken from the top of the smear zone demonstrated the ability to mineralize ^{14}C-phenanthrene, utilizing oxygen from the headspace. NA and BV columns did not show any significant reduction in the total soil TPH as compared to BS column which demonstrated a zero-order removal rate of 35.5 mg rTPH kg^{-1} contaminated soil d^{-1}. On the basis of respiration analysis, biodegradation rates of 0.33 and 0.49 mg TPH kg^{-1} contaminated soil d^{-1} were achieved in BV and BS columns, respectively.

INTRODUCTION

Many Department of Defense (DoD) installations, similar to the Fort Drum facility, are continuously faced with the challenge of how to dispose petroleum hydrocarbon contaminated soils. Total petroleum hydrocarbon (TPH) includes gasoline, diesel, aviation fuels, lubricating oils, as well as heavy oils and grease. Sources of contamination include spills and leaks at fueling stations, above and under ground fuel storage tanks (UST), maintenance sheds, waste oil pits, and vehicle wash areas. These contamination problems are not unique to DoD installations; but are encountered by local governments as well as private sector because of spills and leaks in facilities such as gas stations, farming operations, or garages. As of April 1995, over 287,000 USTs were confirmed to be leaking petroleum hydrocarbons into the surrounding environment (US EPA, 1995). Although petroleum spills may not cause immediate human health concerns, removal is needed to avoid future problems like contamination of groundwater aquifer.

There are many technologies available to combat petroleum hydrocarbon contamination in soils. The Dig and haul method of disposing the contaminated soils into landfills can be extremely costly and environmentally unfriendly. Other physical technologies like incineration and thermal desorption are expensive and difficult to implement due to regulatory issues and public acceptance. Bioremediation, utilizing intrinsic potential of indigenous microorganism, is one of the cost-effective and environmentally acceptable options. A few of the more widely practiced bioremediation technologies include landfarming, bioventing, and heap-pile bioremediation. TPHs are a rich source of organic matter, and microorganisms use these as a source of carbon and energy to fuel metabolic processes. Bacteria and fungi are the main agents responsible for the biodegradation of petroleum products, but some molds, yeasts, cyanobacteria, and green algae have shown the ability to oxidize certain components. Hydrocarbon biodegradation generally proceeds more rapidly under aerobic conditions (Brock et al., 1994), and field investigations have demonstrated that BTEX biodegradation in groundwater can be limited by oxygen (Aelion and Bradley, 1991).

The main objective of this research study was to investigate the alternative bioremediation technologies on the basis of intrinsic TPH degradation capability of native subsurface microbial consortia for follow on full-scale remediation activities. Additional study also evaluated the parameters to enhance the TPH biodegradation potential of resident microorganisms.

Site Description. Fort Drum Military installation is located in upstate New York, approximately 10 miles northeast of Watertown. Area 1595 is one of the nine fuel storage and dispensing areas located centrally in the Gasoline Alley. Previously, two 25,000 and one 12,000 gallon underground storage tanks (USTs) containing diesel fuel and unleaded gasoline were buried at Area 1595. The three USTs were replaced in 1975, and a one-inch hole was discovered in one of the USTs. Fueling activities were discontinued in 1994. Hydrocarbon contamination remains in the subsurface in the immediate vicinity, and down gradient of the former UST locations. A separate-phase product layer above the water table has been reported in six wells since 1995. Movement of the product layer, caused by seasonal fluctuations of the water table, is believed to have created a "smear zone" of petroleum contaminated soil, which serves as a source of dissolved-phase contaminants to ground water moving through the area. High concentrations of BTEX have been reported in surface water samples collected from the Area 1595 Creek, immediately downstream of the impoundment. High concentrations of several polycyclic hydrocarbons (PAHs) have been reported in sediment samples collected from the same area.

MATERIALS AND METHODS

Soil Cores and Groundwater. An initial soil core used for chemical and microbiological characterization was extracted from Area 1595. An additional three cores used in the bench-scale column study were extracted within a 10 feet radius of the initial core. Soil cores spanned a continuous depth to about 15 feet below the ground surface reaching several feet below the groundwater table. Soil was collected in acetate liners approximately 5 feet in length. Groundwater samples were extracted, in conjunction with initial subsurface core.

Experimental Setup. Microcosm setup consisted of 250 mL respirometry flasks (Reliance Glass). Aliquots of soil were placed into the flask and challenged with radio-labeled acetate and phenanthrene tracers. Flasks were equipped with center wells that contained 2 mL 1N potassium hydroxide solution for trapping $^{14}CO_2$ resulting from mineralization of ^{14}C-labeled acetate or ^{14}C-labeled phenanthrene.

Bench-scale column setup comprised of 6 ft glass columns with an inside diameter of 3.25 in (Figure 1). Soil columns were kept in a walk-in cooler at 10°C, the average yearly aquifer temperature for the Fort Drum area. To accommodate the 15 feet deep soil cores two columns were connected in series with ¼ inch stainless steel tubing. Sample ports were located at about 1 foot intervals along the length of the column. The lower two ports (1-10 and 1-23), used for taking water samples, were fitted with two-way valves and packed with glass fiber. To simulate the saturated zone of the contaminated site, groundwater from the site was added through port 1-0 at the bottom of the first column to a height of approximately 32 inches. At each sampling event, all free water was removed from the column and replaced with contaminated ground water from the site. In the soil column simulating bioventing air was forced through port 1-34, about 2 inches above the saturated zone. Air was forced in through port 1-0 located in the lower end cap of soil column representing biosparging. Airflow through the BV and BS columns was initiated at 4 mL min^{-1} and maintained at this rate for 5 weeks. During the final 4 weeks airflow was reduced to EPA recommendation of 1 mL min^{-1} that corresponds to an estimated residence time of approximately 60 h in soil. No air was forced into columns simulating natural attenuation in the aquifer. The ends of each column were closed with a 3.5 inch Teflon plug fitted with a tubing connector and screened with a 3 inch diffuser stone.

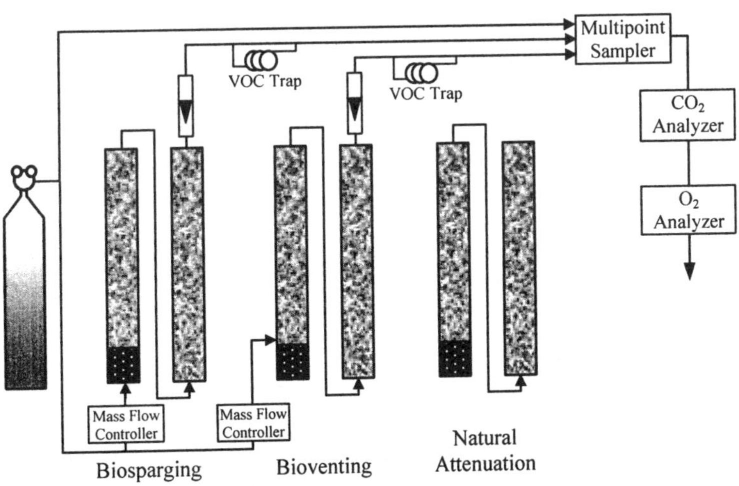

FIGURE 1. Schematics of bench-scale column setup.

Soil and water samples were taken at 2 week interval followings a 7 day equilibration period (Table 1) for chemical and microbiological analysis. TPH quantification was performed by using a HP-6890 gas chromatograph (GC) equipped with an SPB-5 capillary column. Phospholipid fatty acids (PLFA) were identified and quantified on a HP-5973 mass selective detector interfaced to a HP-6890 GC equipped with a J&W DB-5ms capillary column. Water samples were analyzed for VOCs using EPA Method SW846-8260A. Reproducibility of the GC analyses averaged a standard error of 9% while replicate analyses of soil extracts averaged a standard error of 15%.

TABLE 1. Sampling Schedule

Days after start (d)		**0**	**7***	**15**	**29**	**44**	**58**	**70**
Airflow in BV & BS (mL/min)		**0**	**0**	**4**	**4**	**4**	**1**	**1**
Port Tag	**Column depth (in)**							
2-TOP	0		S	S				S
2-58	7		S	S	S	S	S	S
2-46	19		S	S	S	S	S	S
2-34	31		S	S	S	S	S	S
2-23	42		S	S	S	S	S	S
2-10	55		S	S	S	S	S	S
2-0	65		S					S
1-TOP	65		S	S				S
1-58	72		S	S	S	S	S	S
1-46	84		S	S	S	S	S	S
1-34	99		S	S	S	S	S	S
1-23	107	W#	S,W	S,W	S,W	S,W	S,W	S,W
1-10	120	W#	S,W	S,W	S,W	S,W	S,W	S,W
1-0	130	W#	S,W	S,W	S,W	S,W	S,W	S,W

S = Soil sampling; W = Water sampling

* Represents end of the equilibration period

Water obtained from site groundwater and represents initial water concentration for each treatment

Exit air from the BV and BS columns was collected in Tedlar bags. Gas analysis was done by sequentially passing the collected air through a photoacoustic infrared multi-gas analyzer, and a fuel-cell type oxygen detector. The cumulative consumption of O_2 and production of CO_2 were calculated from airflow rates and the compositions of inlet and exit gases, using Equations 1 and 2, respectively.

$$\Delta O_2 = \frac{Q}{22{,}400}\left[O_2^{in} - O_2^{out}\left(\frac{1 - O_2^{in} - CO_2^{in}}{1 - O_2^{out} - CO_2^{out}}\right)\right]\Delta t \cdot 1{,}440 \quad (1)$$

$$\Delta CO_2 = \frac{Q}{22{,}400}\left[CO_2^{out} - CO_2^{in}\left(\frac{1 - O_2^{in} - CO_2^{in}}{1 - O_2^{out} - CO_2^{out}}\right)\right]\Delta t \cdot 1{,}440 \quad (2)$$

where ΔO_2 is the O_2 consumption (mole); ΔCO_2 is the CO_2 production (mole); Q is the airflow rate (mL min^{-1}); O_2 and CO_2 are the molar fractions of oxygen and carbon dioxide, respectively, in inlet and exit air; and Δt is the elapsed time (d).

RESULTS AND DISCUSSIONS

Chemical and Microbiological Characteristics. The results of chemical and biological analysis of the initial soil core are summarized in Figure 2. TPH concentrations were low near the ground surface and increased with depth. Contaminant concentration rapidly decreased to levels similar to surface soils near aquifer levels (~10.5 feet), but peaked to the highest value at 11.5 feet. Low TPH concentration at 10.5 feet bounded on top and bottom by higher contaminant concentrations suggest possible contaminant-microbial interaction. The soil TPH values reported are for the recoverable TPH (rTPH) and do not include the unrecoverable fraction. The recovery was found to be 58±5%. Microbial biomass in the soil core decreased by an order of magnitude in the first three feet below the surface. There was a significant increase in subsurface microbial biomass at 8-10 feet depth, near the top of the TPH contamination zone, which indicated a contaminant influence on the microbial growth.

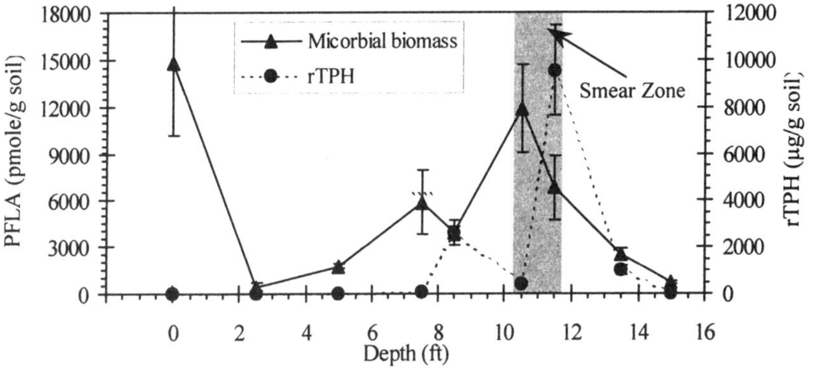

FIGURE 2. Vertical profile of rTPH and biomass concentration in soil core.

Mineralization of ^{14}C-acetate showed the unequivocal evidence of microbial respiration. Microorganisms from the top of the smear zone (10.5 feet) were capable of mineralizing about 37% of ^{14}C-acetate to $^{14}CO_2$ in 12 days. Significant mineralization of ^{14}C-phenanthrene was observed in control (without nutrient amendments), with soil from the top of the smear zone under saturated conditions, utilizing headspace O_2. These results are in agreement to the findings reported by Barker et al., (1987) and Ridgway et al., (1990) that O_2 is an effective electron acceptor in field studies for biodegradation of petroleum hydrocarbons. In addition, to the mineralization of phenanthrene, 20% greater reduction in the total rTPH was observed in the control flask as compared to the sterile control.

Contaminant Biodegradation in Soil Phase. Initial TPH concentrations in the three soil cores (NA, BV, and BS columns) are presented in Figure 3. In each column relatively high TPH concentrations were found in the bottom half of the cores where the concentrations peaked and then decreased again as the saturated zone was approached. There was not any significant change in the overall TPH concentration profile in the NA and BV columns over the 10 week evaluation period. In the NA column total rTPH was reduced from 104.6 to 86.3 g at the end of test, while in the BV column rTPH was reduced from 69.2 to 63.1 g at the completion of 9 weeks of bioventing. Biosparging resulted in a significant reduction in the TPH concentration. Total rTPH in the BS column was reduced from its initial value of 75.9 to 39.3 g. Zero-order (concentration independent) removal rate calculated from the initial and final rTPH levels in the biosparging column was 35.5 mg rTPH kg^{-1} contaminated soil d^{-1} based on total mass of 16.4 kg contaminated soil treated.

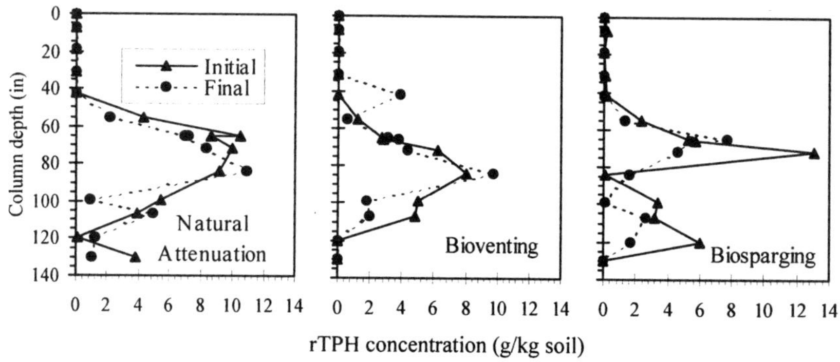

FIGURE 3. Initial and final rTPH profile in NA, BV, and BS columns.

The heterogeneity of the cores was also evident in terms of microbial analysis (Figure 4). Corresponding to insignificant changes in the TPH concentration, no significant changes in the biomass concentration were observed in the NA and BV columns. However, microbial biomass in the BS column showed a significant reduction in most areas of the column as a result of introducing air in the saturated zone.

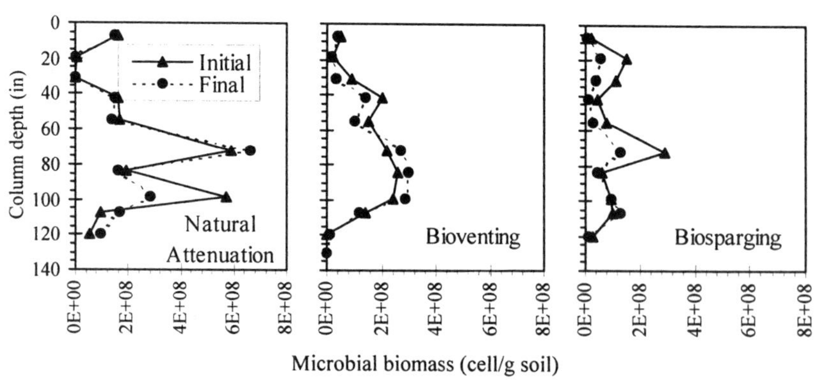

FIGURE 4. Initial and final biomass profile in NA, BV, and BS columns.

Contaminant Biodegradation in Aqueous Phase. TPH concentration in water samples from the bottom, middle, and top of the saturated zone of the NA, BV, and BS columns is illustrated in Figure 5. In the NA column, except for the anomalous results for 4th sampling event, there was not any significant change in the TPH concentrations over a 10-week period. On the 4th sampling, water samples were drawn from the ports at a higher flow rate that probably resulted in significant volatilization of the contaminant from the sample. Aqueous TPH concentration in the top of the saturated zone in the BV and BS columns increased by a factor of approximately seven during the equilibration period. These data suggest a redistribution of hydrocarbons between the soil and aqueous phases, however, this redistribution did not change the concentration of TPH in the soil significantly because there was about two orders of magnitude more mass of TPH in the soil as in the aqueous phase. In the BS column the aqueous phase TPH concentrations at the top of the saturated zone decreased dramatically in sample events following the equilibration period, which indicates a rate of aqueous phase TPH degradation exceeding the rate of TPH desorption from the soil. A similar reduction in aqueous TPH concentrations was not observed in the middle and lower levels of the saturated zone, possibly as a result of lower population of viable biomass in these zones.

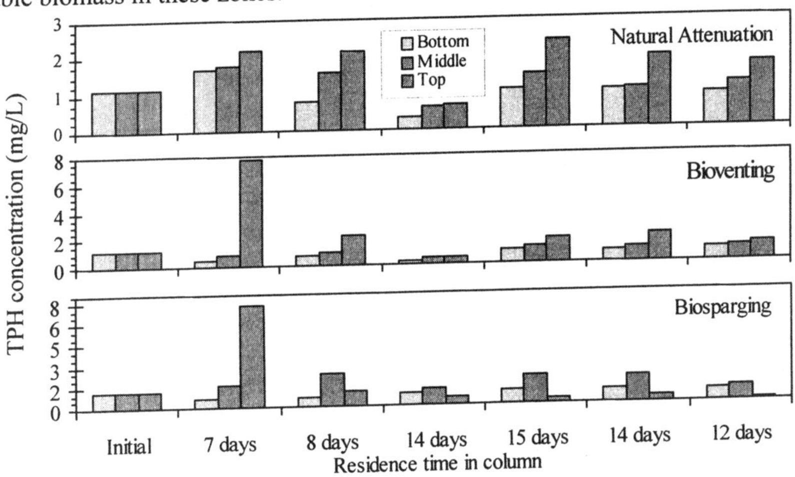

FIGURE 5. TPH concentration in aqueous phase in NA, BV, and BS columns.

Respiration Gas Analysis. As air passed through the columns, concentration of CO_2 increased with a corresponding decrease in O_2 concentration as a result of microbial respiration. The cumulative plots of O_2 consumption and CO_2 production are shown in Figure 6. Using the well-defined relationship (3.2 g O_2 g^{-1} HC) between aerobic HC metabolization and O_2 consumption (Cookson, 1995), the cumulative O_2 consumption was converted to cumulative TPH biodegraded. Over the 9 weeks, 1.77 g TPH (corresponding to 0.177 mole O_2 consumed) was degraded in the BV column as compared to 1.6 g TPH (equivalent to 0.16 mole O_2 consumed) degraded in the BS column. As shown in Figure 6, O_2 consumption and CO_2 production stabilized between day 42 to 62, resulting in a steady O_2 consumption and CO_2 production rate of 0.57 and 0.45 mmole d^{-1} in the BV column; and 0.80 and 0.54 mmole d^{-1} in the BS column.

Based on the mass of contaminated soil, the estimated zero-order removal rate between day 42 to 62 in the BV column (17.3 kg soil) was 0.33 mg TPH kg^{-1} contaminated soil d^{-1} (corresponding to 0.57 mmole O_2 consumed d^{-1}) and 0.49 mg TPH kg^{-1} contaminated soil d^{-1} in BS column (16.4 kg soil). Comparison of the exit gas data under two different aeration rates (Figure 6) suggested little benefit from feeding air at a higher rate than the rate recommended by EPA.

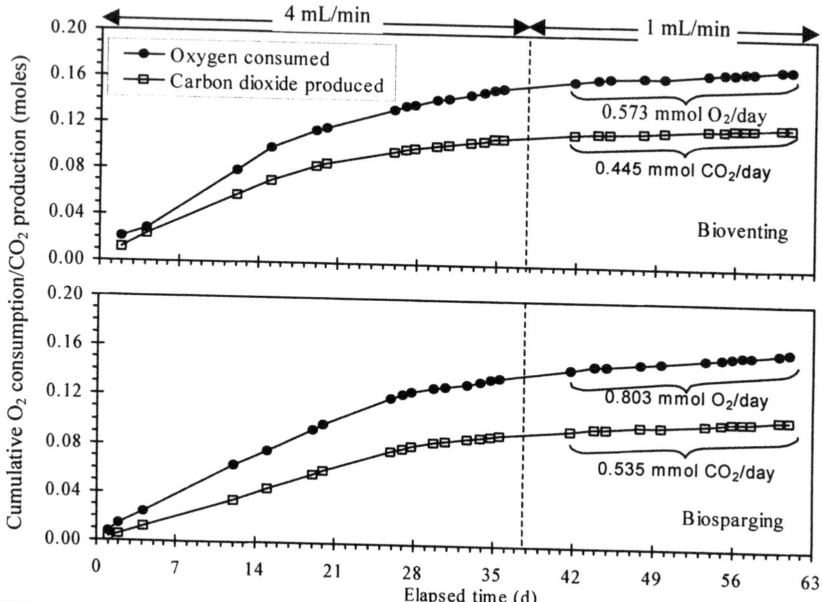

FIGURE 6. Cumulative O_2 consumption and CO_2 production in BV and BS columns.

REFERENCES

Aelion, M., and P.M. Bradley. 1991. "Aerobic Biodegradation Potential of Subsurface Microorganisms from a Jet Fuel Contaminated Aquifer". *Appl. Microbiol. 57*: 57-63.

Barker, J.F., G.C. Patrick, and D. Major. 1987. "Natural Attenuation of Aromatic Hydrocarbons in a Shallow Aquifer". *Ground Water Monitor Rev. 7*: 64-71.

Brock, T.D., M.T. Madigan, J.M. Martinko, and J. Parker. 1994. *Biology of Microorganisms.* 7th Edition. Englewood Cliffs, NJ.

Cookson, J.T. Jr. 1995. *Bioremediation Engineering, Design and Application.* McGraw-Hill: Washington, D.C.

Ridgway, H.F., J. Safarik, D. Phipps, P. Carl, and D. Clark. 1990. "Identification and Catabolic Activity of Well Derived Gasoline Degrading Bacteria from a Contaminated Aquifer". *Appl. Environ. Microbiol. 56*: 3565-3578.

US EPA. 1995. *Straight Talk on Tanks*. United States Environmental Protection Agency; EPA 510-K-95-003, July 1995.

PERFORMANCE ASSESSMENT OF THE PHASED REMEDIATION OF A FORMER GAS MANUFACTURING PLANT

Adrian R. G. Shields (Komex, Bristol, UK)
Antony Jones, Jeremy Mason, Steve Smith (King's College London, UK)
Adrian Butler, Howard Wheater, Nigel Bell (Imperial College London, UK)

ABSTRACT: The remediation of a former manufactured gas plant (MGP) in the north of England is the subject of both a commercial remediation project and an ongoing collaborative research programme. A risk-based remediation strategy has been adopted, which involved an initial risk assessment to identify appropriate cleanup criteria and removal of primary sources. This was followed by remedial technology pilot testing, and the installation and monitoring of a full scale in-situ air sparging system. This system is currently operational. Periodic measurements of dissolved organic and inorganic components, and highly detailed time series of temperature, Eh, pH, dissolved oxygen and piezometry have been obtained for the sparge system. Naphthalene degradative activities have been assessed through the application of Filters for the Recovery of Organisms from Groundwater Samples (FROGS). Integration and interpretation of the data is undertaken using a biogeochemical groundwater model. Significant reductions in dissolved phase hydrocarbons are being achieved within and downgradient of the aeration zone. Integration of laboratory challenge tests for hydrocarbon mineralisation rates with in-situ measurements from the sparge system have demonstrated the response of the subsurface environment to the changes in redox conditions following the introduction of the sparge barrier. These results also provide guidance on the long-term effectiveness of the sparge barrier.

INTRODUCTION

The site was operated as an MGP between the 1850s and 1960s. Following this period, the works was decommissioned and demolished, and has lain vacant since. The aquifer beneath the site is an unconfined, fine grained sand with a depth to groundwater of approximately 1m. Groundwater flow is generally from west to east across the site, although a marked divergence in flow direction is also evident, with a north-easterly flow in the north of the site and a south-easterly flow in the south of the site. Although major contaminant sources have been removed (tar tanks and gasholder bases), groundwater contamination remains. Principal contaminants include phenols, BTEX (benzene, toluene, ethylbenzene and xylenes) compounds and light fraction PAHs (polynuclear aromatic hydrocarbons). Figure 1 shows a site plan including the locations of pilot scale and full scale equipment, together with key monitoring points.

Objectives. Research work at the site is ongoing. Interests include the development, installation and testing of an air sparge treatment curtain (Komex), the development of techniques for identification of aerobic and anaerobic mineralisation rates (King's College London), and the development of a variably saturated groundwater flow model with multi-species contaminant transport capabilities, incorporating coupled biochemical and geochemical reactions (Imperial College). The research and commercial work will ultimately be combined in the development of a refined risk assessment, which will be used to identify whether any additional remedial work is required at the site.

FIGURE 1. Site plan and key monitoring piezometers

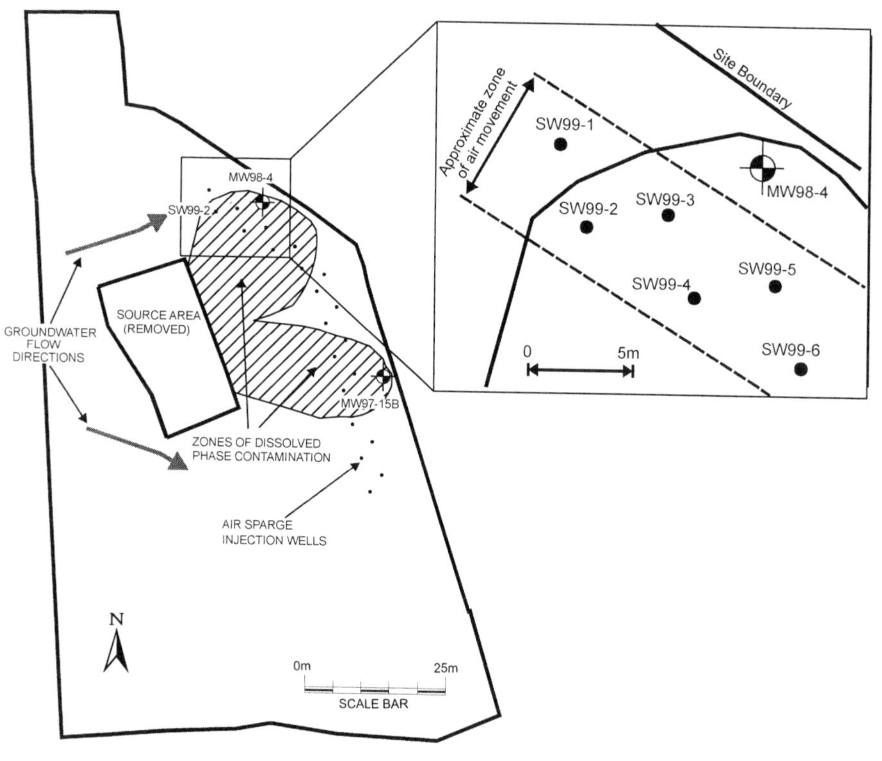

REMEDIATION WORKS

Following contaminant source removal works (Figure 1), pilot scale testing of both air sparging and oxygen releasing technologies was carried out to identify the most effective technology to apply during later remediation phases at the site (Shields et al, 2000). This work demonstrated that air sparging was the more effective technology, and a full-scale air sparging treatment curtain was designed and implemented. The system was commissioned in August 1999 and is currently operational. The system comprises twenty-two vertical air injection points (Figure 1), with overlapping zones of influence, creating an aeration curtain designed to mitigate off-site dissolved phase contaminant migration.

DATA ACQUISITION

Performance data were acquired as outlined below:

Groundwater Analytical Data. An array of groundwater monitoring piezometers located within both the northern and southern groundwater contamination plumes, in the zone of aeration, and downgradient of this zone are used to monitor groundwater quality. Water samples were analysed by contract laboratory for organic and inorganic components. Figure 1 shows the key monitoring locations discussed below.

Field Monitored Parameters. A YSI 600XLM multi-parameter water quality probe was used to take discrete measurements of measure pH, conductivity, redox potential, temperature and dissolved oxygen from piezometers across the site. The probe was also deployed long-term in monitoring well MW98-4, logging the above parameters on an hourly basis.

Biodegradative Activities. Monitoring of the temporal variation in naphthalene biodegradative activity was achieved through the application of Filters for the Recovery of Organisms from Groundwater Samples (FROGS). FROGS (Jones, 2000) enable the sampling of representative biofilms from groundwater wells permitting the subsequent quantification of biodegradative activity and extraction of nucleic acids. The presence and quantification of expression, of genes involved in the degradation of contaminants can then be made.

RESULTS AND DISCUSSION

Time series data acquired from repeated water sampling show that significant reductions in dissolved phase hydrocarbons are being achieved and maintained within, and downgradient of the aeration zone, for naphthalene, phenol and BTEX compounds (Figures 2 and 3). Within the zone of air movement, such as at SW99-2 (Figure 2), mass removal rates are relatively rapid (up to 6 $\mu gL^{-1}hr^{-1}$ for benzene), over the first 2 weeks of the project. During this time, significant concentrations of hydrocarbon vapour were recorded in the SVE offgas, confirming volatilization as a major mass removal mechanism. However, downgradient of the zone of air movement, such as atMW97-15B (Figure 3) and MW98-4, negligible air movement precludes significant mass removal by volatilization, yet contaminant reduction is still observed. The evidence for mass removal via biodegradation is discussed below.

Figure 4 shows time series data of dissolved oxygen concentrations at MW98-4, downgradient of the air sparge curtain, recorded on an hourly basis by the downhole probe. On startup of the air sparge system, dissolved oxygen

FIGURE 2. Variations in key organic contaminants during air sparging, SW99-2.

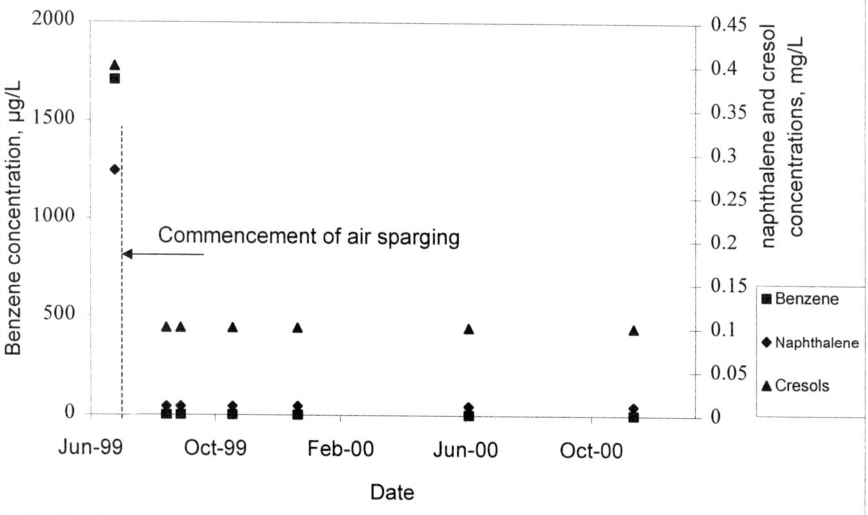

FIGURE 3. Variations in key organic contaminants during air sparging, MW97-15B.

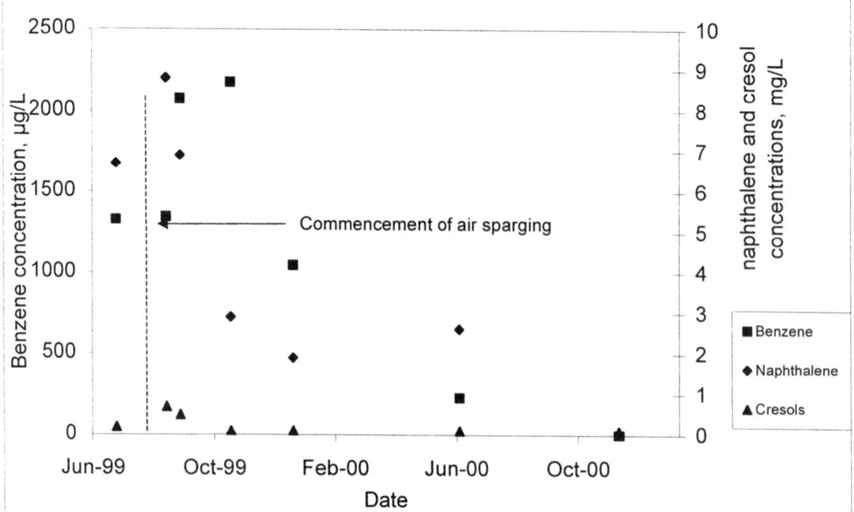

FIGURE 4. Dissolved oxygen concentration, MW98-4.

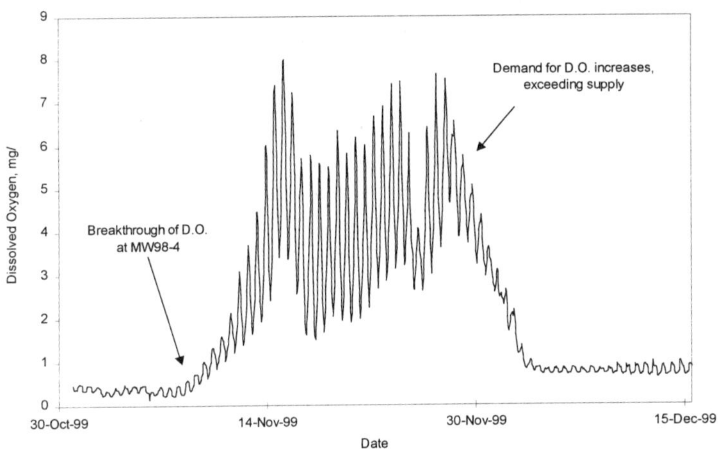

concentrations were persistently below 1 mgL^{-1}. Dissolved oxygen remained depressed for approximately 80 days, until early November 2000. After this point a gradual increase up to 6 - 8 mgL^{-1} was observed. The 'sawtooth' nature of the trace is a reflection of the pulsed air injection cycles made by the air sparge system. The 80 day delay to the onset of downgradient aerobic conditions reflects the groundwater travel time from the edge of the air sparge well zone of influence to MW98-4, as shown below.

$$t = \frac{\eta_e (s_2 - s_1)}{Ki \times 3600 \times 24} \qquad (1)$$

Where:

t = travel time of oxygenated groundwater to MW98-4, days.
K = Hydraulic conductivity of MW98-4; 1 x 10^{-6} ms^{-1} (Komex, 1998)
i = hydraulic gradient; 0.03 (Komex, 1997)
η_e = Effective porosity of aquifer; 0.2 (Komex, 1997)
s_1 = Radius of influence, air sparge wells; 3.5m (Shields, 2000)
s_2 = Distance from nearest sparge well to MW98-4; 4.5m

The calculation yields a value for t of 77 days, corresponding closely with the observed dissolved oxygen breakthrough.

The decline in dissolved oxygen is attributed to an increase in subsurface oxygen demand associated with aerobic biodegradation. Previous work at the site has shown that a diverse population of aerobic bacteria are present in the subsurface, including naphthalene and phenanthrene degraders (Jones, 1999) and BTEX degraders (work not yet published). A comparison of the naphthalene mineralisation rates identified for these hydrocarbons corresponds favourably with laboratory analysed BOD (biochemical oxygen demand) data. A maximum naphthalene mineralisation rate of 2.41 $\mu gL^{-1}hr^{-1}$ was recorded for MW98-4, through analysis of FROG material. BOD analyses for groundwater samples from MW98-4 varied between 58 and 125 $\mu gL^{-1}hr^{-1}$. Although the naphthalene mineralisation rate therefore represents only 2 - 4 % of the BOD, this is considered to be a realistic proportion, given the presence of significant concentrations of more readily biodegradable species, such as BTEX compounds, and taking natural respiration (biodegradation of natural organic compounds such as humic acids, biodegradation of dead biomass).

SUMMARY AND CONCLUSIONS

A full scale air sparging system has been designed and implemented, and is operating successfully. The corroboration of hydrogeological, chemical and microbiological data has elevated the confidence which can be ascribed to these data. This in turn provides a sound basis for model development work based upon the site dataset currently being carried out. The model, which combines variably saturated groundwater flow with multi-species contaminant transport incorporating coupled biochemical and geochemical reactions, is being used in conjunction with the site data to develop a framework for assessing the long-term impacts the remediation strategy (Butler et al., 1999). In order to achieve this, the effects of source term variability, contaminant cosolubility, competitive demands for nutrients and terminal electron acceptors are also included in the model's structure.

FURTHER WORK

The combined research programme remains ongoing. The comprehensive modelling framework will ultimately be used to refine the preliminary site risk assessment, which will then be used to update the site cleanup criteria. Remedial planning for the site can then be reviewed and scaled up or down accordingly.

ACKNOWLEDGEMENTS

This work has been supported by the UK Engineering and Physical Science Research Council (EPSRC). Additional support has been provided by Komex Europe, Lattice Property and TES Bretby Laboratories.

REFERENCES

Butler, A., H. Gao, and H.S. Wheater. 1999. "Modelling Controls on Natural Attenuation at Contaminated Gasworks Sites." *2nd NNAGS Conference on Natural Attenuation, June 1999, Sheffield, UK.*

Jones, A.D.G., H.C. Fenwick, R.D. Henkler, A.L. Glover, and J.R. Mason. 2000. "Assessment of Constraints to Bioremediation in Contaminated Sites". *Goldschmidt 2000 Symposium, 2000, Oxford, UK.*

Jones, A.D.G., J. Mason, S. Smith, H.S. Wheater, A.B. Butler, J.N.B Bell, H. Gao, A.R.G. Shields, P.E. Hardisty, and S. Wallace. 1999. "Hydro-biologic Controls on Transport and Remediation of Organic Pollutants." In A.Leeson and B. Alleman (Eds.), *The Fifth International In Situ and On-Site Bioremediation Symposium. San Diego, California, April 1999.* Battelle Press, Columbus, Ohio.

Komex, 1997. "Environmental Assessment of Former Gasworks" (confidential report).

Komex, 1998. "Phase I Remediation Works of Former Gasworks" (confidential report).

Shields, A.R.G, Hardisty P.E., Butler, A.B., Wheater, H.S., Jones, A.D.G, Mason, J., Wallace, S., 2000. "Remediation of a Former Gas Manufacturing Plant using a Research-Supported In-Situ Air Sparging Programme." In *Consoil 2000: Seventh International FZK/TNO conference on Contaminated Soil. Leipzig, Germany, 2000.* Thomas Telford, London.

SOIL VAPOR EXTRACTION AND BIOVENTING OF CONTAMINATED LANDFILL MIXED WASTE

M. Talaat Balba, Darlene Coons, Cindy Lin, Ian Richardson, Sue Scrocchi, and ***Alan Weston*** (Conestoga-Rovers & Associates)

ABSTRACT: Disposal of chemical waste at a landfill facility in Ohio resulted in release of chemicals into the groundwater. Fluctuation of the groundwater table extended the contamination throughout the landfill. Waste excavated at the Site as part of source removal activities needed to be treated to acceptable levels so it could be disposed. The waste consisted of a non-homogenous mixture of municipal and industrial waste, including rubber strips, putrescible garbage, wood, metal, and other solids. The compounds of concern (COCs) in the waste included benzene, toluene, ethylbenzene, and xylenes (BTEX), trichloroethene (TCE), tetrachloroethene (PCE), and vinyl chloride monomer (VCM). Soil vapor extraction (SVE) followed by bioventing as a polishing step was identified as a potential treatment alternative for the remediation of this waste. A bench-scale treatability study (Phase I of USEPA remedial action) was conducted to assess whether this technique could be applied to effectively treat the waste and reduce the contaminant concentrations. The treatability study was conducted in phases: SVE to remove the TCE, PCE, and VCM, followed by bioventing to enhance biodegradation of the BTEX compounds. The SVE treatment was 99.3 percent effective in removing the TCE. The toxicity characteristic leaching procedure (TCLP) regulatory limit for TCE is 0.5 mg/L. All of the rubber and waste samples were below this limit with greater than a 99 percent confidence level after 6 weeks. Bioventing resulted in greater than 78 percent removal of the BTEX from the soil and greater than 99 percent removal from the rubber material. Based on the successful results of the treatability studies, a pilot test is being conducted.

INTRODUCTION

As part of Remedial Investigation/Feasibility Study (RI/FS) activities being performed at a landfill site in Ohio, Conestoga-Rovers & Associates (CRA) has been conducting drum and buried container removal activities at the Site since November/December 1998. Excavated wastes are stockpiled in on-site staging pads pending characterization. Waste materials that pass Toxicity Characteristic Leaching Procedure (TCLP) criteria are used as backfill at the site. Any waste that fails the TCLP criteria is held in stockpiles pending final disposition. The compounds detected above TCLP criteria in the waste include benzene, trichloroethene (TCE), tetrachloroethene (PCE), and vinyl chloride monomer (VCM). TCE is the primary constituent that fails the TCLP criteria. The waste is a mixture of municipal and industrial waste, including rubber strips, putrescible garbage, wood, metal, and other solids.

In October 1999, the USEPA granted approval to stage the waste failing TCLP criteria in on-site stockpiles for up to 2 years to allow for evaluation of on-site treatment. Vapor extraction was identified as a potential treatment alternative for this waste.

VE is a proven technology for the treatment of soil impacted with a wide range of VOCs, particularly chlorinated solvents. The technology relies principally on creating a vacuum to vaporize and remove compounds that have a low molecular weight and high Henry's Law constant (greater than 0.01 dimensionless units). Henry's Law constants for the compounds of concern indicate that they are strippable at typical laboratory temperatures between 15 and 20°C (see Table 1).

TABLE 1. Henry's Law Constants of COCs.

Compound	15°C	20°C
Benzene	0.16	0.19
Toluene	0.21	0.23
Ethylbenzene	0.19	0.25
Xylene	0.19*	0.24*
TCE	0.28	0.35
PCE	0.47	0.59
VCM	0.71	0.90

Note:
* Average of values for o, p, and m-xylene.

Objective. Although VE has been used successfully at numerous sites, the performance of the process on mixed soil/debris has not been adequately evaluated. Therefore, CRA conducted a bench scale treatability study (Phase I of USEPA remedial action) to assess whether VE can be applied to treat the excavated soil/debris mixture from this Ohio site failing the TCLP criteria. The main objective of the laboratory treatability study was to determine the feasibility of using VE process to reduce the TCE, PCE, and VCM concentrations in the waste to below TCLP limits and to determine the rate of reduction of VOC concentration over time. Bioventing was then applied as a polishing step to determine its effectiveness in completing the reduction in the concentrations of BTEX in the waste.

MATERIALS AND METHODS

Waste Characterization. An approximately 100-gallon composite sample of rubber material and other waste material was collected from a freshly exposed area of excavated waste material. Upon confirmation that the material did contain VOCs above TCLP limits, the 100-gallon composite sample was shipped in sealed containers to the CRA Laboratory in Niagara Falls. Upon arrival at the laboratory, CRA sized the larger material down to allow placement in the treatment vessel, and the rubber strips were cut into approximate 6-inch strips.

Experimental Design. The treatability study was performed in a 27-gallon stainless steel vessel connected out of the side at the bottom to a vacuum/pressure pump. The pump was used to pull air through the waste. A pictographic view of the treatment vessels is provided in Figure 1.

FIGURE 1. Treatment vessels.

The composite waste material was placed in the vessel on top of a stainless steel mesh screen above a loop of copper tubing with air inlet holes. The inlet holes were spaced around the loop to provide uniform air withdrawal from the base of the material. The copper tubing sat on a thick layer of gravel/stones with a glass bead layer covered with glass wool between the copper tubing and the steel screen. The composite waste material was placed on top of the screen and was manually compacted. The waste was covered with a lid including an opening to admit influent ambient airflow. Fresh air flowed through the opening in the stainless steel lid, through the waste, across the mesh, and out of the vessel through the copper tubing.

A control stainless steel cylinder vessel of the same dimensions was fabricated and provided with an air inlet hole in the lid, but with no exit hole. It was filled with waste material and allowed to stand next to the active vessel throughout the test.

Testing Protocol. The treatability study was conducted for a period of 10 weeks; 6 weeks of VE and 6 weeks of bioventing. The VE system was initiated by activation of the vacuum pump. The vacuum pump was operated continuously and a flow regulation valve was adjusted to maintain a flow of approximately 30 pore volumes per day during the test. (This flow was selected based on the observed excellent permeability of the composite waste material. However, if this rate can not be achieved in the field, the treatment may require slightly longer duration to achieve the treatment goals).

A gas GC sampling loop was connected in parallel with the main evacuating line before the pump. The gas loop was sampled and analyzed to assess the VOC concentrations in the vapor phase and the amounts removed from the waste by the VE treatment. The data were used as a tool to assist in the evaluation of process performance. A sampling port for Summa canisters was also provided. The Summa canister data were used to confirm that VOCs were extracted by the treatment.

A schematic of the treatability study apparatus is shown in Figure 2 below.

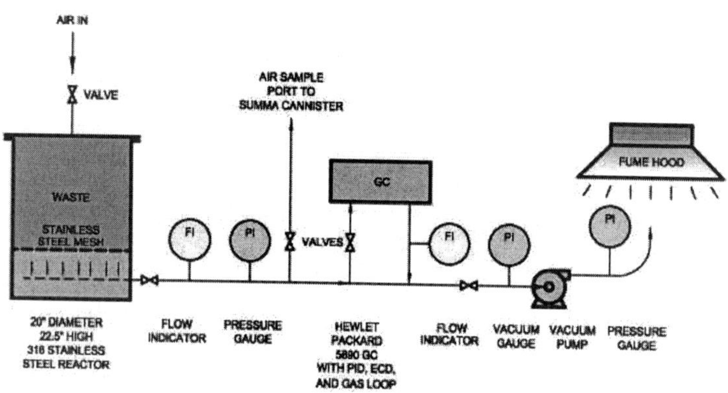

FIGURE 2. Schematic of Apparatus

During the study, samples from the treated rubber material and from the other waste material were collected and analyzed. The following data were also recorded:

i) flow rate from the extraction point (daily);
ii) pressure at the extraction point (daily);
iii) ambient and soil temperature (daily); and
iv) moisture content in the waste (weekly).

After 6 weeks of VE treatment, the system was switched to a "bioventing" system to determine if enhancement of nutrient conditions would enhance performance and VOC removal. The system was inoculated with microbes and nitrogen and phosphorus inorganic fertilizer, and the system was monitored for 6 weeks.

SAMPLING AND ANALYSIS

Random samples of the rubber and waste material were collected and analyzed prior to starting the study to establish baseline conditions. The samples were analyzed for total VOCs, TCLP VOCs, and moisture content using SW-846 methods. During the study, random samples of the rubber and waste material were collected from the treatment vessel and analyzed for total VOCs and moisture content at T=2, 4, and 6 weeks.

During VE, the gas loop was sampled and analyzed for specific VOCs (VCM, TCE, PCE, and benzene) at half-hour intervals during the first 24 hours, at one-hour intervals for the remainder of the first week, and then at four-hour intervals for the remainder of the test. The gas loop samples were analyzed by Method SW 846-8021B against a certified 1 part per million (ppm) gas standard containing TCE, PCE, VC, and benzene in nitrogen.

Summa canister samples were collected at T=0 and 2 hours and at T=2 weeks and analyzed for VOCs by Method TO-14. The effluent from the pump was sampled after the 6 weeks using a Tedlar bag so that the results could be compared with the gas loop data. TCE and air were injected into Tedlar bags to produce air samples with TCE concentrations of 0.91, 9.1, and 91 parts per million volume (PPMV). These samples were analyzed using the gas loop GC as a QC check on the analytical standard.

At the end of the VE phase and prior to bioventing, soil samples were collected and analyzed for pH, moisture content, nitrogen, phosphorus, and microbial counts. Samples were then collected again at T=2, 4, and 6 weeks and also analyzed for these parameters to ensure the presence of adequate nutrient levels for microbial activity.

RESULTS AND DISCUSSION

The data showed that TCE was removed efficiently from the material over the course of the treatment. The cumulative removal of TCE over time was calculated and plotted, as shown in Figure 3 below. The largest amounts of TCE were removed at the beginning of the treatment, with decreasing amounts removed as the treatment progressed.

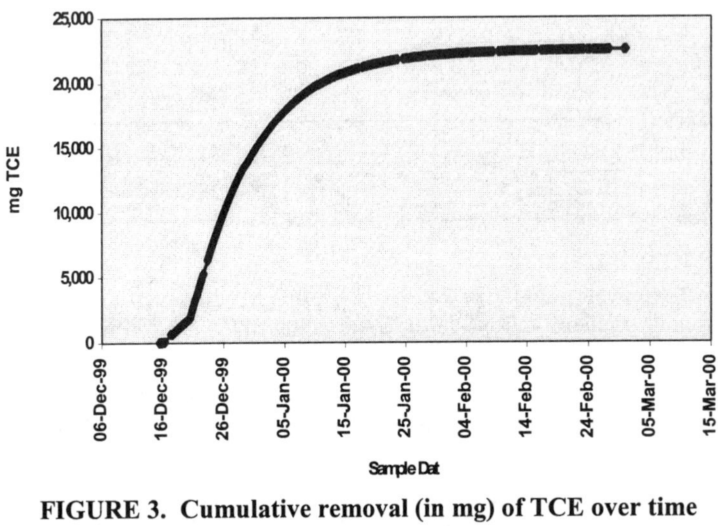

FIGURE 3. Cumulative removal (in mg) of TCE over time

The mean TCLP concentration of TCE in the rubber material was reduced from 3.9 to 0.018 mg/L and in the remaining waste it was reduced from 0.81 to 0.015 mg/L. Based on mass removed, the mass of TCE estimated in the waste prior to VE treatment was 28g. The mass of TCE in the waste after VE treatment was estimated at 0.19 gm, suggesting that the VE treatment was 99.3 percent effective in removing the TCE.

The TCLP regulatory limit for TCE is 0.5 mg/L. All of the rubber and waste samples collected after 6 weeks of VE treatment were below the limit with greater than a 99 percent confidence level.

The bioventing system maintained a pH within the range of 8.8 to 9.3. Moisture content in the soil averaged around 13.5 percent during the bioventing treatment. Nitrogen and phosphorus were maintained at levels sufficient for biodegradation. Total microbial counts were in the range of 10^6 CFUs/g (colony forming units per gram). The BTEX-specific degrading microbial population ranged between 10^4 and 10^5 CFUs/g. The specific BTEX microbial counts were assessed using solid media amended with BTEX as the sole source of carbon and energy.

Analytical results of bioventing indicated that enhancing the system with BTEX-specific microbial cultures did increase the degradation of BTEX. Greater than 78 percent removal of BTEX in soil and greater than 99 percent removal of BTEX in the rubber was observed after 6 weeks. The results are summarized in Table 2.

TABLE 2. Bioventing analytical results

Compound (mg/Kg)	SOIL T=0	SOIL T=6 weeks	SOIL % Loss	RUBBER T=0	RUBBER T=6 weeks	RUBBER % Loss
Benzene	ND 2.0	ND 2.0	-	ND 20	ND 2.0	-
Toluene	4.9	0.85J	83	ND 2	0.04J	>99
Ethylbenzene	20	3.3	84	435	0.32J	>99
Xylenes (total)	87	19	78	3115	3.6	>99

Notes:
J = Associated value is estimated.
ND = Non-detect at associated value.

CONCLUSION

In bench-scale studies, six weeks of VE treatment reduced TCE concentrations in the soil and waste to levels below the TCLP limit of 0.5 mg/L with greater than a 99 percent confidence level. Polishing using bioventing further enhanced degradation of BTEX compounds in both the soil and rubber waste. A field pilot test is now being considered to evaluate the treatment under field conditions.

ACKNOWLEDGEMENTS

CRA would like to acknowledge Michelle Simon of the USEPA and Linda Cumming of Battelle for their valuable input to this project.

REFERENCES

USEPA. 1998. *Remediation Case Studies: Debris and Surface Cleaning Technologies, and Other Miscellaneous Technologies. Vol. 13.* EPA 542-R-98-017.

USEPA. 1991. *Soil Vapor Extraction Technology Reference Handbook.* EPA 540/2-91/003.

PRECISE CONTAMINANT DISTRIBUTION DELINATION USING MEMBRANE INTERFACE PROBE (MIP) TECHNOLOGY

Alex Nadolishny (Nedatek, Inc., Brandon, FL)
Donald Ochs (Regenesis, Cinnaminson, NJ)
Paul Fleischmann (ZEBRA Environmental Corp., Lynbrook, NY)

ABSTRACT: The Membrane Interface Probe (MIP), a real-time subsurface logging tool that provides continuous readout of relative VOC concentrations and electrical conductivity of the soil has been utilized to obtain a comprehensive set of data regarding the distribution of jet fuel in the subsurface prior to the injection of Oxygen Release Compound (ORC), an oxygen source used for bioremediation. The subsurface data was used to achieve precise placement of the ORC.

INTRODUCTION

A number of direct-sensing tools have been developed recently in an attempt to expedite the process of subsurface data collection, while minimizing the amounts of investigation-derived waste. The technologies include a number of Cone Penetrometer Technology (CPT)-deployed systems such as Laser-Induced Fluorescence (LIF), Rapid Optical Screening Tool (ROST) and Fuel Fluorescence Detector (FFD), and the less expensive percussion-resistant systems, such as Direct Sensing Soil Conductivity (DSSC) and the Membrane Interface Probe (MIP). The percussion-resistant tools allow tool advancement with common percussion soil sampling machines such as Geoprobe, EarthProbe, and others. The availability of these systems makes it possible to apply direct-sensing tools on medium-scale projects with great efficiency.

The application of enhanced bioremediation technologies, by adding electron donors or electron acceptors to the subsurface, requires that detailed site information is available prior to the product placement activities. The information includes: distribution of contaminants in the subsurface, presence/absence of NAPL and soil stratigraphy. The use of the direct sensing tools allows this information to be provided in real time, and with very high resolution.

Objective. The objective of this project was to utilize the Membrane Interface Probe (MIP) technology to obtain a comprehensive set of data regarding the distribution of the jet fuel in the subsurface prior to the injection of Oxygen Release Compound (ORC). The subsurface data were used to achieve precise placement of the ORC.

Site Description. The site is located between a runway and a taxiway of a former Air Force Base. A number of Jet A fuel spills occurred over an extended period of time at a fuel pumping station and along the product piping.

The soils in the vadose zone are represented by fine to medium sands, with silty units present below the water table. It was determined that the static water table has been fluctuating by as much as 15 feet (4.5 m) during several dry years prior to this investigation. The majority of the logging locations were situated on a grassy area between the runway and the taxiways, with several locations situated on a paved area.

METHODS AND EQUIPMENT

A custom-built Membrane Interface Probe (MIP) system was provided by ZEBRA Environmental Corporation. The probe was advanced into the soil using a Geoprobe 5400 series probe unit. The same unit was later utilized to perform the injection of the Oxygen Release Compound (ORC).

The Membrane Interface Probe. The MIP is a direct-sensing screening tool that provides real-time data on relative VOC concentrations and electrical conductivity of the soils.

The system utilizes a proprietary Teflon TFE membrane placed on a heating plate. Nitrogen carrier gas sweeps the inner side of the membrane, bringing the contaminants transported across the membrane to the detectors mounted in the carrier vehicle.

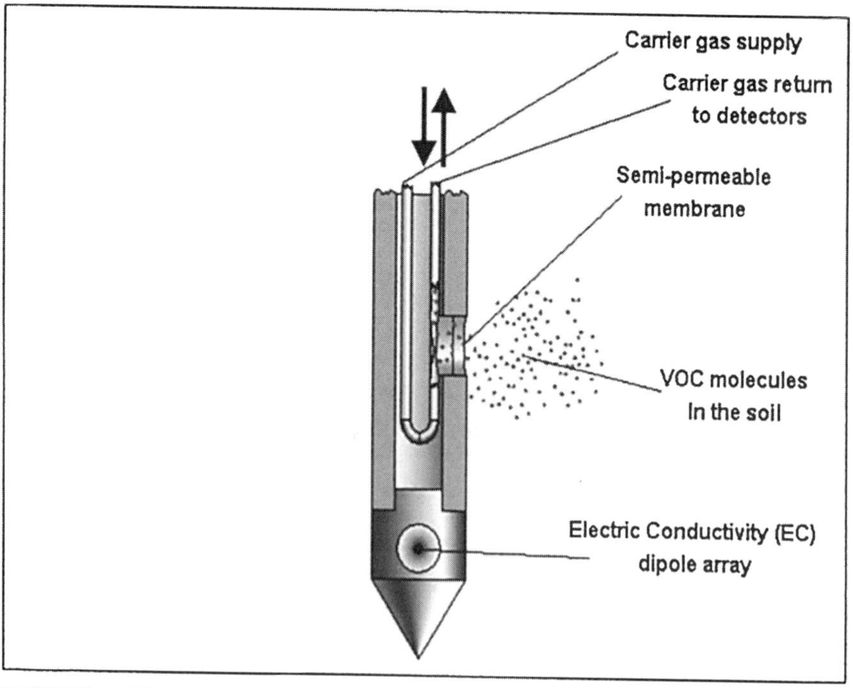

FIGURE 1. The downhole portion of the Membrane Interface Probe system schematic. (The image is courtesy of EA Engineering of Newburgh, NY)

The carrier gas pressure is maintained at 4 to 8 psi on the inner side of the membrane. This prevents the water from breaking through the membrane by maintaining a pressure gradient across the membrane. The presence of the gradient, however, does not interfere with the transfer of the VOC molecules across the membrane in a direction opposite to the pressure gradient. The reason is the mechanism of the VOC transfer: the VOC molecules are not transported by the flow of the gas diffusing through the membrane pores (the flow is towards the outside of the probe); instead, they get absorbed into the hydrophobic matrix of the membrane and get desorbed on the other side of the membrane, where they get picked up by the carrier gas flow. The movement of the molecules results from a concentration gradient instead of pressure gradient. The heating of the membrane increases the rate of the transfer, increases vapor pressure for VOCs present in the soil adjacent to the membrane, and volatilizes some of the compounds with low vapor pressures at the ambient temperatures. The area affected by the heater is limited to several cubic inches immediately adjacent to the membrane. Due to extremely high depth resolution of the MIP equipment (theoretically as high as 1 inch) and effects of the heat and soil disturbance, replicate measurements cannot be obtained with the MIP within the same borehole.

The carrier gas was analyzed utilizing OI Analytical Model 4450 Tandem PID/FID detectors. An Electron Capture Detector (ECD) also was utilized during screening for halogenated hydrocarbons.

QA/QC procedures were limited to monitoring all the physical parameters of the equipment (heater operation, carrier gas pressure, conductivity system continuity and isolation tests), and sensitivity checks performed prior to each logging location. The sensitivity checks were performed by the means of introducing an exact volume of a standard water solution of BTEX constituents to the membrane at operating temperature, and measuring resulting detector response and gas trip time. Adjustments to the gas flow, or if necessary, probe/membrane replacement were performed if the measured detector response was less than 85% of the standard response magnitude.

The MIP is a screening tool and does not provide data on exact concentrations of contaminants in the subsurface. The detector output was recorded in uV (microVolts), and then a correlation was established between the MIP data and sample analysis results. The MIP data typically show correlation coefficient of .65 - .85 when compared to analytical results. The focus of this particular application was on obtaining information about distribution of the contamination, and detailed correlation study was not performed.

The information recorded by the system is displayed in real time on a computer screen (see FIGURE), and it is recorded in ASCII format on a computer disc. Upon completion of a project, the data can be used to print out the logs and to prepare vertical cross-sections (see RESULTS) or aerial view maps.

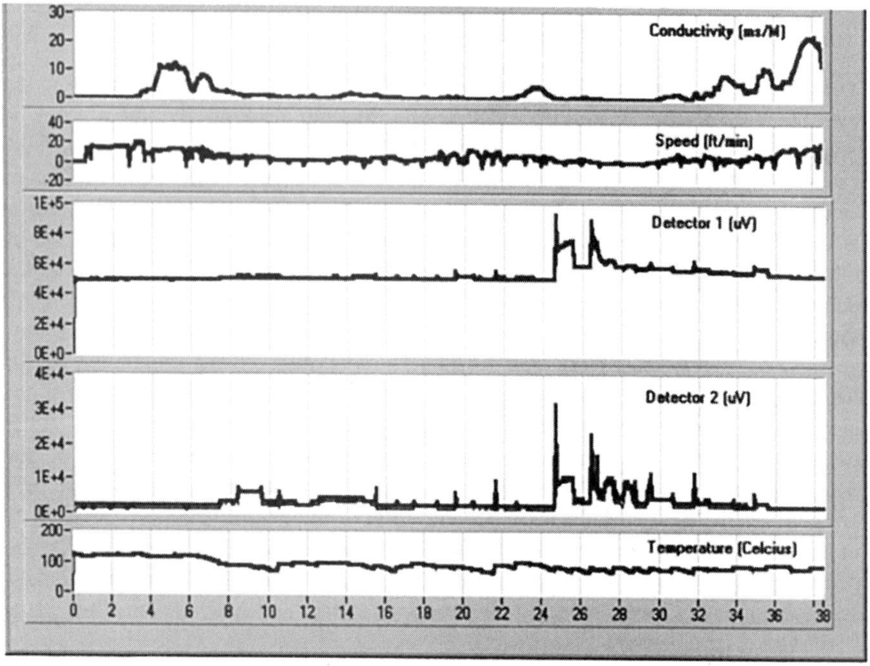

FIGURE 2. A screen shot of the MIP system output.

RESULTS

Prior to the MIP logging, an extensive smear zone within the groundwater fluctuation range was expected. Accordingly, plans were made to inject ORC evenly within the interval of 11- to 48- feet (3.3 to 14.4 m) below grade.

A total of 11 locations were logged during two days on site. The locations were situated along two lines dissecting the area of the investigation. A typical MIP logfile from the site is shown as FIGURE below.

The data collected on site indicated that the bulk of contamination became trapped in the deeper, silty layers, and was not brought up to existing water table after the rise of the static water table level. The MIP data were confirmed by random discrete soil sampling.

The conclusions were utilized to make modifications to the original work plan. The modifications included a change of the ORC injection interval to correspond with the contaminant distribution. The data obtained with the MIP system were used to create a graphical representation of the distribution of the Jet A fuel contamination in the subsurface (see FIGURE 4). The cross-sections were prepared using Surfer 6 software (Golden Software, Golden, CO). The cross-sections, along with the individual logs, were provided as deliverables (results) to the personnel responsible for making a decision regarding the placement of the injection points and depth of the injection.

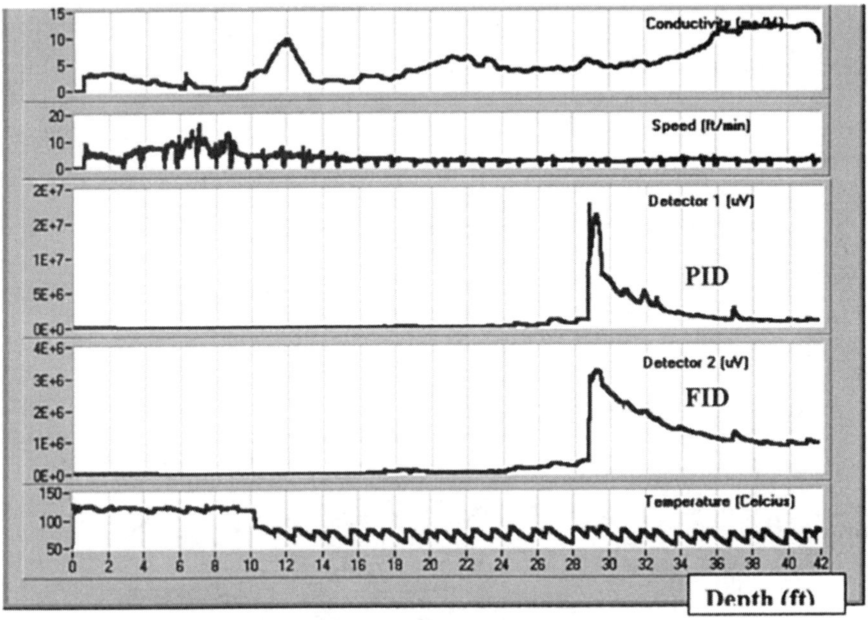

FIGURE 3. A MIP log file recorded on site. The static water table was observed at 10.5 ft (3.15 m) below grade surface.

CONCLUSION

The use of a rapid site assessment tool such as the MIP allowed the client to obtain crucial contaminant distribution information during only two days immediately prior to the injection of the ORC. The data obtained with the high-resolution MIP system provided a distribution model dramatically different from the previous one based on data collected from the sampling of monitoring wells. The analysis of the data resulted in changes to the injection work plan, particularly in regards to the total depth of borings and injection interval. It is believed that without such a precise distribution model, significant amounts of the ORC would have been wasted. The MIP logging was performed utilizing the same Geoprobe equipment that was later used for the ORC injection, and was accomplished in one mobilization with the injection crew. This provided the client with a significant cost reduction.

The Membrane Interface Probe (MIP) has been developed by Geoprobe Systems, Salina, KS. ORC is a registered trademark of Regenesis, San Clemente, CA.

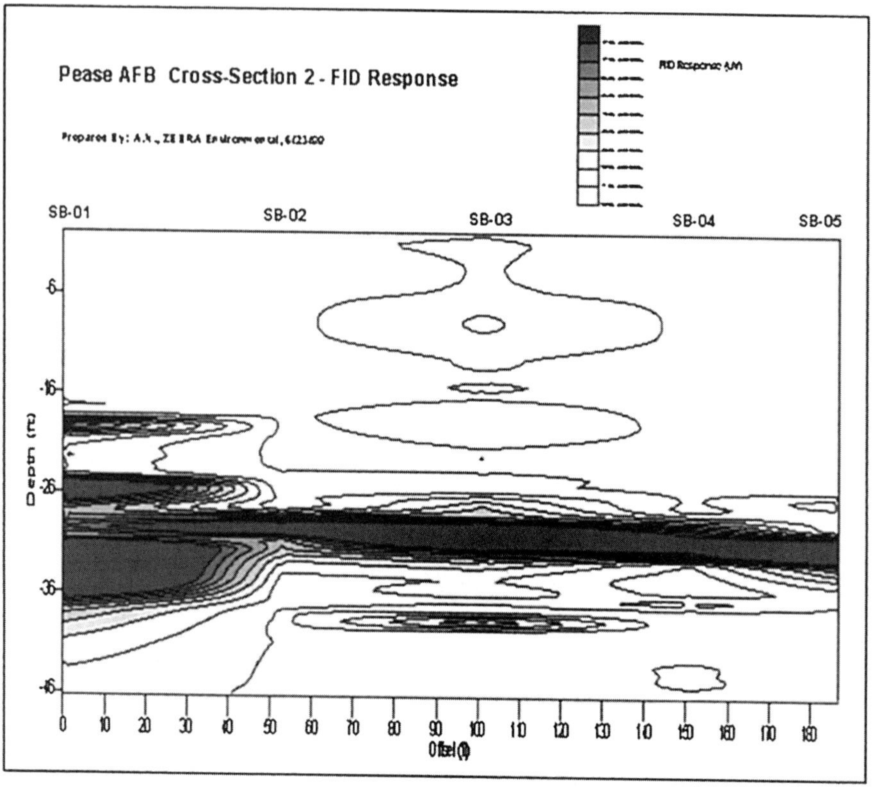

FIGURE 4. A vertical cross-section of the site representing the distribution of the contamination, based on the MIP system response (FID data used).

REFERENCES

Christy, T.M., May 1996. "Drivable Permeable Membrane Sensor For the Detection of Volatile Compounds in Soil", presented at the 1996 National Outdoor Action Conference, Las Vegas, Nevada.

Calderone, G.M., P.A. Conde, A. Nadolishny, and M. Evans. 2000. Using Membrane Interface Probe for Delineation of Chlorinated Solvent Plume. Paper presented at the EnviroExpo, Session on Innovative Technologies for Site Remediation, Boston, Massachusetts. 11 May.

P. Conde and G. Calderone, 2001. Comparison of Membrane Interface Probe Technology with Traditional Direct-Push Site Investigation Techniques. AEHS Annual Conference on Contaminated Soils, Sediments and Water 19-22 March 2001. San Diego, California.

COMETABOLIC AIR SPARGING FIELD DEMONSTRATION WITH PROPANE TO REMEDIATE TRICHLOROETHENE AND *cis*-DICHLOROETHENE

Adisorn Tovanabootr (*Post Mortem*), ***Lewis Semprini***, Mark E. Dolan, and Mohammad Azizian (Oregon State University, Corvallis)
Victor S. Magar, Dennis DeBacker, Andrea Leeson (Battelle, Columbus, OH)
Capt. David Kempisty (Air Force Research Laboratory, Tyndall AFB, FL)

ABSTRACT: A cometabolic air sparging (CAS) demonstration was conducted at McClellan Air Force Base (AFB), California, to treat chlorinated aliphatic hydrocarbons (CAHs) in groundwater using propane as the cometabolic substrate. A propane-biostimulated zone was sparged with a propane/air mixture and a control zone was sparged with air alone. Propane-utilizers were effectively stimulated in the saturated zone with repeated intermediate sparging of propane and air. Propane delivery, however, was not uniform, with propane mainly observed in down-gradient observation wells. Trichloroethene (TCE), *cis*-dichloroethene (c-DCE), and dissolved oxygen (DO) concentration levels decreased in proportion with propane usage, with c-DCE decreasing more rapidly than TCE. The more rapid removal of c-DCE indicated biotransformation and not just physical removal by stripping. Propane utilization rates and rates of CAH removal slowed after three to four months of repeated propane additions, which coincided with the depletion of nitrogen (as nitrate). Ammonia was then added to the propane/air mixture as a nitrogen source. After a six-month period between propane additions, rapid propane-utilization was observed. Nitrate was present due to groundwater flow into the treatment zone and/or by the oxidation of the previously injected ammonia. In the propane-stimulated zone, c-DCE concentrations decreased below the detection limit (1 μg/L), and TCE concentrations ranged from less than 5 μg/L to 30 μg/L, representing removals of 90 to 97%. In the air sparged control zone, TCE was removed at only two monitoring locations nearest the sparge-well, to concentrations of 15 μg/L and 60 μg/L. The responses indicate that stripping as well as biological treatment were responsible for the removal of contaminants in the biostimulated zone, with bistimulation enhancing removals to lower contaminant levels.

INTRODUCTION

Aerobic cometabolism has potential for the remediation of groundwater contaminated with chlorinated aliphatic hydrocarbons (CAHs) such as trichloroethene (TCE) and 1,2-*cis*-dichloroethene (c-DCE) (McCarty and Semprini, 1994). In-situ remediation technologies, such as air sparging, can potentially remediate sites much less expensively and more effectively than conventional ex-situ technologies. In-situ air sparging remediates groundwater through a combination of volatilization and enhanced biodegradation of contaminants (Marley and Bruell, 1995). Cometabolic air sparging (CAS) is an innovative form of conventional air sparging designed to remediate CAH-contaminated groundwater and potentially vadose zone contamination. CAS involves the sparging of air and a gaseous cometabolic substrate, such as methane, into an aquifer and the overlying vadose zone (Travis and Rosenberg, 1997). Potential gaseous cometabolic substrates include methane, propane, butane, propylene, and possibly toluene, which have all been used as growth substrates to promote cometabolic transformation of CAHs.

A demonstration of CAS treatment was performed in the subsurface at the McClellan Air Force Base, California (AFB). The subsurface at McClellan AFB was contaminated with a mixture of CAHs including primarily TCE, c-DCE, and minor amounts of chloroform (CF), 1,1-dichloroethene (1,1-DCE) and 1,1-dichloroethane (1,1-DCA). In the CAS demonstration, propane was added as the cometabolic substrate. Propane was selected because previous microcosm studies performed with groundwater and aquifer solids from McClellan AFB showed indigenous propane-utilizers were effective at transforming CAH mixtures of TCE, CF, and 1,1,1-trichloroethane (1,1,1-TCA) (Tovanabootr and Semprini, 1998).

The objective of this demonstration was to evaluate the effectiveness of CAS for remediating CAH contamination in groundwater and the vadose zone using indigenous propane-utilizing microorganisms. Air and propane were delivered to the subsurface and the temporal and spatial concentration distributions of dissolved oxygen, propane, and the CAHs of interest in the groundwater and vadose zone were determined. Propane and oxygen uptake rates and CAH transformation rates were determined at monitoring locations in both the aquifer and the vadose zone. The effectiveness of sparging with and without propane addition was determined to compare cometabolic CAH biodegradation versus CAH air stripping from the groundwater. This was accomplished by having two treatment zones, one that received air alone and one that receive both propane and air. Nitrogen (as nitrate) was also tracked as a nutrient that could become limiting with the prolonged utilization of propane. Presented here are the results from the treatment of the saturated zone.

MATERIALS AND METHODS

The experimental test site was located in Operable Unit A (OU A) at the southern end of the Base in an area with relatively high CAH groundwater concentrations. TCE and c-DCE concentrations exceeded 500 μg/L in this area. A deep vadose zone exists at the site, with the contaminated groundwater level being approximately 105 ft below the ground surface (bgs). One sparge well received air only (A-zone), and one received a mixture of propane and air (C-zone). The two sparge wells were located approximately 100 ft apart. Propane gas was fed to one sparge well, and an air compressor was used to deliver air to both wells. A soil vapor extraction system (SVE) was installed, but was never used in the experiments. Six multi-level monitoring points for soil-gas and groundwater sampling surrounded each sparge well.

The distance from the sparge well to the inner and outer multilevel monitoring wells was 7.5 ft. and 15 ft., respectively. Monitoring points in the cometabolic treatment zone receiving propane were designated MW-C1 through C6, while those in the control test plot (received air only) were designated MW-A1 through A6. Both test plots were operated equally in relation to the flow rate and the time of sparging.

The treatment system was operated as a passive treatment system, with sparging occurring over a short time duration, followed by long periods without sparging. Propane and air were pulse injected into the saturated zone once or twice a week for duration of 5 to 10 hr. A mixture of air containing propane (2%) was injected at an airflow rate of 5 scfm. Propane concentrations were increased to 4% during the tests. The control test plot was subjected to the same treatment; however, only air was injected.

The sparging resulted in propane and air being added to both the saturated and the vadose zones. The tests were performed in a series of steps as follows:

- *Day 0-36:* Collection of background concentration measurements
- *Day 36:* Propane injection initiated at a concentration of approximately 2% in air
- *Day 36-76:* Propane concentrations in the saturated and vadose zone were monitored without air sparging
- *Day 80-120:* Weekly injection of propane and air
- *Day 120-160:* Frequency of propane and air sparging was increased to twice a week in an attempt to accelerate the bioremediation of TCE and c-DCE
- *Day 160-188:* No propane or air was added so that propane oxidation rates could be verified and compared to earlier rates, and to determine if chlorinated solvent transformation would continue without additional propane
- *Day 188-220:* Ammonia was added with propane and air at a concentration of about 0.01 % after it was determined that nitrogen was limiting in the treatment zone
- *Day 220-290:* No sparging of propane or air
- *Day 300-400:* Collection of background concentration measurements
- *Day 400-550:* Propane injection at a concentration of 1% in air
- *Day 450-500:* Ethylene addition activity tests
- *Day 500-550:* Methane addition as a cometabolic substrate
- *Day 570-580:* No sparing of propane, methane, or air.

RESULTS AND DISCUSSION

Prior to sparging, higher groundwater concentrations of TCE and c-DCE were observed in the C-zone (100 to 1,000 μg/L), compared to the A-zone (60 to 400 μg/L). After sparging propane and air into the C-zone, a fairly long lag period of 20 to 60 days occurred before propane utilization was evident. Shown in Figure 1 are groundwater concentrations contours after the first addition of propane and air to the C zone and air only into the A zone. The contours are based on averaged groundwater concentrations at the 113-ft and 117 ft monitoring locations. In the C-zone, propane and air were mainly distributed to monitoring

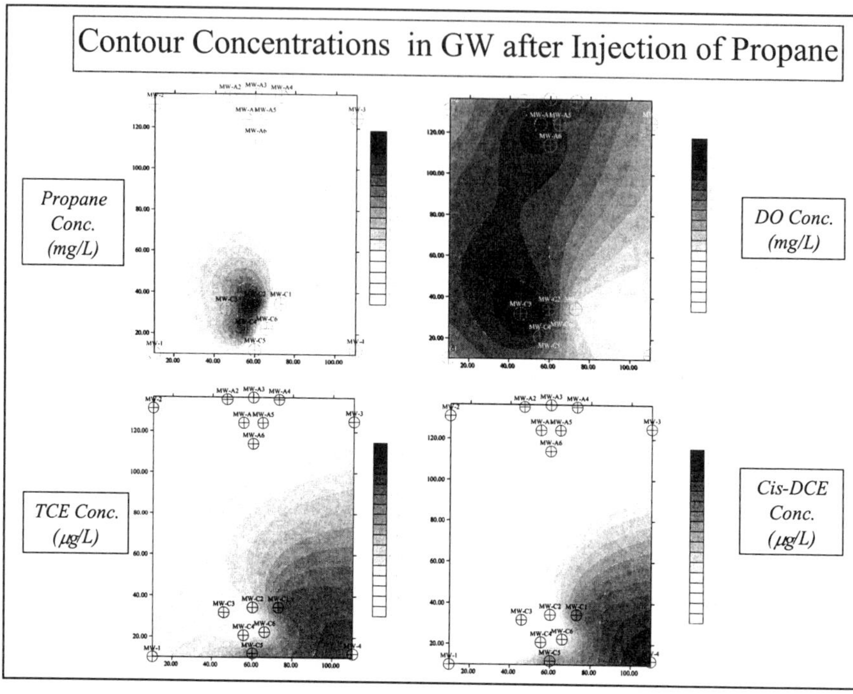

FIGURE 1. Contours of propane, DO, *cis*-DCE, and TCE after the first sparging with propane. The A-zone and C-zone are in the upper and lower portions of the diagram, respectively.

wells MW-C2, MW-C3, and MW-C4, downgradient of the sparge well. The predictions of distant propane concentrations are likely a result of the contour analysis, since monitoring wells were not located in the center of the demonstration zone. The propane concentrations in the saturated zone varied greatly, and were correlated with the dissolved oxygen concentrations, indicating some areas were more effectively sparged than others.

Groundwater concentrations of propane at the MW-C2, MW-C3, MW-C4 locations averaged about 0.3 to 1.2 mg/L after the sparge events, while dissolved oxygen concentrations were in the range of 6 to 7 mg/L. In contrast wells MW-C1, MW-C5, and MW-C6 had propane concentrations of less than 0.1 mg/L, and dissolved oxygen concentrations in the range of 1.6 to 2.5 mg/L. The results show the heterogeneous distribution of propane and dissolved oxygen that resulted from the sparging. The highest concentrations occurred at the groundwater monitoring locations closest to the sparge well (7.5 ft), and were near their solubility limits for the partial pressures of the sparge gas. The TCE and *cis*-DCE concentrations show a gradient in concentration across the C-zone, with much higher concentrations in the upgradient locations of MW-C1, MW-C5, and MW-C6. Thus, there was potential for transport of high concentrations of TCE and *cis*-DCE into the downgradient locations where propane and air were being delivered.

The A-zone had much lower CAH concentrations initially, with *cis*-DCE and TCE in the concentration range of 100 to 300 µg/L. After sparging, the DO concentration in the groundwater ranged from 6 to 7 mg/L. Background DO levels were uniformly higher in the A-zone than the C-zone, so differences in DO due to sparging were less apparent.

The rates of propane utilization also varied at the different monitoring locations. After successive propane additions, the rate of propane utilization increased as the microbial population was stimulated. Lag periods for propane utilization ranged from about 20 days to 50 days.

Figure 2 shows the results of a contour analysis of the groundwater concentration in the test zone after several sparge events, with the successive consumption of propane and DO. Propane was completely consumed in the C-zone. The decrease in the DO levels (2 to 4 mg/L) in the C-zone, in response to

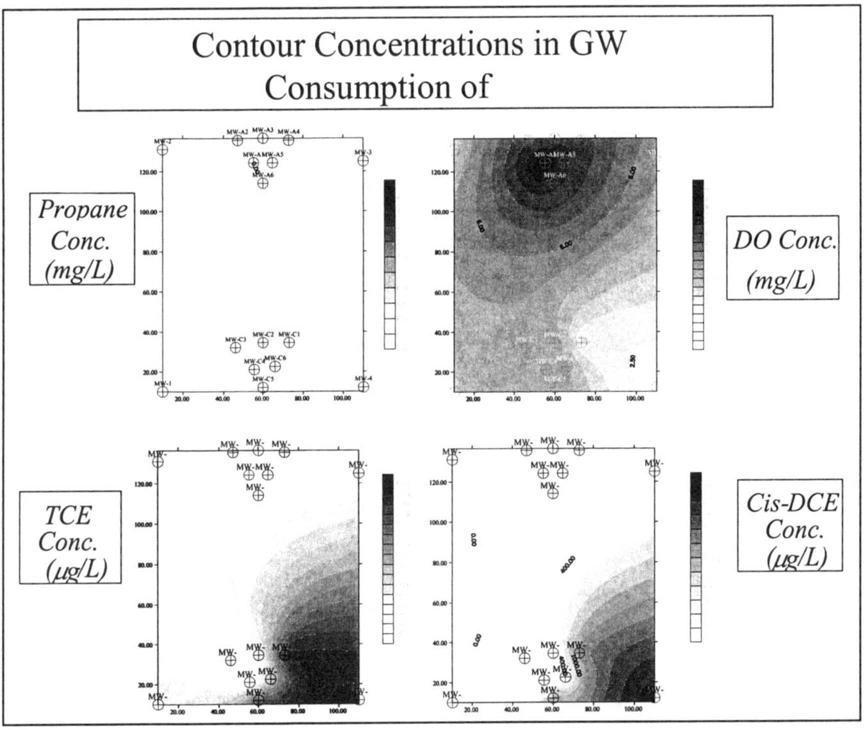

FIGURE 2. Contours of propane, DO, *cis*-DCE, and TCE after the consumption of several propane additions. The A-zone and C-zone are in the upper and lower portions of the diagram, respectively

the utilization of oxygen, was apparent. Low dissolved oxygen concentrations and high TCE and *cis*-DCE concentrations remained in the upgradient portion of the C-zone that was not effectively sparged. The portion of the C-zone that was affected by sparging and propane utilization had reduced *cis*-DCE and TCE concentrations, with *cis*-DCE reduced to a greater extent than TCE. In the A zone, some reduction in *cis*-DCE and TCE also occurred, but to a lesser extent on a relative concentration basis than was achieved in the biostimulated C-zone.

Temporal histories of propane, DO, *cis*-DCE, and TCE groundwater concentrations at monitoring well MW-C3-117 for the entire demonstration test are shown in Figure 3. This monitoring well was 15-ft downgradient from the sparge well. The pulsing in DO and propane concentrations indicates the successive sparge events with propane. Propane added on day 36 gradually decreased after 25 days. Sparging at 80 days resulted in an increase in propane concentration, which was rapidly consumed along with DO. Sparging twice a

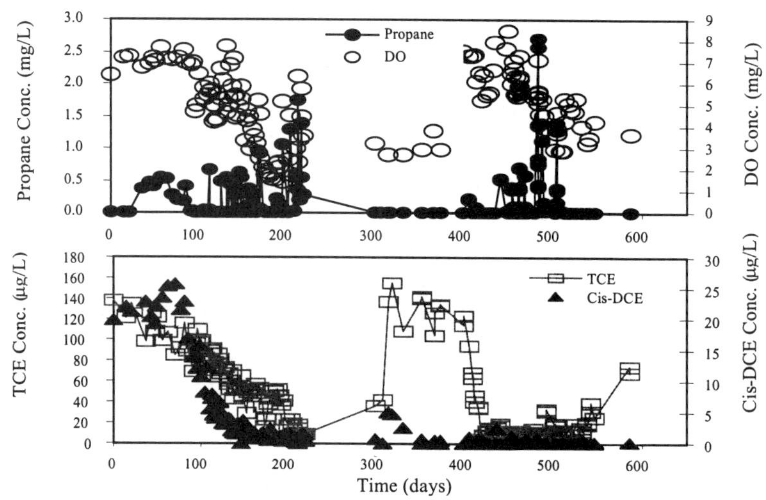

FIGURE 3. Concentration histories of propane, DO, *cis*-DCE, and TCE at groundwater monitoring well MW-C3-117.

week, up to 160 days, was followed by rapid consumption of propane and DO. Attempts to sparge higher concentrations of propane (4%) into the treatment zone (days 140 to 210) resulted in higher propane concentrations in the saturated zone, but decreased DO levels.

A prolonged period of observation was conducted after high propane addition, without additional sparging (160 to 188 days). Propane was gradually reduced to below detection levels, followed by a corresponding gradual decrease in DO concentrations to about 0.6 mg/L. A 1-week period was needed to consume propane to low concentrations, which was needed to remove competitive inhibition effects on CAH transformation. Thus, sparging twice a week was not likely effective towards CAH treatment. Over the period of 190 to 210 days, propane concentration began to increase in the treatment zone, which coincided with a decrease in nitrate concentration in the groundwater. Ammonia was added at a very low concentration (0.01 %) to the sparge gas (over days 188 to 210) in an attempt to provide a nitrogen source to the treatment zone. Some enhancement in propane utilization rates resulted in response to the ammonia addition.

cis-DCE concentrations also begins to decrease more rapidly than TCE with the successive utilization of propane. This result is consistent with the results of laboratory microcosm studies with aquifer solids and groundwater from the treatment zone that were stimulated on propane (Dolan and Semprini, 1999). During the first 210 days of treatment, *cis*-DCE concentrations decreased to less than 1 μg/L in the treatment zone. TCE concentrations also gradually decreased to below 10 μg/L by the end of the test in well MW-C3-117. At latter times when propane consumption rates decreased due to nitrogen limitations, TCE concentrations increased, and were subsequently reduced to low levels after ammonia was added to the sparge gas.

Similar observations occurred at the other monitoring locations where propane was delivered. The biological transformation of *cis*-DCE and TCE was indicated by: 1) the more rapid removal of *cis*-DCE compared to TCE; 2) the removal of *cis*-DCE and TCE to low concentration levels; 3) the transient responses of the decrease in *cis*-DCE and TCE to the initiation of propane consumption; 4) their increase in concentration as propane utilization rates slowed; and 5) their decrease again with ammonia addition.

During the period from 200 to 400 days, no propane or air were sparged to the treatment zones. TCE concentrations at MW-C3-117 increased to levels existing in the treatment zone prior to biostimulation, while DO levels were maintained at about 3 mg/L. *cis*-DCE concentrations did not increase to background levels during the rest period. The reason for this is not known, but it is possible that biotic removal of *cis*-DCE continued for a prolonged period after propane addition was stopped. The initiation of propane addition around day 400 resulted in a rapid decrease in TCE concentration. Propane was added at about 1 % during this period to maintain high DO levels in the treatment zone. During the second season of testing, *cis*-DCE concentrations in the treatment zone generally were maintained below 1 μg/L and TCE concentrations were generally maintained between 5 to 30 μg/L. The reasons why TCE concentrations could not be decreased further are not known. No ammonia was added during the second season. Increases in nitrate resulting from groundwater flow into the treatment zone and from ammonia nitrification to nitrate were enough to keep nitrogen from becoming limiting. Propane addition was stopped at about 500 days. TCE concentration began to increase in response to the stopping of propane addition, while *cis*-DCE concentrations remained low.

Results from the corresponding A well, MW-A3-117 that received only air, are shown in Figure 4. The sparge events were maintained at the same interval and flowrate as the C-zone. Little reduction in *cis*-DCE and TCE concentrations were observed at this location during either season of field-testing. Oxygen was maintained at a fairly constant level during the testing, at around 6 to 7 mg/L, slightly lower than the maximum achieved in MW-C3-117, indicating that sparging was less effective at this location. The results do show very limited removal due to sparging. Two monitoring locations 7.5 ft from the A-zone sparge well showed decreases in *cis*-DCE and TCE concentrations at similar rates, indicating stripping from the groundwater was the main removal mechanism.

The results of this field demonstration show propane to be an effective cometabolic substrate for the treatment of CAH mixtures. More effective treatment of the CAHs was achieved compared to sparging with air alone. The sparging process was shown to be affected by aquifer heterogeneities, with propane and air delivered in a non-uniform pattern. Thus the ability to use sparging to deliver gaseous cometabolic substrates to the saturated zone must be evaluated for each site. If cometabolic sparging can be effectively applied, it potentially represents a low cost means of remediating CAH contaminated groundwater.

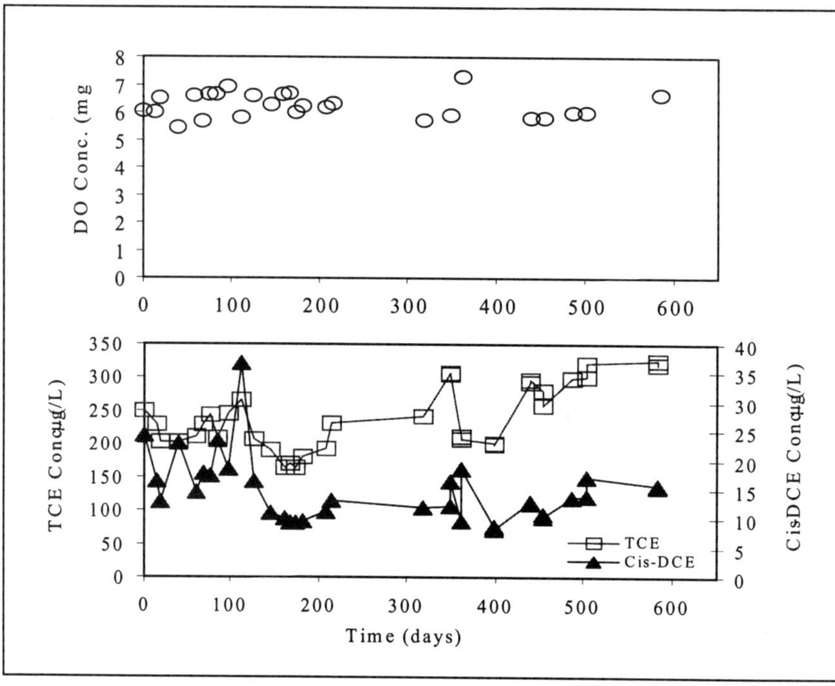

FIGURE 4. Concentration histories of DO, *cis*-DCE, and TCE at well groundwater monitoring well MW-A3-117.

ACKNOWLEDGMENTS

The field demonstration was funded by the Environmental Security Technology Certification Program (ESTCP) under the project title *Use of Cometabolic Air Sparging to Remediate Chloroethene-Contaminated Groundwater Aquifers.* The authors wish to thank Chris Coonfare and Sarah McCall (Battelle), and Derek Rogers (OSU) for their invaluable assistance in this effort.

REFERENCES

Dolan. M.E. and L. Semprini. 1999. "Aerobic Cometabolic TCE Transformation in Propane-fed Aquifer Microcosms," Abstracts from the 4th International Symposium on Subsurface Microbiology, Vail, CO, August 22-27.

Marley, M. and C.J. Bruell. 1995. *In Situ Air Sparging: evaluation of Petroleum Industry Sites and Considerations for Applicability, Design and Operation.* American Petroleum Institute Publication No. 4609.

McCarty, P.L. and L. Semprini, "Ground-water Treatment of Chlorinated Solvent in: Groundwater Clean-up Through Bioremediation," in *Handbook of Bioremediation*, Lewis Publishers Inc., Chelsea, MI, pp 87-116 (1993).

Tovanabootr, A. and L. Semprini. 1998."Comparison of TCE transformation abilities by methane and propane-utilizers." *Bioremediation Journal* 2:105-124.

Travis, B.J. and N.D. Rosenberg. 1997. "Modeling In Situ Bioremediation of TCE at Savannah River: Effects of Product Toxicity and Microbial Interactions on TCE Degradation." *Environ. Sci. Technol.* 31:3093-3102.

VADOSE ZONE BIODEGRADATION POTENTIAL DURING COMETABOLIC AIR SPARGING OF CHLOROETHENE-CONTAMINATED GROUNDWATER

Regina M. Lynch, Sarah McCall, Victor S. Magar, Andrea Leeson
(Battelle Environmental Restoration Department, Columbus, OH USA)
Mark Dolan, Lewis Semprini, and Mohammad Azizian
(Oregon State University, Corvallis, OR)
Capt. Dave Kempisty
(U.S. Air Force Research Laboratory (AFRL/MLQE), Tyndall AFB, FL)

ABSTRACT: A field study was conducted to investigate the effectiveness of cometabolic air sparging for remediation of groundwater contaminated with chlorinated solvents. The study was conducted for 18 months at McClellan AFB, CA, and propane was selected as the primary substrate. Intermittent weekly groundwater sparging with 1 to 4 % propane in air was conducted to stimulate propane biodegradation and simultaneous cometabolic biodegradation of primarily trichloroethene (TCE) and *cis*-dichloroethene (c-DCE). While the saturated zone showed significant reductions in contaminant concentrations using propane as the primary growth substrate, the vadose zone remained unresponsive after 500 days even though significant concentrations of propane were reaching the vadose zone. Subsequent groundwater sparging with methane appeared to stimulate propane, TCE, and c-DCE degradation in the vadose zone, suggesting that these compounds were degraded cometabolically by methanotrophs. This paper focuses on the results of the vadose zone cometabolic treatment process.

INTRODUCTION

Chlorinated aliphatic hydrocarbons (CAHs) have been used for years as industrial cleaners and degreasers and comprise some of the most common groundwater contaminants found at federal installations. The United States Air Force (USAF) has identified more than 900 sites that are contaminated with chlorinated solvents in excess of drinking-water standards and that may require cleanup (United States Environmental Protection Agency [U.S. EPA] and USAF, 1993). Similar cleanup situations are faced by industry, other government agencies, and up to 85% of Superfund sites.

The Environmental Security Technology Certification Program (ESTCP) supported an 18-month field study to investigate the cometabolic air sparging (CAS) process at McClellan Air Force Base (MAFB), California. The target CAH compounds for the CAS technology included a variety of chlorinated solvents, including trichloroethene (TCE), dichloroethene (DCE) isomers, vinyl chloride (VC), trichloroethane (TCA), and lower chlorinated ethane isomers. CAS is similar to conventional air sparging, but instead of adding air alone into an aquifer, CAS includes addition of a gaseous organic growth substrate to promote the in situ cometabolic degradation of CAH compounds. The CAS concept combines knowledge of cometabolic degradation using bacteria that grow on gaseous substrates such as propane, methane, and butane with the more conventional air sparging technology.

While successful demonstrations have been reported using aerobic cometabolism with methane as the primary growth substrate (Travis and Rosenberg, 1997), cometabolic CAH degradation by propane-degrading bacteria is unique to this study.

OBJECTIVES

TCE and *cis*-dichloroethene (*c*-DCE) were the major target contaminants in the groundwater and vadose zone at MAFB, and primary focus was given to these compounds for process optimization. Remaining contaminants were monitored as secondary compounds to assess the impact of the treatment process on multiple groundwater and soil gas contaminants. The demonstration was conducted in two phases, each lasting six to nine months. Phase 1 focused on CAH cometabolism in the saturated zone; Phase 2 focused on propane degradation and CAH cometabolism in the vadose zone and optimization of the groundwater remediation process. During each operational phase, two sparge wells were operated, one located in an active zone and one in a control zone. The active zone was fed air plus propane and the control zone was fed air only. CAH, propane, and oxygen concentrations were monitored in the surrounding saturated and vadose zones, along with standard geochemical parameters such as pH, conductivity, and temperature in the saturated zone and relative humidity in the vadose zone.

This paper focuses on the results of Phase 2, describing the effectiveness of CAH remediation in the vadose zone. A companion paper describes groundwater remediation results (Tovanabootr et al., 2001).

SITE DESCRIPTION

MAFB is located 7 miles north of Sacramento, CA. In July 1987, MAFB was placed on the U.S. EPA's National Priorities List. This demonstration was conducted at Operable Unit A, the location of a former demolition and scrap material burning and burial pit that was operational from 1964 to 1969. Materials that were burned or buried included residues, scrap material, fuels and oils, solvents, semi volatile organic compounds and volatile organic compounds. In 1969, the activities ceased and the pit was filled. In the vadose zone, background levels of TCE and *c*-DCE ranged from 500 to 800 μg/L in the active zone and 300 to 600 μg/L in the control zone. The groundwater begins at 108 ft bgs, and the soil consists largely of fine-grained materials such as silt, clay, and sandy silt underlain by medium-grained materials such as silty or clayey sand or gravel.

MATERIALS AND METHODS

The CAS demonstration consisted of two sparge wells, each installed to a depth of 12 ft below the groundwater table, or approximately 120 ft below ground surface (bgs). The sparge wells were installed approximately 100 ft apart to compare the effectiveness of air sparging with and without the cometabolic growth substrate. Six monitoring points were installed surrounding each sparge well for collection of soil gas and groundwater at discrete depths. The soil-gas monitoring points were used to measure soil-gas concentrations of carbon dioxide, oxygen, propane, and CAHs. Monitoring points in the vadose zone consisted of small-diameter ¼-inch nylon tubing to depths of 30, 85, 95, and 105 ft bgs. They terminated with 6-inch-long, 1-inch-diameter, 20-slot PVC screens. Installation of vadose zone monitoring clusters conformed to MAFB requirements (McClellan AFB, 1999).

The following nomenclature was used to identify the sparge system and monitoring assembly. Wells in the control zone were labeled A and those in the active zone were labeled C. Soil gas points were labeled by location (1 through 6) and depth (30-, 85-, 95-, and 105-ft bgs).

SAMPLING AND ANALYSIS

The soil gas monitoring points at the 95-ft depth were monitored approximately weekly, while those at the 30-ft depth and 85-ft depth were monitored less frequently (approximately monthly) because they were less responsive to sparging activities. All of the 105-ft monitoring points were saturated and could not be sampled, except for one of the monitoring points in the active test plot (C4-SG-105).

The temperature, dew point, and relative humidity of the soil gas were analyzed using a Control Company Digital Hygrometer/Thermometer Model 35519-041, equipped with a digital display. Gaseous concentrations of carbon dioxide and oxygen were analyzed using a GasTech series gas monitor Model 205, equipped with a digital display. Meters were calibrated each day of use according to manufacturer specifications.

Soil gas was extracted from the vadose zone with a diaphragm soil-gas sampling pump with Teflon™ seals and diaphragm. A sufficient amount of soil gas was purged before sample collection and field parameters were measured to exchange the gas in the lines for a minimum of three volumetric displacements. This was done by running the diaphragm pump for 10 minutes before each sampling event to flush the lines. Pre-cleaned Tedlar™ bags were purged three times with the soil-gas of each sample point before sample collection, to ensure representative gas samples. Samples were collected in the Tedlar™ bags and were analyzed on site using a gas chromatograph (GC) equipped with a flame ionization detector (FID) and a photoionization detector (PID) (Hewlett Packard, Model 5890 Series II; GSQ-PLOT column [J&W Scientific, Folsom, CA]) equipped with a 16-port purge-and-trap. The GC and a purge-and-trap unit were housed in a field laboratory trailer provided by the US Air Force.

Propane and the following analytes were targeted for on-site analyses based on the contaminants detected at the site: tetrachloroethene (PCE); TCE; c-DCE; 1,1-DCE; chloroform (CF); and 1,1-dichloroethane (1,1-DCA). The GC was calibrated against six-point propane and CAH standard curves taken from a specialty gas cylinder with a certified mixture of the target contaminants (Scott Specialty Gases, Plumsteadville, PA). Gas samples (5 mL) were drawn from the Tedlar™ sample bags using a hand-held syringe, and were hand injected into the purge-and-trap autosampler for analysis.

SITE OPERATIONS

During Phase 2, the focus of site operations was to stimulate biodegradation of CAHs in the vadose zone. Figure 1 shows that CAH and propane degradation appeared to begin in the vadose zone near the end of Phase 1. Propane accumulated in the vadose zone up to about 5%, and then around Day 125, propane and CAH concentrations began to decrease, indicating that the CAS process was occurring in the vadose zone. After Day 200, the site was left inactive for 73 days. Due to the promising results at MAFB shown in Figure 1, investigation of a second site was abandoned and work continued at MAFB. Regular sparging and sampling events resumed (Phase 2). During Phase 2, the frequency and duration of sparging as well as the amount of propane injected were varied in an attempt to stimulate vadose zone propane and CAH degradation.

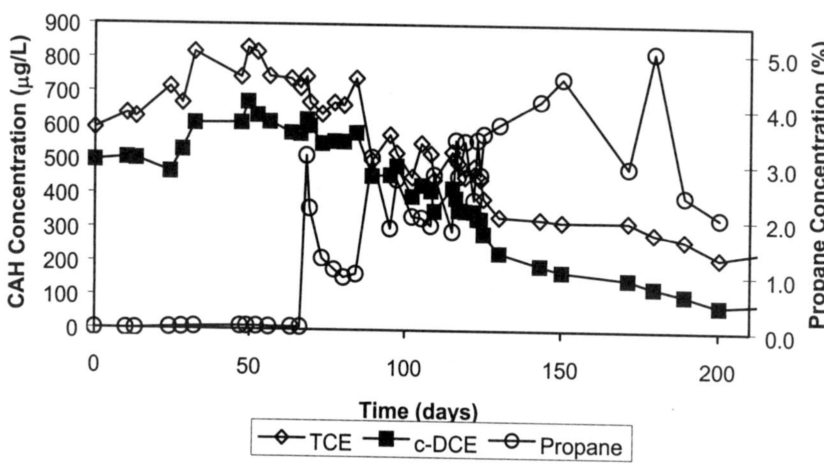

FIGURE 1. Phase 1 active zone soil gas CAH and propane concentrations in C6-SG-105.

Phase 2 was conducted for a total of 219 days between May and December 2000 (i.e., from Days 327 to 546). Throughout Phases 1 and 2, propane and air were sparged intermittently (weekly or biweekly) with 2 to 4% propane at 5 standard cubic feet per minute (scfm) (141.6 Lpm). While high propane and CAH degradation rates were observed in the saturated zone during Phase 1, there was no sign of significant propane or CAH degradation in the vadose zone.

On Day 430, a series of push-pull tests were conducted with nitrous oxide, ammonia, and dinitrogen tetroxide at both the control and the active sites to inject nitrogen directly into the vadose zone. Unfortunately, these results were inconclusive (data not shown). Decreases in oxygen and increases in carbon dioxide were observed, which would have indicated increased biological activity, but no significant decreases in propane, TCE, or c-DCE were observed. Both dinitrogen tetroxide and ammonia were discontinued after dinitrogen tetroxide appeared highly corrosive to instruments and ammonia temporarily elevated the pH to >10 near one groundwater well.

In a final attempt to stimulate biological degradation in the soil gas, propane injections were terminated and 1% methane in air was injected into the active sparge well. Sparging with methane occurred on Days 512, 520, 526, and 546, and then was discontinued to allow for acclimation and growth of methanotrophs.

RESULTS AND DISCUSSION

Vadose zone propane, methane, and CAH concentrations in two active soil gas monitoring points (C1-SG-95 and C3-SG-95) are shown in Figures 2 and 3. Between Days 394 and 492, a slow increase in soil gas propane concentrations that approached the sparge concentration of approximately 1 to 2% was observed in both monitoring points. TCE and *c*-DCE concentrations were also very stable in both soil gas monitoring points. Thus, there was no sign of propane degradation or cometabolic CAH degradation throughout the propane feed period. Similar results were observed in the remaining monitoring points at the site.

On Days 488 and 503, 0.1% ethylene was added to the 4% propane sparge gas. The ethylene was added to determine whether there was degradation activity at the site other than propane degradation. After the addition of ethylene, propane concentrations in monitoring point C1-SG-95 fluctuated significantly, while those in C3-SG-95 remained relatively stable. Ethylene was depleted within one week in both wells (data not shown). There was no sign of corresponding TCE or c-DCE degradation and sparging with ethylene was discontinued.

Methane was used as the cometabolic substrate from Days 512 to 546. Figures 2 and 3 show complete methane degradation after the first sparge event on Day 512. When methane was reinjected on Days 520, 526, and 546, degradation was not as rapid, but by the final sampling event (Day 587), methane concentrations were non-detectable in both C1-SG-95 and C3-SG-95, as well as the remaining monitoring points (not shown). Propane, TCE, and *c*-DCE were depleted concurrently with methane. In monitoring point C3-SG-95, both TCE and c-DCE were reduced almost below their respective detection limits, indicating degradation of these compounds through cometabolism by methanotrophs. The experiment was discontinued on Day 587, so these results could not repeated in the field. Nonetheless, the fact that propane, TCE, and *c*-DCE concentrations were extremely stable for 150 days during Phase 2 until methane was injected and degraded suggests that their degradation was due to cometabolism by methanotrophs. These results also suggest that propane was degraded by methanotrophs. An alternative though unlikely explanation is that while propane was degraded by propane-degrading bacteria coincidentally with methane, degradation of TCE and *c*-DCE was due to cometabolism by propane-degrading bacteria; this possible hypothesis also could not be verified in the field.

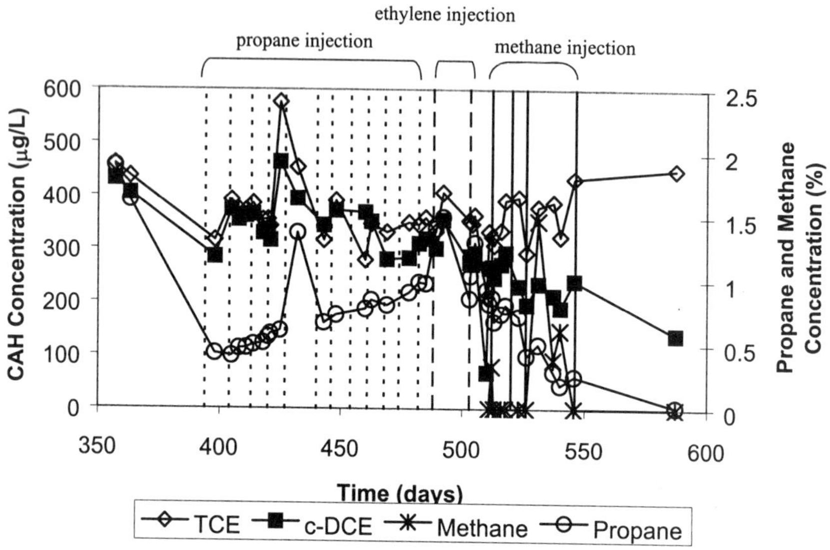

FIGURE 2. Phase 2 active zone soil gas CAH, propane, and methane concentrations in C1-SG-95.

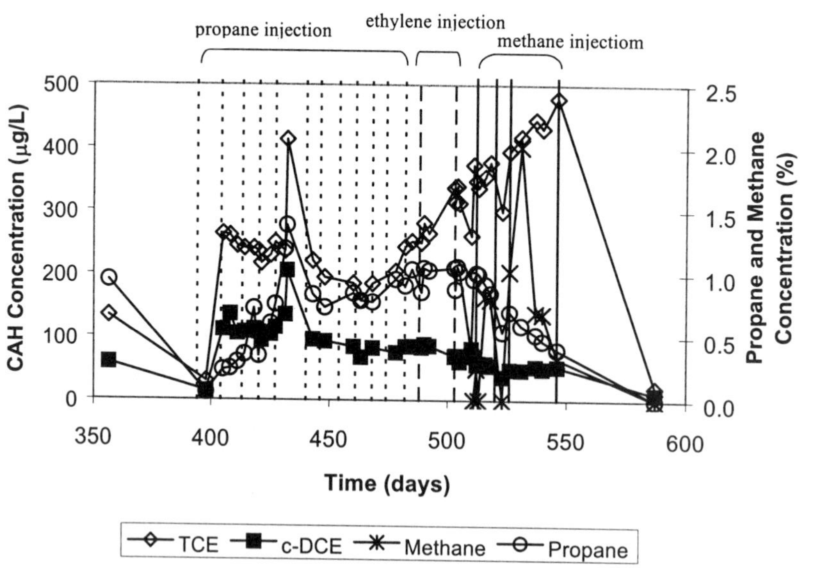

FIGURE 3. Phase 2 active zone soil gas CAH, propane, and methane concentrations in C3-SG-95.

CONCLUSIONS AND LESSONS LEARNED

After over 500 days of exposure to propane in soil gas (approximately 1 to 4%), there was no evidence of propane degradation in the vadose zone and no evidence of CAH cometabolism by propane-degrading bacteria. Methane was injected into the vadose zone to stimulate methanotrophic CAH cometabolism. The first signs of propane degradation and cometabolic CAH degradation coincided with the introduction of methane into the vadose zone, suggesting that their degradation activity was due to cometabolism by methanotrophs. Insufficient time was devoted to the methane injection experiment, and the cometabolic CAH degradation activity could not be verified by reinjecting methane into the aquifer.

There are three possible explanations of why propane degradation was not observed in the vadose zone at MAFB: 1) the necessary bacteria may not have been present in the vadose zone; 2) conditions may not have been suitable for growth of propane degrading bacteria; and 3) CAH concentrations or the combination of CAHs in the vadose zone may have been toxic to bacteria. The fact that propane-degrading bacteria were present in the saturated zone (Tovanabootr et al., 2001) indicates that propane-degrading bacteria were present at the site. It is unlikely, but possible, that propane degraders were present in the saturated zone and not the vadose zone. The fact that ethylene (data not shown) and methane were degraded suggests that the contaminants were not toxic to the bacteria that degraded these compounds. However, they may still have been toxic to propane-degrading bacteria. Thus, the most likely explanation was that environmental conditions were not suitable for propane degraders in the vadose zone. Laboratory microcosm studies and results of the saturated zone showed that nitrogen limitations resulted in significantly decreased propane degradation and a corresponding loss of CAH cometabolism (Dolan, 2000; Tovanabootr et al., 2000). Nitrate concentrations in the saturated zone at the onset of the experiment were approximately 5 mg/L. Such low nitrogen concentrations also were likely to have been present in the porewater of the vadose zone, although nitrogen was not measured in the vadose zone. If this were the case, it is possible that nitrogen was limiting to propane degradation in the vadose zone. Alternatively, low moisture content in the vadose zone also could have limited propane degradation. However, humidity measurements showed that the relative humidity in the vadose zone exceeded 80%.

The observation that methane was degraded in the vadose zone could be explained by the fact that methanotrophs more readily fix nitrogen than propane-degrading bacteria. Thus, methane degradation may not have been nitrogen limited.

The failure of the CAS process to stimulate cometabolic degradation of CAHs in the vadose zone using propane means that multiple growth substrates would be required for complete CAH degradation in the saturated and vadose zones. The cost benefit of CAS over conventional air sparging comes from the lower costs of introducing propane and/or methane to the saturated and vadose zones compared with having to employ a soil vapor extraction (SVE) system equipped with aboveground off-gas treatment.

REFERENCES

Dolan, M.E. 2000. Personal communication.

McClellan Air Force Base. 1999. Work Implementation Plan, Passive Soil Vapor Extraction. Report prepared for McClellan AFB by URS Griener. McClellan AFB, California. January.

Tovanabootr, A., M.E. Dolan, M. Azizian, L. Semprini, V.S. Magar, D. DeBacker, A. Leeson, and D. Kempisty. 2001. "Cometabolic Air Sparging Field Demonstration with Propane to Remediate Trichloroethylene and *cis*-Dichloroethylene." In A. Leeson, R.E. Hinchee, P.C. Johnson, and L. Semprini (Eds.). *In Situ Aeration and Aerobic Remediation*, Battelle Press: Columbus, OH (in press).

Tovanabootr, A., M.E. Dolan, L. Semprini, V.S. Magar, and A. Leeson. 2000. "Cometabolic Air Sparging Field Demonstration with Propane to Remediate a Chloroethene and Chloroethane Co-contaminated Aquifer." *Battelle International Conference on Chlorinated and Recalcitrant Compounds*. Monterey, CA.

Travis, B.J. and N.D. Rosenberg. 1997. "Modeling *In Situ* Bioremediation of TCE at Savannah River: Effects of Product Toxicity and Microbial Interactions on TCE Degradation." *Environ. Sci. Technol. 31*:3093-3102.

United States Environmental Protection Agency and USAF. 1993. *Remediation Technologies Matrix and Reference Guide.*

MODELING BIOSPARGING AND NATURAL ATTENUATION OF CHLORINATED ETHENES AT SRS

Philip A. Weeber (GeoTrans, Roswell, Georgia)
Peter F. Andersen (GeoTrans, Roswell, Georgia)
Bryan J. Travis (Los Alamos National Laboratory, Los Alamos, New Mexico)
David C. Noffsinger (Westinghouse Savannah River Co., Aiken, South Carolina)

ABSTRACT: An in situ biosparging system, consisting of two horizontal injection wells, has operated since October 1999 at the U.S. Department of Energy (DOE) Savannah River Site (SRS). The remediation activities were designed to enhance the biological activity within distinct trichloroethene (TCE) and vinyl chloride (VC) plumes. The first horizontal injection well (SLH-1) targets the TCE plume by injecting air, methane, and nutrients to enhance cometabolic degradation. The second horizontal injection well (SLH-2) injects air to promote aerobic biooxidation of the VC. Downgradient of the injection wells, remediation of the groundwater continues through the processes of natural attenuation. By conjunctively using two numerical models, chlorinated degradation was simulated including (1) the cometabolic (methanotrophic) degradation of TCE, (2) the biostimulation of VC, and (3) the natural attenuation of both TCE and VC. The combined use of near-well and regional models coupled with this unique set of performance data allows for improved understanding and application of this technology.

INTRODUCTION

The Sanitary Landfill (SLF) is a 70-acre (28.3 hectare) unlined landfill located at the U.S. DOE Savannah River Site (SRS). The SLF received wastes from 1974 to 1994 and was subsequently capped in 1997. Groundwater beneath the SLF is contaminated with a number of constituents. The most widespread contaminants that exhibit plume-like characteristics are trichloroethene (TCE) and vinyl chloride (VC). This contamination is being remediated through the use of two horizontal injection wells. The southern biosparging system (SLH-1) stimulates cometabolic (methanotrophic) TCE biodegradation with the pulsed injection of air, methane, and microbial nutrients (triethyl phosphate and nitrous oxide). The western biosparging system (SLH-2) injects air to stimulate in situ aerobic biooxidation of VC. The two horizontal wells were installed in 1997 and 1998 and began operation in October 1999. Figure 1 depicts the location of the two horizontal wells and contaminant concentrations in June 2000, after 8 months of biosparging activities.

June 2000 isopleths for TCE represent a significant decline in concentrations to levels below remediation goals (June 1997 isopleths shown in Durant et al., 2000). June 2000 isopleths for VC show that along the VC injection well, concentrations are observed to decline as groundwater passes through the bioremediation zone. Data does not exist within the predicted radius of influence of either biosparging well to accurately depict near-well concentrations. VC concentrations underneath the

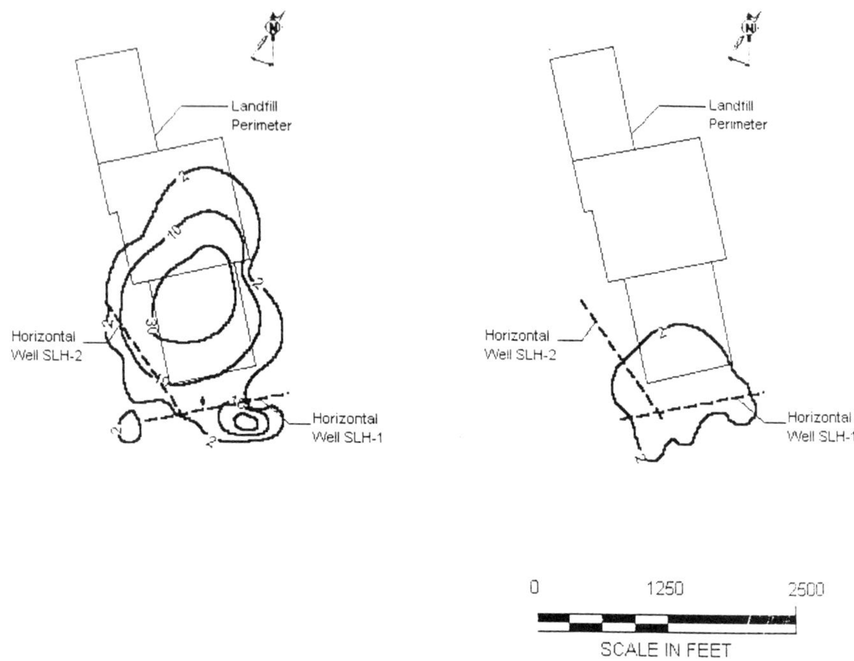

FIGURE 1. Sanitary Landfill site map depicting injection well locations and June 2000 TCE and VC plume configurations.

southern portion of the capped landfill have significantly increased since 1997. Also, VC has been detected south of SLH-1, the TCE biosparge well.

APPROACH

In 1998, Phase I work predicted performance of the biosparging systems and developed an operational plan (Durant et al., 2000). These predictive simulations utilized the codes TRAMPP (Travis and Rosenberg, 1997) and RT3D (Clement et al., 1998). TRAMPP has the capability to simulate a variety of processes in three dimensions, including solute air-water partitioning, Monod kinetics, cometabolism, rate-limited sorption, and nutrient limitation. TRAMPP also allows specification of a pulsed injection schedule. The complex biosparging performance for both TCE and VC was individually demonstrated within the radius of well influence.

Phase II groundwater modeling evaluated the operational performance of the biosparging systems. Quarterly water quality data from 90 days and 180 days after system start-up were analyzed. This unique collection of data allowed comparison between field data and the TRAMPP simulations for cometabolic degradation of TCE and the aerobic biooxidation of VC. RT3D natural attenuation modeling was then compared to data collected at a downgradient seepline. The combined use of near-well and regional models coupled with performance data allowed for improved understanding and application of this technology. RT3D simulations then were used to collectively model the subsequent natural attenuation of TCE, DCE, and VC.

Conceptual Model for Natural Attenuation. Natural attenuation of TCE, DCE, and VC is actively occurring at the SLF. Dissolved organic carbon introduced by leachate from the SLF has caused the development of anaerobic conditions in the groundwater beneath much of the SLF (WSRC, 1997). The combined anaerobic conditions and abundance of organic carbon have created favorable conditions for reductive dechlorination of TCE. The accumulation of VC in the SLF groundwater provides direct evidence that TCE has been reductively biodegraded. An accumulation of chloride provides additional evidence of the in situ destruction of the chlorinated compounds. Although the biosparging systems are presently introducing oxygen into the subsurface at the downgradient edge of the SLF, data suggests that the core of the plume is still anaerobic directly underneath the landfill.

Aerobic conditions occur within and downgradient of the sparging wells. Since reductive dechlorination of TCE is inhibited by oxygen, it is assumed that TCE degradation does not occur significantly in the aerobic portions of the plume (outside the capped areas of the landfill). Biodegradation of DCE and VC, however, can be accelerated by the presence of oxygen (Bradley and Chapelle, 2000; Verce et al., 2000). Thus, aerobic biodegradation is likely a significant natural attenuation mechanism for residual DCE and VC occurring downgradient of the sparge wells.

Site specific natural attenuation rates were estimated from quarterly mass estimates. The GIS program TINMASS (GeoTrans, 1996) was used to quantify contaminant mass-in-place for various points in time during the 1990s. RT3D simulations, using these estimated rates, predictively modeled the natural attenuation of TCE, DCE, and VC in the possible case that the biosparging system was turned off.

Conceptual Model for Biosparging Performance. Engineered in-situ bioremediation began in October 1999. The two wells are operated on a bi-weekly injection schedule. The 48-hour injection regimen for the TCE system consists of 24 hours of air injection, followed by 8 hours of air, methane, triethyl phosphate (TEP), and nitrous oxide, followed by 16 hours of air injection. Typical injections into the TCE plume consist of 240 standard cubic feet per minute (scfm) of air, and during the eight hours of nutrient injection, methane at 0.41% percent by volume, nitrous oxide at 0.008% by volume, and TEP at 0.003% by volume. The VC injection well is operated on the same bi-weekly schedule with 48 hours of air injection at an average of 270 scfm.

Biodegradation was modeled separately for TCE and VC, using a two-dimensional cross section in TRAMPP. TRAMPP uses a sophisticated variable order (Monod) model for describing methane and TCE cometabolism, and VC oxidation. Nutrient consumption and protozoa grazing are also described by similar variable order expressions in TRAMPP. The parameters used in these models are rarely, if ever, measured in the field. In many cases, only laboratory estimates of these parameters are available, and few published values are available. Given the paucity of data for these Monod parameters, the possible range of values for each parameter is not known; nor is the error involved with the parameter estimates known. For the purposes of the modeling presented here, we used published, peer-reviewed values for the biokinetic parameters, and provided references for the articles from which these values were obtained (refer to Table 1 in Durant et al., 2000). The site-specific nutrient injection schedule was also incorporated into the model.

RESULTS AND DISCUSSION

Biosparging model simulations were performed using the historical injection schedule and the biokinetic parameter estimations/assumptions noted above. Model results were compared to October 1999, March 2000, and June 2000 data from monitor wells located along the two-dimensional model transects. Biosparging modeling predicted a 60 foot radius of influence, both upgradient and downgradient, in which bioremediation would take place. The monitoring wells, for both the TCE and the VC system, are located outside the predicted radius of influence of the injection well. Multiple monitoring wells within this radius of influence would be required for a quantitative model calibration. Therefore, model results were compared to the downgradient trends (i.e. removal percentage observed in downgradient wells).

TCE Biodegradation Model Results. For the TCE system, historical data shows that concentrations in upgradient wells decrease to between 20% and 33% of their initial levels. This decrease occurs without the influence of the biosparging injection well. The biosparging model did not incorporate this upgradient trend. Downgradient wells, outside the radius of influence, show June 2000 concentrations that are 15% to 50% of October 1999 measurements. Biological parameters (i.e. TCE utilization, methane utilization, and nutrient half saturation) were adjusted within the model to replicate the decrease in concentrations observed in the downgradient monitoring wells. These adjustments were order of magnitude changes in biological parameters to assess a qualitative representation of the observed system.

Figure 2 shows the predicted model results for TCE in June 2000. Within the radius of influence of the biosparging well, TCE concentrations are predicted to be less than two parts per billion. Downgradient of the well, a plume remnant exists that was advected past the biological treatment zone before being remediated. The predicted concentrations agree qualitatively with observed conditions at the SLF.

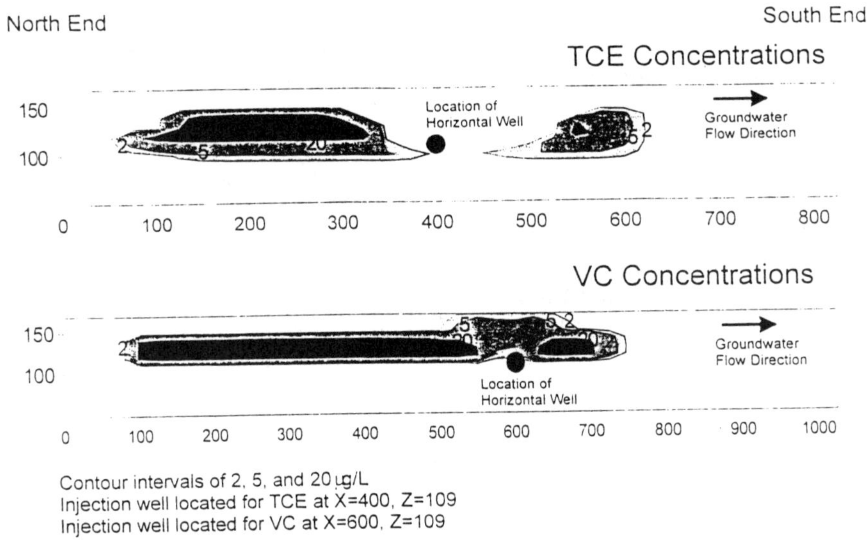

Figure 2. TCE and VC TRAMPP Model Results for June 2000.

VC Biodegradation Model Results. For the VC system, historical data shows increasing upgradient concentrations. VC concentrations increased between 38% and 145%. No monitoring wells exist within the radius of influence of the horizontal injection well. Monitor wells directly downgradient of the VC injection well showed a gradual decline to non-detect concentrations for VC. The biosparging model was iteratively revised, with an emphasis on the biosparging zone, to replicate VC concentration trends as observed at the downgradient observation points. Again, these adjustments were order of magnitude changes in biological parameters to assess a qualitative representation of the observed system.

Figure 2 shows the predicted VC concentrations after 180 days of biooxidation. It is apparent that VC concentrations are reduced through biological treatment, but the contaminant persists at concentrations above remediation goals. The magnitude of predicted downgradient concentrations matches the observed downgradient concentrations with the two part per billion isopleth approximately 200 feet (61 meters) from the biosparging well.

Natural Attenuation Rate Estimation. Natural attenuation was approximated using first-order rate estimates for TCE, DCE, and VC in the SLF groundwater. These rate estimates were obtained by quantitative evaluation of the decreasing trend in contaminant mass-in-place over time. The computed mass of TCE in the SLF groundwater declined by over 50% between 1992 and 1997; much of this decline can be attributed to biodegradation. By linearly fitting chlorinated mass data, the apparent first-order natural attenuation rate for TCE is 4×10^{-4} d^{-1}. The apparent whole-plume mass of DCE also declined significantly during the 1990s, with an estimated first-order decay rate of 1.5×10^{-3} d^{-1}. The estimated rate of DCE degradation in the anaerobic region was only slightly slower (1.2×10^{-3} d^{-1}). It was not possible to estimate the rate of VC attenuation throughout the anaerobic region since there was a net accumulation of VC during the 1990s. For the aerobic region of the plume, the VC attenuation rate was estimated to be 8×10^{-4} d^{-1}. These attenuation rate estimates serve as model parameter values for the prediction of plume fate after the biosparging systems have been turned off.

Natural Attenuation Model Results. RT3D is used to collectively predict the natural attenuation of TCE, DCE, and VC after the biosparging system is turned off and the domain has rebounded to steady state conditions. The initial contaminant mass specified for the natural attenuation simulations represents the mass in the groundwater system after the biosparging wells have ceased operation. The longer the biosparging system operates and aerobic and anaerobic degradation processes occur, the lower the anticipated initial mass within the system. As a conservative assumption, the initial contaminant mass specified for the RT3D simulations was set equal to the contaminant distributions from June 2000.

The initial TCE concentrations are below remediation goals. Simulations show the transport and degradation of these low concentrations and ultimately predict the removal of TCE within 100 days. Likewise, DCE is not a contaminant of concern at the SLF and initial conditions are below remediation goals. However, DCE is tracked as the intermediate between TCE and VC. Simulations predict low levels of DCE in areas that previously contained TCE; these levels reduce to below one part per billion before 200 days.

Vinyl chloride is the most recalcitrant compound at the SLF. Beginning with June 2000 concentrations, RT3D modeling predicts that VC concentrations will persist in the groundwater at concentrations above the remediation goal of one part per billion. Figure 3 depicts the simulated VC plume configuration after 100, 200 and 400 days. After 400 days, the core of the VC plume from underneath the landfill is predicted to decrease to approximately 15% of initial concentrations. The advection of the plume remnant proceeds to a wetland boundary where VC is then discharged to an adjacent river. Natural attenuation alone is predicted to be an inadequate measure for preventing the off-site migration of VC.

Data also suggests that VC is being produced underneath the landfill. An estimated time for VC remediation cannot be computed until VC production curtails and concentrations begin to stabilize and decline over time. The increase in VC concentrations underneath the landfill in relation to the physical placement of the biosparging well is the largest limitation to the overall effectiveness of the treatment program.

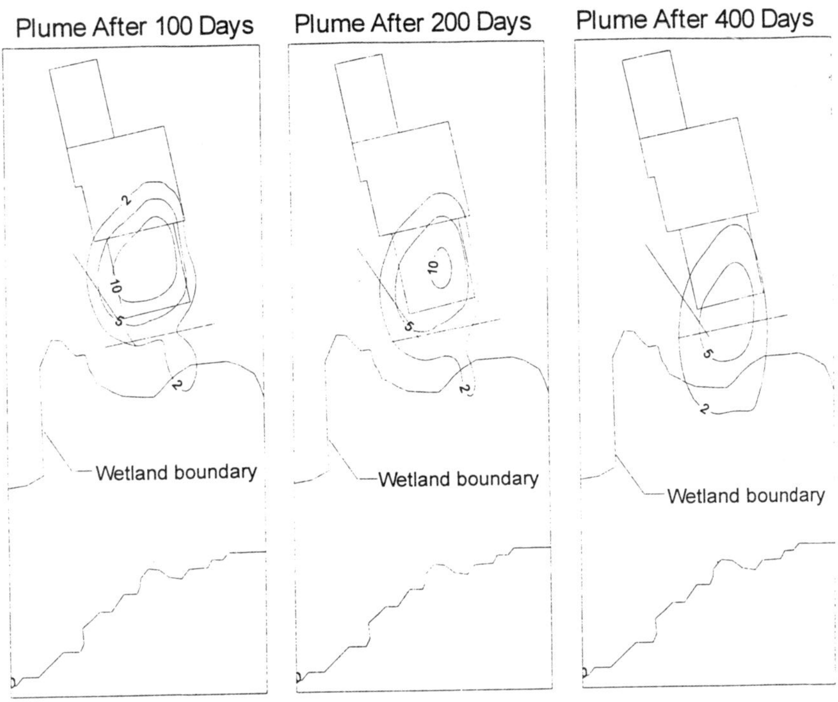

Concentration in micrograms per liter.

Figure 3. RT3D Natural Attenuation Model Results for VC.

CONCLUSIONS

Biosparging and natural attenuation have been successfully combined to remediate TCE and portions of the VC plume at the SLF. The remediation has been enhanced through the operation of two horizontal injection wells. Site specific groundwater quality data has allowed approximation of first-order natural attenuation rates. TRAMPP simulations of cometabolic degradation of TCE and the aerobic biooxidation of VC are demonstrated to be reasonable approximations of system performance. Natural attenuation predictive simulations have shown that VC will exceed remediation goals if biosparging activities were to cease.

Both the TRAMPP and RT3D simulations have several severe limitations. Perhaps the greatest limitation is the uncertainty associated with the source term. For the purpose of these simulations, it was assumed that there is no NAPL phase present at the SLF, and that all the TCE and VC is dissolved in groundwater. This assumption eliminates the possibility that remediation times could be constrained by rate-limited dissolution from a nonaqueous phase. If NAPLs are subsequently detected at the SLF, this assumption would be proven invalid, and new model simulations would be required to improve cleanup time estimates. None of the data collected to date suggests that a source is present. In fact, natural biodegradation appears to have destroyed much of the remaining TCE, and a clear declining trend is apparent. TRAMPP simulations are further limited by a lack of monitoring well data within the radius of influence. A quantitative model calibration would require several points within each active biosparging zone.

The natural attenuation simulations with RT3D are limited as the biodegradation parameter estimates are somewhat uncertain. It should be noted that in this work, the biodegradation rate was approximated by the natural attenuation rate. This approximation, which is commonly employed at chlorinated solvent sites, introduces error in that the two processes are not exactly identical.

Future work at the SLF includes the optimization of the VC injection schedule and continued monitoring of system performance.

ACKNOWLEDGMENTS

The authors thank Neal Durant for his aid in the development of the biodegradation and natural attenuation models and for his contributions to this work.

REFERENCES

Bradley, P. M. and F. H. Chapelle. 2000. "Aerobic Mineralization of Dichloroethene as Sole Carbon Substrate." *Environmental Science and Technology. 40: 2084-2086.*

Clement, T. P., Y. Sun, B. S. Hooker, and J. N. Petersen. 1998. "Modeling Multispecies Reactive Transport in Ground Water." *Groundwater Monit. Rev. Remediation*. Spring: 79-92.

Durant, N. D., P. A. Weeber, P. F. Andersen, D. G. Jackson, and B. J. Travis. 2000. "Predicting Enhanced Bioremediation Performance at the SRS Sanitary Landfill." In G. B. Wickramanayake, A. R. Gavaskar, B. C. Alleman, and V. S. Magar (eds.), *Bioremediation and Phytoremediation of Chlorinated and Recalcitrant Compounds*. C2-4: 429-436. Battelle Press, Columbus, OH.

GeoTrans. 1996. *Application of GIS Technology to Evaluate Intrinsic Remediation.* Prepared for the U.S. EPA Robert S. Kerr Laboratory.

Travis, B. J., and N. D. Rosenberg. 1997. "Modeling In Situ Bioremediation of TCE at Savannah River: Effects of Product Toxicity and Microbial Interactions on TCE Degradation." *Environmental Science and Technology*. 31: 3093-3102.

Verce, M. F., R. L. Ulrich, and D. L. Freedman. 2000. "Characterization of an Isolate that uses Vinyl Chloride as a Growth Substrate Under Aerobic Conditions." *Applied Environmental Microbiology*. 66:3535-3542.

WSRC. 1997. *Intrinsic Bioremediation of Landfills Interim Report*. Westinghouse Savannah River Company Report WSRC-RP-97-17, Rev. No.: 0.

EFFICIENT TCE MINERALIZATION IN A NOVEL METHANOTROPHIC BIOREACTOR SYSTEM

B. Ramsay (Polyferm Canada Inc., Harrowsmith, ON, Canada)
D. Karamanev (University of Western Ontario, London, ON, Canada)
S. Pierre and C. Lafontaine (École Polytechnique de Montréal, Montréal, PQ, Canada)
T. Sakamoto and J. Ramsay (Queen's University, Kingston, ON, Canada)

ABSTRACT: The use of immobilized soil bioreactors (ISBRs) operated under methanotrophic conditions to mineralize trichoroethylene (TCE) was investigated. ISBRs with associated biofilters designed to operate under methanotrophic conditions were constructed at the 2 L and 2000 L scale. TCE degradation rates of up to 230 mg L^{-1} d^{-1} for an initial TCE concentration of 18 mg L^{-1} and a 0.04 d residence time were achieved in a 2 L ISBR. The maximal TCE degradation was 97% at a 0.91 d residence time. The methanotrophic immobilized soil bioreactor showed a TCE degradation rate of up to 7 times higher than the best results reported in the literature for methanotrophic reactors at a similar outlet concentration, with 323% greater removal efficiency for low TCE outlet concentrations. At high TCE outlet concentrations, degradation rates were up to 9 times higher than the best results reported in the literature with a 307% higher TCE removal efficiency. Since the methanotrophic ISBR emits small amounts of gas, methanotrophic TCE biodegradation was also studied in a gas-solid bioreactor (biofilter). Up to 63% TCE removal efficiency was achieved at a gas flow of 2.3 L min^{-1} (reactor residence time of 1.62 minutes) and TCE inlet concentration of 1.37 mg L^{-1} and a methane inlet concentration of 3.05% v/v. Degradation rates of up to 233 g TCE m^{-3} reactor h^{-1} were obtained at a TCE inlet concentration of 20.45 mg L^{-1}, a CH_4 concentration of 2.74% v/v and a gaseous flow rate of 2.4 L min^{-1}. Radiolabelled microcosm experiments verified that TCE was mineralized in both the ISBR and biofilter. A computer-controlled 2000 L ISBR and associated biofilter are now undergoing testing in a methanotrophic regime. They are designed to replace "advanced oxidation" reactors in "pump-and-treat" TCE bioremediation processes.

INTRODUCTION

Trichloroethylene (TCE) is a common groundwater contaminant. Remediation of sites contaminated with this dense non-aqueous phase liquid (DNAPL) is normally by a "pump-and-treat" process. Separation techniques such as air stripping, activated carbon or pervaporation may be used as the treatment step, or a reactor may be employed. Although much less frequently employed than the aforementioned separation techniques, the most commonly used reactors are the "advanced oxidation" type, relying on UV light and oxidants such as H_2O_2 and O_3,. Biological oxidation can also be used but is not yet competitive with "advanced oxidation" processes due to high cost and relative unreliability. Under anaerobic conditions, many microorganisms can dechlorinate TCE and related compounds. However, this often leads to the formation of highly toxic vinyl chloride. Although vinyl chloride can be efficiently biodegraded in a subsequent aerobic process, regulators worry that some may be released into the environment if the aerobic step is not well controlled.

Microorganisms cannot use TCE as their sole source of carbon and energy. No enzymes are known to be specific for TCE but some microbial oxidases for other substrates, such as phenol and ammonia, are nonspecific enough to act on TCE. The organisms possessing these enzymes must be grown on a substrate other than TCE but degrade TCE via cometabolism. The soluble (cytoplasmic) form of methane monooxgenase (MMO_s) has the highest known enzymatic TCE oxidation rate. Unfortunately bacteria possessing MMO_s have very low specific growth rates. Another difficulty in designing bioreactors for TCE degradation is that intermediate products including free radicals irreversibly damage MMOs. Use of a mixed culture may result in some scavenging of these intermediates perhaps offering a degree of protection to the TCE-oxidizers.

Based on the factors mentioned above, the ideal bioreactor design for a TCE-contaminated groundwater pump-and-treat process should contain large amounts of MMO_s-producing organisms in mixed culture. Some form of immobilization would be required due to the low specific growth rate of MMO_s-producing organisms. The immobilized soil bioreactor (ISBR) provides these conditions at a lower cost than competitors such as membrane bioreactors. In an ISBR process, soil (or enriched soil) is loaded into the reactor where it quickly becomes entrapped in porous textile supports. The ISBR is an air-lift design. When applied to pentachlorophenol degradation, it had a volumetric activity in the range of orders of magnitude greater than any other bioreactor yet reported (Karamanev et al. 1998). If similar results can be obtained for TCE mineralization, the ISBR could be competitive with advanced oxidation reactors. One potential pitfall is the escape of TCE in the gas phase due to its volatile nature. The ISBR was redesigned, and coupled to a biofilter to address this possibility. This paper contains some of our findings on the application of the redesigned ISBR and biofilter to TCE biomineralization.

MATERIALS AND METHODS

Laboratory-Scale Methanotrophic ISBR. The process incorporated three distinct units (Fig. 1): a unit for TCE, air, and mineral salts medium addition, a methanator for the addition of CH_4, and the ISBR itself. A geotextile (Texel 400, Texel, St-Elzear, Can.) (H = 36.4 cm, L = 7.8 cm t = 0.5 cm) was used to immobilize the soil (2.55 g). The soil was collected from the aerobic-anaerobic interface of a swamp (Lac des Atocats, Mount St-Bruno, Que, Can.). All the tubing inside the reactor was stainless steel. Liquid phase agitation was achieved by recirculating the headspace gas using a peristaltic pump and Teflon™ tubing.

Laboratory-Scale Methanotrophic Biofilter. A stainless steel biofilter with an internal diameter of 0.097 m and 0.5 m in height was used for biofiltration (Fig. 2). It was divided into three sections. Under each section, there was a grid (9 mesh per inch) to keep the filtration media in place. The air stream, was humidified and enriched with methane. The CH_4 and air flows were controlled to obtain the desired CH_4 concentration. Gaseous TCE was generated by evaporation of liquid TCE. The biofilter was loaded with the filtration medium and supplied with CH_4 (4% v/v) for two weeks to develop a methanotrophic bacterial population after being inoculated with a TCE-degrading consortia obtained from the immobilized soil bioreactor .

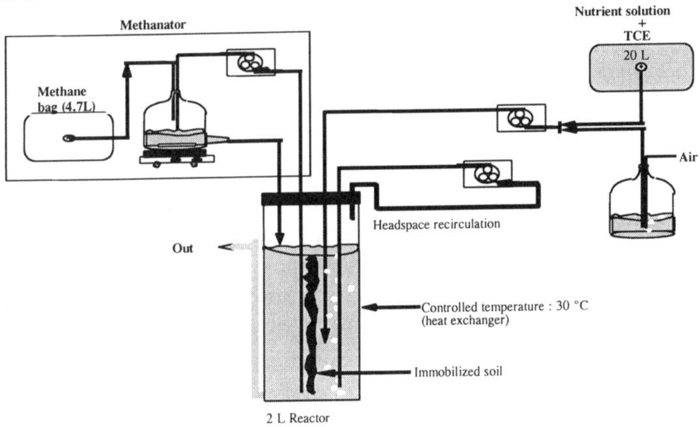

FIGURE 1. Laboratory-scale methanotrophic ISBR for TCE degradation.

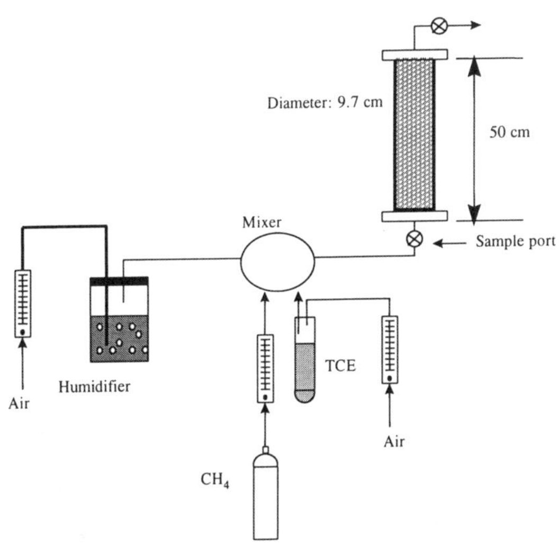

FIGURE 2. Laboratory-scale methanotrophic bioreactor for gas-phase TCE degradation.

The filtration media was a mixture of compost and perlite (50% v/v) (Les Tourbières Premier Ltd, Rivière-du-Loup, Que., Can.). This media had already been used successfully for the biodegradation of dichloromethane (Ergas et al., 1994). The compost used is an organic compost made of peat moss, plant residues and sterilized manure (Les Tourbières Premier Ltd). It contains more than 50% of organic matter and is pH-balanced.

Analytical Procedures. O_2, CH_4, and TCE analysis, as well as ^{14}C-TCE mineralization were conducted as described elsewhere (Lafontaine, 1997; Pierre, 1997).

RESULTS AND DISCUSSION

TCE Mass Balance in an Abiotic Reactor. The original ISBR was designed for the degradation of relatively nonvolatile compounds such as pentachlorophenol so it did not incorporate headspace gas recirculation. Since aeration was required for mixing and metabolism, there was a constant gas stream leaving the reactor. However, under these conditions a significant portion of the TCE (about 25%) was transferred to the gas phase due to volatilization (Fig. 3). Changes were made to the ISBR to avoid abiotic loss of TCE. Aerated water was added to the reactor in a separate stream than that of the contaminated water. This resulted in very similar inlet and outlet TCE concentrations except when the reactor was opened to add the geotextile support (Fig. 4). This system (Fig. 1) was used for all TCE biodegradation experiments.

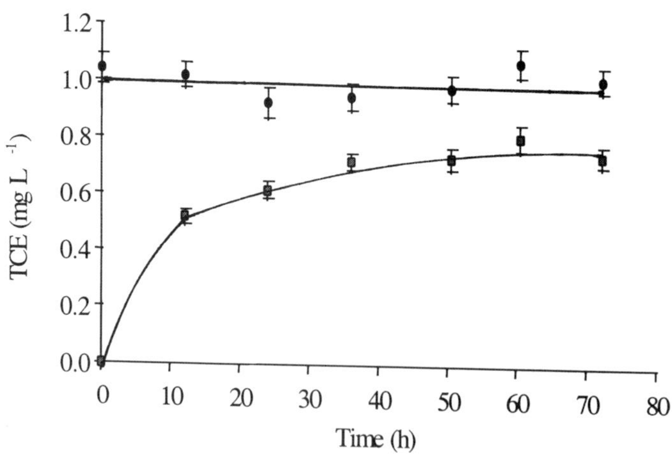

FIGURE 3. Abiotic TCE (1 mg L^{-1}) addition in an ISBR with 0.5 L air min^{-1} aeration (●)[TCE] inlet, (■) [TCE] outlet. τ : 6 hours; temperature : 30 °C.

TCE Degradation in the ISBR. TCE was fed to the ISBR at various concentrations, methanation rates, and hydraulic residence times. In general, the TCE degradation rate increased linearly with the amount of TCE fed (1st order kinetics) and with the residence time. In contrast, when the CH_4 addition rate was diminished from 0.063 to 0.024 mg L^{-1} min^{-1}, TCE degradation increased about 16%, probably due to competitive inhibition. At the end of this experiment which lasted 8 days, the CH_4 supply was lowered and only the TCE concentration in the feed and the residence time were varied. For a given residence time, TCE degradation rate varied linearly as a function of the feeding TCE concentration. TCE degradation rates of up to 230 mg L^{-1} d^{-1} for an initial TCE concentration of 18 mg L^{-1} and a residence time of 0.04 d were measured. Maximum TCE removal efficiency was 97% at a residence time of 0.91 d.

TCE Degradation in the Biofilter. As in the ISBR, the TCE degradation kinetics varied directly with TCE concentration and hydraulic residence time (Table 1) and inversely with the CH_4 concentration (data not shown). The highest activity was achieved with an inlet concentration of 20.5 mg L^{-1} TCE (out = 14.6 mg L^{-1} TCE) which resulted in a degradation rate of 233 g TCE m^{-3} reactor h^{-1}. There was some loss of activity over time probably due to the toxic effect of TCE intermediates on the microbial consortium.

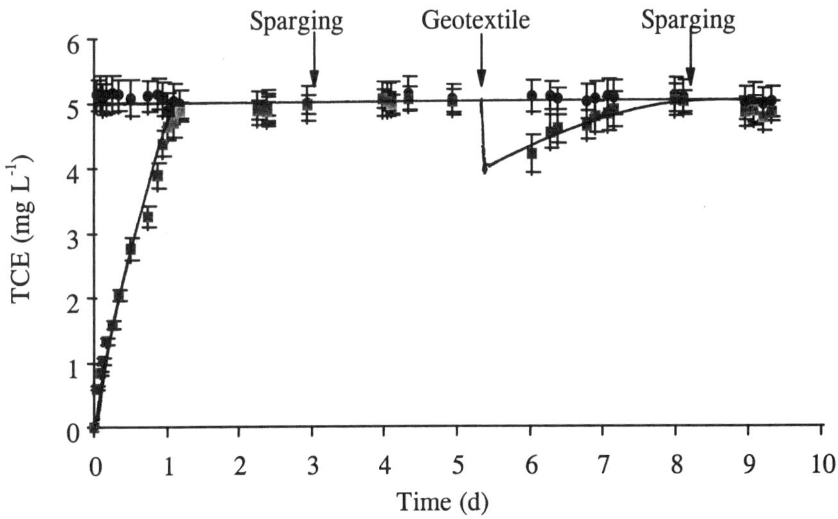

FIGURE 4. Abiotic addition of TCE (5 mg L^{-1}) to ISBR with headspace gas recycle and (M)[TCE] inlet, (O) [TCE] outlet. τ :6 h; Temp:30 °C; headspace gas recirculation:0.5 L min^{-1}. (↓ sparging indicates onset of gas recirculation period) (↓ geotextile indicates geotextile support addition with no agitation).

TABLE 1. Operation of a 3.7 L biofilter at different inlet TCE concentration.

Inlet [TCE] (mg L^{-1})	1.37	3.15	5.85	20.45	41.06
% TCE degradation	62.77	21.59	49.4	28.7	8.86
Inlet [CH_4] (% v/v)	3.05	3.65	3.82	2.74	2.87
Flow rate (L min^{-1})	2.288	2.404	2.248	2.442	2.583
Residence time	1.617	1.539	1.649	1.515	1.432
TCE degradation rate (g TCE m^{-3} reactor h^{-1})	31.95	26.56	105.20	232.76	152.54

TABLE 2. Comparison of the results of TCE biodegradation in a methanotrophic immobilized soil reactor with other published reports of reactors used for TCE degradation.

Reference	$[TCE]_{in}$ (mg L^{-1})	$[TCE]_{out}$ (mg L^{-1})	Residence time (d)	Degradation rate (mg L^{-1} d^{-1})	Efficiency (% TCE removal)	k_{TCE} (d^{-1})
Arvin 1991	1.2	0.924	0.2	1.38	23	1.49
Lanzaronne and McCarty 1990	0.0025	0.0185	0.02	0.33	26	17.8
Aziz et al. 1995*	0.097	0.075	0.0065	3.38	22.7	45.1
Lab-scale ISBR	2.26	0.061	0.91	2.42	97.3	39.7
Fennel et al. 1993	11.3	8.6	0.11	24.5	23.9	2.85
Lab-scale ISBR	18.12	8.46	0.042	230	53.3	27.2

*pure culture of *M. trichosporium* OB3b

TCE Mineralization. While the reactors showed high TCE degradation rates, further tests were required to show whether mineralization was occurring. Microcosms containing aqueous samples from the ISBR or solid-state samples from the biofilter both showed mineralization of [^{14}C]-TCE. The amounts and rates of TCE mineralization depended on the TCE and methane concentrations but up to 91% of the initial TCE carbon was oxidized to CO_2 in some microcosms indicating that complete mineralization was possible.

Pilot-Scale ISBR. A 2000 L ISBR with associated biofilter (Fig. 5) was constructed and is now undergoing testing. It is designed for the treatment of groundwater containing volatile contaminants and can be operated in a methanotrophic mode. Headspace gas is recycled through spargers placed under every second textile support panel to achieve an "air-lift" effect. Methane and oxygen concentrations in the ISBR and biofilter are maintained at the desired levels via mass flow controllers and based on on-line gas analysis data.

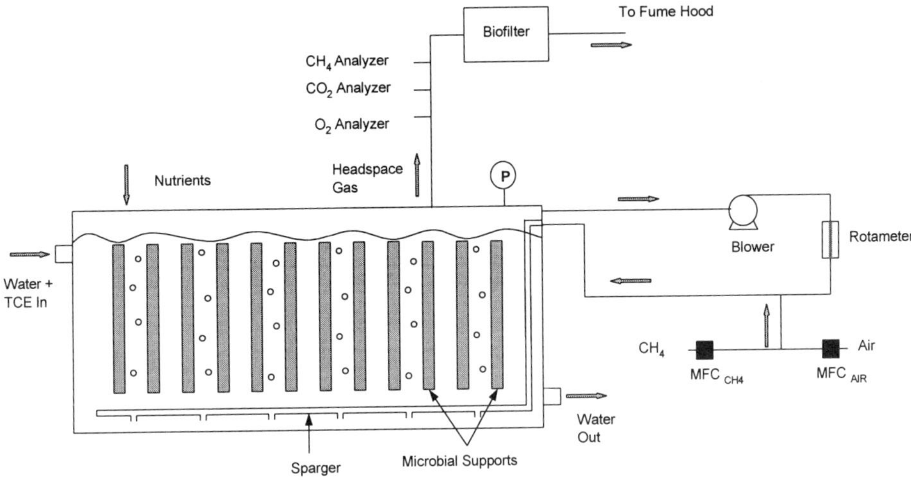

FIGURE 5. Pilot-scale (2000 L) methanotrophic ISBR.

CONCLUSIONS

The 2 L ISBR was found to be more efficient than other bioreactor designs except for those employing axenic culture (Table 2). At low concentrations, TCE degradation rates are normally dependant on the concentration of TCE in the reactor (first order kinetics). Compared with the fixed film reactor of Arvin (1991) and the soil column of Lanzaronne and McCarty, (1990), for low TCE outlet concentrations (i.e. the concentration of TCE in the reactor), the methanotrophic immobilized soil bioreactor showed a TCE degradation rate of 1.75 to 7.3 times higher, with an efficiency 323% and 274% greater and a higher first order kinetic constant. For very high TCE outlet concentrations, the degradation rates were up to 9.4 times higher than the best results reported for the fixed film reactor of Fennel et al., 1993 with a 307% higher efficiency. Aziz et al. (1995) reported data with a higher 1st order degradation constant but employed a hollow fiber membrane reactor with a pure culture of *Methylosinus trichosporium* OB3b which would likely be impractical for field use. The 1st order degradation constant (based on TCE concentration) of the 2 L methanotrophic ISBR was 1 to 2 orders of magnitude lower than that of advanced oxidation reactors. This means that an ISBR must be 10 to 100 times larger than an advanced oxidation reactor. However, the ISBR would still be substantially less expensive to construct. Although a thorough study has not been conducted, it appears that operation costs would also be substantially lower and that the reactor is stable enough for practical application. Optimization of the 2000 L ISBR should lead to increased efficiency but testing must be conducted under actual pump-and-treat conditions. Its performance at typical groundwater temperatures is of particular concern, although excellent results were obtained for PCP degradation when the temperature was shifted from 23 to 9°C (Karamanev et al., 1999). After several days of adaptation, the PCP degradation rate had actually increased.

There is very little literature data available for methanotrophic biofilters. However, they appear to be promising as devices to polish exit air from aqueous phase bioreactors or in combination with strippers for the degradation of certain volatile groundwater such as TCE or materials that are less highly halogenated.

ACKNOWLEDGEMENTS

This work was supported by the Natural Sciences and Engineering Research Council of Canada (NSERC), ALCAN, Petro-Canada, Hydro-Québec, Analex Co., Browning-Ferris Industries, SNC-Lavalin and Centre Québecois de Valorisation de la Biomasse (CQVB) and by Polyferm Canada.

REFERENCES

Arvin, E. 1991. "Biodegradation kinetics of chlorinated aliphatic hydrocarbons with methane-oxidizing bacteria in an aerobic fixed biofilm reactor." *Water Res. 25*:873-881.

Aziz, C. E., M.W. Fitch, L. K. Linquist, J. G. Pressman, G. Georgiou, and G. E. Speitel Jr. 1995. "Methanotrophic biodegradation of trichloroethylene in a hollow fiber membrane bioreactor". *Environ. Sci.Technol. 29*:2574-2583.

Ergas, S. J., Kinney, K., Fuller, M.E. et Scow, K.M. 1994. "Characterization of a compost biofiltration system degrading dichloromethane". *Biotechnol. Bioeng. 44*: 1048-1054.

Fennell, D. E., Y. M. Nelson, S. E. Underhill, T. E. White, and W.J. Jewell. 1993. "TCE degradation in a methanotrophic attached-film bioreactor". *Biotechnol. Bioeng. 42*:859-872.

Karamanev, D. G. and R. Samson. 1998. "Kinetics of the biodegradation pentachlorophenol by the biofilm developed in immobilized soil bioreactor". *Environ. Sci. Technol. 32*:994-998.

Karamanev, D. G., B. A. Ramsay, and M.-C. Aly-Hassan. 1999. "Design and experimental study of immobilized soil bioreactor in in-well biodegradation of pentachlorophenol". *Can. J. Chem. Eng. 77*:936-941.

Lafontaine, C. 1997. "Biofiltration en phase gazeuse du trichloroéthylène sous conditions méthanotrophes". MESc thesis, Ecole Polytechnique de Montréal. 165pp.

Lanzarone, N. A. and P. L. McCarty. 1990. "Column studies on methanotrophic degradation of trichloroethylene and 1,2-dichloroethane". Ground Water *28*: 910-919.

Pierre, S. 1997. "Comportement d'un réacteur à sol immobilisé en conditions méthanotrophes pour la biodégradation du trichloroéthylène". MESc thesis, Ecole Polytechnique de Montréal. 111pp.

AEROBIC COMETABOLIC TRANSFORMATION OF TRICHLOROETHYLENE AND *cis*-DICHLOROETHYLENE IN PROPANE FED AQUIFER MICROCOSMS

Brian Timmins, Mark Dolan, and Lewis Semprini
(Oregon State University, Corvallis, Oregon)

ABSTRACT: A 15-microcosm matrix was created to determine the potential for using propane-utilizing microorganisms to aerobically cometabolize chlorinated aliphatic hydrocarbons (CAH) at McClellan Air Force Base, California. Microcosms were constructed using aquifer materials and groundwater that were aseptically obtained from the treatment location where cometabolic air sparging (CAS) tests were conducted using propane as the primary substrate. Trichloroethylene (TCE) and 1,2-*cis*-dichloroethylene (c-DCE) were tested at concentrations near those found at the field site. Unlike the field demonstration, the lag time to onset of propane utilization took only 9-12 days of incubation. Upon stimulation with propane both CAHs were readily transformed. Nitrogen limitations similar to those observed in the field demonstration were observed in microcosms prepared with site groundwater and no nitrogen amendments. Propane utilization rates decreased and TCE transformation ceased when nitrate became limiting in the microcosms. Both nitrate and ammonia were found to restore propane oxidation rates and to cause resumption of CAH transformation. Microcosm transformation yields of 0.035 μg c-DCE/μg propane and 0.053 μg TCE/μg propane were obtained when nitrogen was not a limiting factor.

INTRODUCTION

The objective of this study was to use microcosms to better understand factors affecting cometabolic air sparging (CAS) that was being performed at McClellan Air Force Base (AFB) to bioremediate trichloroethylene (TCE) and 1,2-*cis*-dichloroethylene (c-DCE) subsurface contamination. The AFB site has a deep vadose zone extending 108 ft (32.9 m) below ground surface (bgs), and chlorinated aliphatic hydrocarbon (CAH) concentrations above 500 μg/L in the saturated zone (Tovanabootr et al., 2000). A mixture of air containing propane (2%) was injected at an airflow rate of 5 standard cubic feet per minute (scfm) once or twice a week for a duration of 5-10 hours over a period of 80 days (Tovanabootr et al., 2000). Removal of CAHs was observed in the saturated zone where effective propane and oxygen delivery occurred.

Microcosms provide a good screening method to determine if the indigenous microorganisms from a site can catalyze specific transformations of interest. They are relatively inexpensive to construct, easy to use, allow versatility of amendments, and often provide reproducible results. Microcosms have been widely used to determine the biodegradability of organic contaminants at the laboratory scale, and to determine the impact of site-specific physical, chemical, and hydrogeologic conditions (Bedient et al., 1994). Previous microcosm studies with subsurface solids and groundwater from another site at McClellan showed propane utilizers effectively transformed mixtures of TCE, 1,1,1-trichloroethane, and chloroform (Tovanabootr and Semprini, 1998).

The main objectives for the microcosm studies were to:

1) Determine whether indigenous microcosms that utilize propane were present at the site.
2) Estimate the lag time required to stimulate propane-utilizing microorganisms.
3) Evaluate how effectively the indigenous microorganisms transform CAHs.
4) Quantify transformation yields (µg TCE or µg c-DCE/ µg propane).
5) Estimate the nitrogen requirements of the indigenous microorganisms and test the effects of ammonia as a nitrogen source.
6) Determine microbial kinetic parameters.

Results will be presented for the first five objectives.

MATERIALS AND METHODS

Soil and Groundwater Collection. All aquifer material and groundwater used to construct the microcosms was obtained directly from the saturated zone of the CAS demonstration site. Aquifer solids were collected in six by 1.5-inch brass sleeves from the depth interval of 112-113 ft bgs (34.1-34.4 m) during the installation of three of the site monitoring wells. The solids were aseptically removed from the sleeves, wet sieved (No. 4 sieve) to remove large particles, and then mixed together before being placed in the microcosms. Groundwater was obtained from an existing monitoring well at the site and sparged with nitrogen gas to remove CAHs before being placed in the microcosms.

Microcosm Preparation. Materials used to construct the microcosms were autoclaved (140 C for 60 minutes) to prevent microbial contamination. Fifteen-150 mL bottles were filled with aquifer material (40 g) and groundwater (75 mL) under a laminar flow hood, leaving 60 mL of air-filled headspace. The microcosms were sealed with butyl rubber septum caps, inverted, and incubated on a shaker table (200 rpm) at 20 C. Propane was added to the microcosms via headspace injection of >98% pure propane (Alltech Associates Inc, Deerfield, IL). TCE and c-DCE (>99.9% pure, Fisher Scientific, Pittsburgh, PA) were added via water-saturated solutions injected into the liquid phase of the microcosm. Nitrate was added using an aqueous solution of $NaNO_3$.

Analytical Methods. Microcosm performance was monitored using headspace analysis for propane, CAHs, and DO. Headspace samples were obtained using a 100-µL gas tight syringe and analyzed on an HP 6890 gas chromatograph equipped with a Photo-Ionization Detector (PID) and a Flame Ionization Detector (FID) in series. Separation was obtained using a 0.53 mm megabore GS-Q capillary column (J&W Scientific, Folsom, CA). The total mass of each compound in a microcosm was calculated assuming equilibrium partitioning (unitless Henry's constants at 20°C of 28.9 for propane, 0.342 for TCE and 0.123 for c-DCE). Headspace oxygen was measured using a HP 6890 gas chromatograph equipped with a thermal conductivity detector (TCD) with argon as a carrier gas. Microcosms were re-equilibrated to atmospheric pressure by injecting oxygen into the headspace before each substrate injection. Nitrate was analyzed using a Dionex 40001 ion chromatograph with a conductivity detector. All analytes were quantified by comparison to external standards.

RESULTS AND DISCUSSION

Microcosm Matrix. Below is a list of the microcosms and amendments employed in the test:

Control 1 (Poisoned) – amended with $HgCl_2$, propane, and CAHs
Control 2 (Autoclaved) – amended with propane and CAHs
Control 3 (No-substrate) – amended with CAHs only
Live 4 through 6 – amended with propane and CAHs
Live 7 through 9 – amended with propane only
Live 10 through 12 – amended with propane, CAHs, and nitrate
Live 13 through 15 – amended with propane and nitrate
(All live microcosms were performed in triplicate to permit statistical analysis.)

Controls. Three controls were operated; Control 1 contained 25 mg/L of mercuric chloride along with propane and CAHs; Control 2 was autoclaved twice over a 48-hour interval before propane and CAHs addition; Control 3 was used as a no-substrate or intrinsic control. All microcosm controls were periodically analyzed for the 101-day duration of the study. Microcosm controls initially contained approximately 790 μg propane, 36 μg c-DCE, and 114 μg TCE. Propane concentrations remained essentially constant in the Controls 1 and 2 over the entire study period. Limited CAH losses were observed in the controls over the 101-day test period. An average TCE loss of 37% (std deviation = 2.2%) was observed in the three control bottles. Controls 1 and 2 averaged a 20% loss of c-DCE where Control 3 showed a 40% c-DCE loss. The slow, continuous CAH losses were most likely due to slow leakage from the bottles or partitioning into the butyl rubber septa. The greater c-DCE loss in the no-substrate control may have been partially due to intrinsic biotic c-DCE transformation. In any case, the CAH losses in the controls were quite small when compared to CAH removal observed in the active microcosms.

Live Microcosms with CAHs. Two groups of triplicate microcosms contained CAHs and propane. The first group contained background levels of nitrate (4.8 mg/L) found in the site groundwater. The second group was amended with nitrate to achieve an aqueous concentration of 125 mg/L as NO_3^-. The microcosms initially contained 820 μg propane, 33 μg c-DCE, and 104 μg TCE. All six propane stimulated microcosms behaved similarly until day 15 (Table 1). Lag times for propane consumption ranged from 9 to 12 days. After propane concentrations decreased, CAH concentrations also began to decline. All microcosms showed c-DCE being cometabolized at a faster rate than TCE, with c-DCE being essentially completely removed while TCE was partially removed. The average transformation yields were 0.036 μg c-DCE/μg propane and 0.056 μg TCE/μg propane. These are conservative estimates since no attempt was made to maximize the transformation yields.

TABLE 1. Comparison of lag times, propane utilization, and CAH transformation in microcosms with and without nitrate amendment.

Microcosm Type	Lag Time (days)	Propane Utilized (μg)	c-DCE Transformed (%)	TCE Transformed (%)
Without nitrate	9-12	1,590	100	72
	9-12	1,570	100	75
	9-12	1,150	100	59
With nitrate	9-12	1,170	100	57
	9-12	1,140	100	58
	9-12	1,500	100	94

On day 15 the non-nitrate amended microcosms began to metabolize propane at a slower rate, and TCE transformation ceased. Nitrate concentrations in the non-amended microcosms were below the detection limit (25 μg/L) when sampled on day 22. Propane utilization continued until day 41 in the absence of detectable nitrate, but at a much slower rate. From day 21 to day 41, no measurable TCE transformation occurred. This same behavior was witnessed in the McClellan AFB demonstration when nitrate concentrations fell below detectable levels, and propane utilization rates slowed (Tovanabootr et. al., 2000).

In the nitrate amended microcosms, propane utilization and CAH transformation rates did not decrease through successive additions of propane. All of the CAHs were transformed by day 18. On day 28, c-DCE and TCE were again added to the nitrate-amended microcosms and all six microcosms were analyzed 24 hours later, by which time propane was completely utilized. Table 2 shows the amounts of propane utilized and CAHs transformed by day 29.

TABLE 2. Amounts of propane utilized and CAHs transformed with the second addition of propane on day 28.

Microcosm Type	Propane Utilized (μg)	c-DCE Transformed (%)	TCE Transformed (%)
Without nitrate	17	*	0
	44	*	0
	47	*	0
With nitrate	402	32	24
	346	32	25
	359	32	34

* Not present in the microcosm

The nitrate-amended microcosms were incubated in the absence of propane until day 33. During this interval TCE concentrations remained constant, with no indication of further TCE transformation after the propane was consumed by day 29. Propane was again added to the nitrate-amended microcosms on day 33, but unlike previous additions, 2 to 3 days were required for the microcosms to utilize the propane. TCE remained constant until propane concentrations were substantially reduced, after which the remaining TCE was cometabolized. It appeared that incubation in the presence of TCE and the absence of propane had a negative affect on the propane-utilizing microbial population.

Live Microcosms without CAHs. Two groups of triplicate microcosms contained propane and no CAHs. The first group contained background levels of nitrate (4.8 mg/L) found in the site groundwater, while the second group was amended with nitrate to achieve an aqueous concentration of 125 mg/L as NO_3^-. The microcosms initially contained approximately 860 μg propane each. Both groups had similar lag times of 9-12 days, and behaved similarly through day 26 of consecutive feedings. After the propane addition on day 26, the rate of propane consumption in the non-nitrate amended microcosms began to decrease (Table 3). Nitrate concentrations in the non-amended microcosms were below the detection limit when sampled on day 31.

TABLE 3. Comparison of propane utilization rates from day 0-33 between microcosms with and without nitrate amendments.

Microcosm Type	Day 0-26 Utilization Rate (μg/hr)	Day 26-27 Utilization Rate (μg/hr)	Day 28-29 Utilization Rate (μg/hr)	Day 31-33 Utilization Rate (μg/hr)
Without nitrate	34.7	9.9	6.5	6.0
	29.7	14.1	9.1	3.1
	21.1	12.4	8.7	6.3
With nitrate	39.2	*	*	*
	21.2	*	*	*
	23.5	*	*	*

* Rate greater than or equal to the initial rate (Day 0-26).

The three microcosms without nitrate amendments were able to utilize an average of 2570 μg propane before showing the effects of nitrogen limitation. This compares with the three non-nitrate amended microcosms that contained CAHs that utilized an average of 1550 μg propane (40% less) before showing the effects of nitrogen limitation. More nitrogen was required per propane utilized in the microcosms that were transforming c-DCE and TCE. Although the cause of this is unknown, the synthesis of new cells or proteins as a consequence of transformation product toxicity may have been responsible for the greater nitrogen/propane requirements.

Ammonia as a Nitrogen Source. Ammonia was tested as a nitrogen source for propane utilizers in the non-nitrate amended microcosms containing CAHs. The microcosms had previously shown decreased propane utilization rates and little or no TCE transformation as a result of nitrogen limitations. Prior to adding ammonia, propane was injected into all three microcosms and propane and TCE concentrations were monitored for 11 days. TCE remained constant while propane was utilized at a very slow rate, similar to what was previously observed. On the 11th day, 0.2 ml of ammonia gas was injected into the headspace of two of the three microcosms. Within two days the rate of propane utilization rapidly increased in the ammonia-amended microcosms and TCE transformation resumed. All residual TCE was transformed and propane utilized within three days of ammonia addition. The microcosm that did not receive any ammonia showed no enhancement in rates. Propane (220 μg), c-DCE (40 μg), and TCE (43 μg) were added six days after the ammonia addition (day 17). Five additions of propane were required over a 13-day period to cometabolize all of the c-DCE and TCE. Again c-DCE was removed sooner than TCE. On the 30th day c-DCE (42 μg) and TCE (45 μg) were added again along with propane (220 μg). The propane utilization rate decreased, and both c-DCE and TCE transformation slowed. Over the next 23 days propane was injected five times with the utilization rate decreasing with each addition. By day 45, approximately 38-67% of the TCE and from 92-100% of the c-DCE had been transformed. After day 45, TCE cometabolism ceased, but c-DCE cometabolism continued at a slow rate. Throughout the experiment the microcosm without ammonia addition showed slow propane utilization and no TCE cometabolism.

Nitrogen Requirements. A comparison of nitrogen requirements and propane utilization was calculated for microcosms using nitrate or ammonia as a nitrogen source in the presence or absence of CAHs and is presented in Table 4.

TABLE 4. Propane and nitrogen utilization in microcosms under different conditions.

Microcosm Type (Propane Amended)	Propane Utilized (μmol)	Nitrogen Utilized (μmol)	Ratio Nitrogen/Propane (μmol/μmol)
Nitrate with CAHs	36	5.8	0.16
	36	5.8	0.16
	34	5.8	0.17
Nitrate w/o CAHs	58	5.8	0.10
	58	5.8	0.10
	58	5.8	0.10
Ammonia with CAHs	41	8.3	0.20
	37	8.3	0.22

The amount of nitrate or ammonia utilized was determined from direct measurement or from the known amount added to the microcosm. The amount of propane utilized was estimated to be the amount used before a significant decrease in propane oxidation rate was observed. The amount of nitrogen required for propane utilization was significantly higher in microcosms with

CAHs present. Ammonia and nitrate appeared to provide equal benefits and resulted in similar nitrogen/propane requirements in the presence of CAHs. More nitrogen may be needed when CAHs are present due to transformation product toxicity requiring synthesis of protein or new cells.

CONCLUSIONS

The results obtained from limited microcosm testing were found to be highly correlative with results obtained during the field demonstration. The indigenous microorganisms present at the McClellan AFB treatment area have the ability to utilize propane and transform c-DCE and TCE. The transformation of c-DCE was more rapid than TCE, which was consistent with observations from the CAS demonstration site. A maximum transformation yield of 0.053 μg TCE/μg propane and 0.035 μg c-DCE/μg propane was observed. A potential limiting factor for using CAS as a form of bioremediation was the amount of bioavailable nitrogen naturally present at the site. Nitrogen limitation was found to occur quickly when propane was used as a carbon source, and once the system became nitrogen limited propane utilization decreased and TCE transformation ceased. The microcosm tests showed ammonia was effective as a nitrogen source and caused an increase in propane utilization rates and CAH transformation. However, the possible long-term effects of using ammonia as the sole nitrogen source, such as increasing the groundwater pH or increasing the nitrogenous oxygen demand, were not investigated here. Nitrogen requirements in the presence of CAHs are larger than the requirements in the absence of CAHs, which is potentially due to CAH transformation product toxicity requiring new cell synthesis and/or maintenance.

REFERENCES

Bedient, P.B., H. S. Rifai, and C. J. Newell. 1994. "Microcosm." *Ground Water Contamination: Transport and Remediation.* PTR Prentice Hall Inc. 211-2.

Tovanabootr, A., M.E. Dolan, L.Semprini, V.S. Magar, A.Leeson, and A. Lighter. 2000. *Cometabolic Air Sparging Field Demonstration with Propane to Remediate a Chloroethene and Chloroethane Co-Contaminated Aquifer.* In: Physical and Thermal Technologies. Eds. G.B Wickramanayake and A. R. Gavaskar. Battelle Press, Columbus, OH. pp. 67- 74.

Tovannabootr, A. and L. Semprini. 1998. "Comparison of Long-Term TCE Transformation Ability of Methane and Propane-Utilizing Microorganisms Stimulated from the McClellan AFB Subsurface," *Bioremediation Journal. 2*(2): 105-124.

MICROBIAL DEGRADATION OF TRICHLOROETHYLENE (TCE) IN AQUEOUS CULTURE

Chi-Ruey Chen (Florida International University, Miami, FL)
Shonali Laha (Florida International University, Miami, FL)

ABSTRACT: Recent evidence suggests that TCE may be treated by biodegradation under both aerobic and anaerobic conditions using co-metabolic pathways. This involves the use of a primary substrate (such as methane or phenol) that induces the production of oxygenase enzymes from microorganisms that nonspecifically oxidize the TCE. In this study, the microbial degradation of TCE by a mixed methane-oxidizing culture is evaluated and compared with that by pure II methane-oxidizing isolate, *Methylosinus trichosporium* OB3b, in the presence of various substrates (i.e., phenol, toluene, and atrazine). The TCE concentration was significantly decreased within 10 days in the presence of pure culture using 10 or 100 mg/L initial TCE concentration. Similar results were obtained in the presence of mixed culture using 10 or 100 mg/L initial TCE concentration. A first order rate law was found to describe TCE degradation by both cultures. Volatilization losses were observed for control tests. Additional chloride measurements using ion chromatography (IC) were attempt to separate TCE losses through volatilization from TCE biodegradation. Significant variation of chloride content was observed for the triplicate microcosms receiving mixed methane-oxidizing culture and 100 mg/l TCE as indicated by standard deviation measure as 76. Similarly for systems receiving 10 mg/l TCE, the pure and mixed culture microcosms indicated chloride levels <10 mg/l.

INTRODUCTION

Trichloroethylene (TCE) is a common contaminant found in groundwater in the United States. TCE is used as an organic solvent in metal degreasing, dry cleaning and a wide range of other industrial applications. Due to its persistence and suspected carcinogenicity, TCE has been designated a priority contaminant by the EPA. The potential health hazard resulting from exposure to TCE has raised public concern with respect to cleanup of aquifers contaminated with this chemical. The maximum contaminant level (MCL) is 5 μg/l for TCE.

Wilson and Wilson (1985) first observed the aerobic cometabolism of TCE [6]. This involves the use of primary substrates such as phenol, methane, methanol, and toluene that induce the formation of the oxygenase enzymes from microorganisms that fortuitously oxidize TCE to TCE-epoxide.

Objective. The current study evaluates the effectiveness of three microbial consortia at degrading TCE in the presence of various substrates. The objectives of this study are to examine the following considerations necessary to confirm the biodegradation possibility of TCE:

(i) To replicate experiments demonstrating TCE biodegradation under aerobic conditions,
(ii) To test the relative effectiveness of phenol, toluene and atrazine as primary substrates in TCE biodegradation,
(iii) To explore the feasibility of using chloride analyses to provide stoichiometric confirmation of TCE biodegradation.

MATERIALS AND METHODS

Stock solutions of TCE–saturated water were prepared in 100-ml volumetric flasks by adding 0.686 ml of pure TCE (certified ACS. reagent; Fisher Scientific, Fair Lawn, New Jersey) and filled with distilled water. This produced an aqueous concentration of 10,000 mg/l TCE. Stock solutions of phenol were prepared in 100-ml volumetric flasks by adding 0.935 ml of phenol (Fisher Scientific) and filling with deionized water producing an aqueous concentration of 10,000 mg/l. Stock solutions of toluene were prepared with 1.15 ml of toluene dissolved in 18.85 ml of methanol due to toluene's limited aqueous solubility. The resulting toluene concentration in methanol was 50,000 mg/l.

Microbial Methods. Methylosinus trichosporium OB3b and methane-oxidizing mixed culture used in these experiments were obtained from the Department of Civil and Environmental Engineering at University of California at Berkeley [13]. The two cultures were grown at 20°C in nitrate mineral salts (NMS) medium containing 11.76 mM $NaNO_3$, 0.98 mM K_2SO_4, 0.15 mM $MgSO_4$, 0.07 mM $CaSO_4{\bullet}2H_2O$, 0.08 mM $FeSO_4{\bullet}H_2O$, 3.9 mM KH_2PO_4, 6.1 mM Na_2HPO_4, 0.1mM H_2SO_4, 0.001 mM KI, 0.002 $ZnSO_4{\bullet}7H_2O$, 0.002 mM $MnSO_4{\bullet}H_2O$, 0.002 mM H_3BO_3, and 0.004 mM Co $(NO_3)_2{\bullet}6H_2O$.

Analytical Methods. TCE concentrations were determined by using Hewlett Packard (HP) 5890 Gas Chromatography (GC) with an automatic sampler (HP 7673A), electron capture detector (ECD), and a DB-5 capillary column (inside diameter, 0.45 mm; length, 30 m; J & W Scientific, Folsom, Ca.). Helium was used as the carrier gas. The helium flow through the column was maintained at 6.8 ml/min. Nitrogen was used as a makeup gas with a flow rate of 32 ml/min. The GC program used an initial oven temperature of 40°C, raised to 80°C at the rate of 5°C per minute, reaching a final temperature of 120°C at the rate of 15°C per minute. The temperature of the injector and electron capture detector (ECD) were maintained at 250°C. The calibration curves were obtained by liquid-liquid extraction sample analysis in vials containing known amounts of TCE. The vials were prepared by adding known amounts, such as 20, 30, 40, 50, 60 μl of TCE-methanol mixture to 44-ml vials containing 38 ml distilled water and 5 ml pentane. The TCE-amended vials were shaken at 300 rpm at the ambient temperature.

Chloride concentrations were measured using Dionex DX-120 Ion Chromatography (IC) with a IonPad AS14 column (inside diameter, 4 mm) and Auto Suppressed Conductivity for detection. Eluent was composed of 3.5 mM sodium carbonate/1.0 mM sodium bicarbonate. Eluent flow rate through the column was maintained at 1.0 ml/min. Injection volume of 100 ml was used.

RESULTS AND DISCUSSION

Microbial enumeration of the rapid growth cultures was performed. The mixed culture had greater numbers of colonies than pure culture OB3b, and using phenol as the primary substrate resulted in greater numbers of colony-forming units (CFUs). In subsequent TCE biodegradation experiments, phenol was selected as the primary substrate. The average cell count obtained for growth cultures receiving phenol, TCE, and pure culture were 9.75×10^{7} CFU/ml, while the average cell count for mixed culture with phenol and TCE was 3.24×10^{9} CFU/ml. For pure culture exposed to 10 mg/l and 100 mg/l TCE almost complete TCE disappearance was observed in the first five days of incubation (Figure 1). The mixed culture was seen to degrade about 80 percent of TCE within 10 days when using an initial TCE concentration of 100 mg/l. For the microcosms receiving 10 mg/l TCE and mixed culture, TCE concentration was observed to decrease significantly but not completely (Figure 2). The poisoned abiotic controls were intended to provide confirmation of biological losses of TCE. The results suggest that much of the TCE degradation observed could be attributed to abiotic losses such as volatilization, sorption to the glass surface, etc (Figure 3). Significant variation of chloride content was observed for the triplicate microcosms receiving mixed methane-oxidizing culture and 100 mg/l TCE as indicated by standard deviation measure as 76. Similarly for systems receiving 10 mg/l TCE, the pure and mixed culture microcosms indicated chloride levels <10 mg/l (Table 1).

CONCLUSION

The pure culture appeared to be more effective at removing TCE, however, owing to TCE losses observed in the abiotic control experiments it cannot be positively determined that the decreasing TCE concentration is attributable to biodegradation. Chloride analyses also failed to confirm the occurrence of TCE biodegradation. It has been suggested that methane or methanol is required for biodegradation of TCE by methanotrophs as it is a cometabolic process [10], and therefore, one of the shortcomings of this study is that methane was not used as a primary substrate. The results suggest it is difficult to distinguish between biological degradation and other losses of TCE. This is likely to pose greater challenges when implementing bioremediation for cleanup of TCE-contaminated sites.

ACKNOWLEDGMENTS

I wish to thank Nahid Golkar and Dr. Mehrzad Mehran for sharing with me their time and vast knowledge throughout the entirety of this project and Dr. Berrin Tansel, Director of the Drinking Water Research Center for the use of the facilities in the Drinking Water Research Laboratory.A special thanks to my advisor and friend, Dr. Shonali Laha, who gave me the opportunity and confidence necessary for the success of this project.

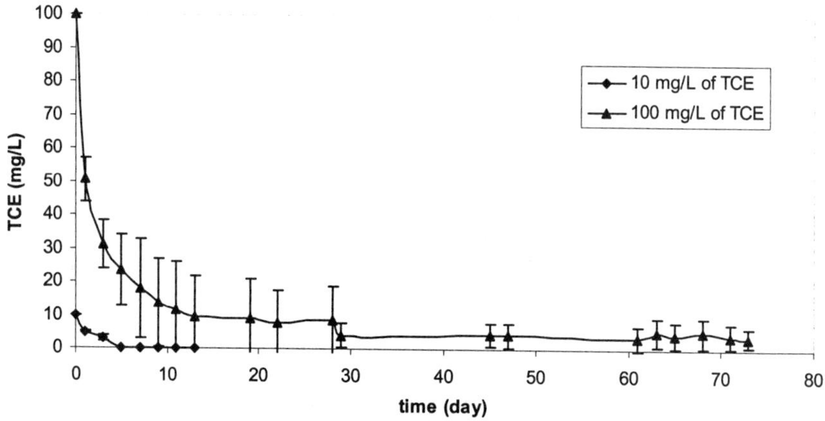

FIGURE 1. TCE biodegradation in flask microcosms incubated at ambient temperature by *Methylosinus trichosporium* OB3b at two initial concentrations.

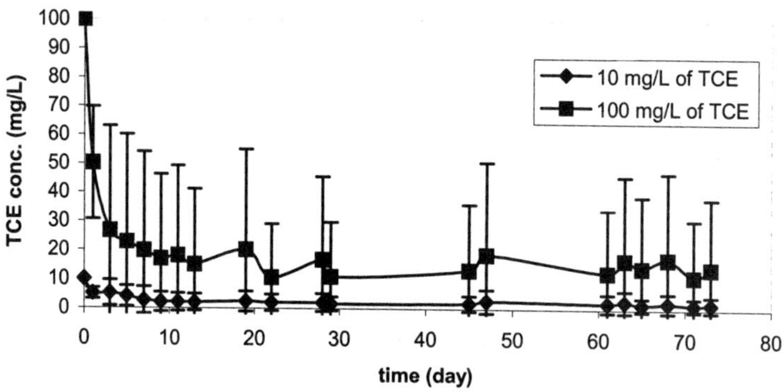

FIGURE 2. TCE biodegradation in flask microcosms incubated at ambient temperature by a mixed methane oxidizing culture at two initial concentrations.

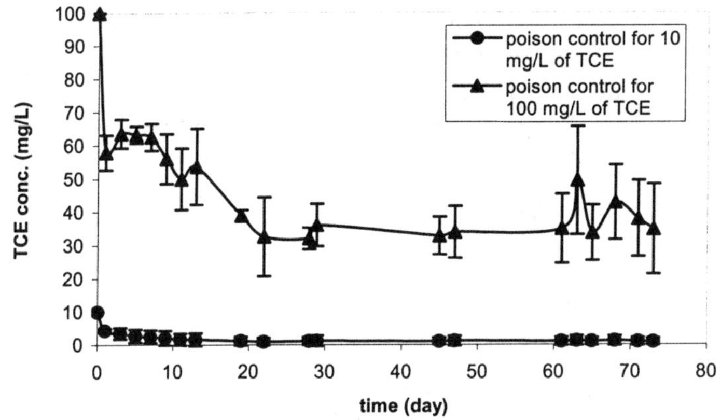

FIGURE 3. TCE losses in flask microcosms incubated at ambient temperature by a mixed methane oxidizing culture with $Hg(NO_3)_2$ at two initial concentrations.

TABLE 1. Results of chloride concentrations in ppm from ion chromatography (IC).

Chloride concentration in microcosm with pure culture OB3b		
0 mg/l of TCE	10 mg/l of TCE	100 mg/l of TCE
4A, 4B, 4C	5A, 5B, 5C	6A, 6B, 6C
Average = 8.27 mg/l	Avg. = 6.76	Avg. = 9.29
Standard Deviation = 8.56	SD = 2.18	SD = 6.69
Chloride concentration in microcosm with mixed methane-oxidizing culture		
0 mg/l of TCE	10 mg/l of TCE	100 mg/l of TCE
7A, 7B, 7C	8A, 8B, 8C	9A, 9B, 9C
Avg. = 2.85	Avg. = 7.21	Avg. = 56.387
SD = 0.30	SD = 2.98	SD = 76.03
Chloride concentration in microcosm with mixed methane-oxidizing culture and Hg $(NO_3)_2$		
0 mg/l of TCE	10 mg/l of TCE	100 mg/l of TCE
10A, 10B, 10C	11A, 11B	12A, 12B, 12C
Avg. = 3.25	Avg. = 4.11	Avg. = 7.40
SD = 0.15	SD = 0.52	SD = 11.71

REFERENCES

Alvarez-Cohen, L., and P. L. McCarty. 1991. "Effects of toxicity, aeration, and reductant supply on trichloroethylene transformation by a mixed methanotrophic culture." *Appl. Environ. Microbiol.* 57:228-235.

Barbee, G. C. 1994. "Fate of chlorinated aliphatic hydrocarbons in the vadose zone and ground water." *Ground Water Monit. Reme.* 11:120-128.

Chang, H. L., and L. Alvarez-Cohen. 1996. "Biodegradation of individual and multiple chlorinated aliphatic hydrocarbons by methane-oxidizing cultures." *Appl. Environ. Microbiol.* 62:3371-3377.

Chu, K. H. 2000. University of California at Berkeley, Department of Civil and Environmental Engineering.

Federal Register. 1989. "National primary and secondary drinking water regulations." *Fed. Reg.* 54:22062-22160.

Haber, C. L., L. N. Allen, S. Zhao, and R. S. Hanson. 1983. "Methylotrophic bacteria: biochemical diversity and genetics." *Science*." 221:1147-1153.

Little, C. D., A. V. Palumbo, S. E. Herbes, M. E. Lidstrom, R. L. Tyndall, and P. J. Gilmer. 1988. "Trichloroethylenen Biodegradation by a methane-oxidizing bacterium." *Appl. Environ. Microbiol.* 54:951-956.

Nielsen, P. H., P. L. Bjerg, P. Nielsen, S. Pernille, and T. H. Christensen. 1996. "In situ and laboratory determined first-order degradation rate constants of specific organic compounds in an aerobic aquifer." *Environ. Sci. Technol.* 30:31-37.

Oldenhuis, R., r. L. J. M. Vink, D. B. Janssen, and B. Witholt. 1989. "Degradation of chlorinated aliphatic hydrocarbons by Methylosinus trichosporium OB3b expressing soluble methane monooxygenase." *Appl. Environ. Microbiol.* 55:2819-2826.

Reddy, K. R., and J. A. Adams. 1998. "System effects of benzene removal from saturated soils and ground water using air sparging." *J. Environ. Eng.* 124:288-299.

Schollhorn, A., C. Savary, G. Stucki, and D.W. Hanselmann. 1997. "Comparison of different substrates for the fast reductive dechlorination of trichloroethene under groundwater conditions." *Water.Res.* 31:1275-1282.

Tsien, H. C., G. A. Brusseau, R. S. Hanson, and l. P. Wackett. 1989. "Biodegradation of Trichloroethylene by Methylosinus trichosporium OB3b." *Appl. Environ. Microbiol.* 55:3155-3161.

Wilson, J. T., and B. H. Wilson. 1985. "Biotransformation of trichloroethylene in soil." *Appl. Environ. Microbiol.* 49:242-243.

BIODEGRADATION OF ETHYLENE DIBROMIDE THROUGH COMETABOLIC MECHANISMS

Kristen E. Hartzell, Christy D. Burton, Andrea Leeson
(Battelle, Columbus, Ohio)

ABSTRACT: Three laboratory-scale batch reactor studies were conducted to determine whether soil microbes could degrade ethylene dibromide (EDB) through cometabolic mechanisms. The cometabolic growth substrates investigated were propane, methane, and natural gas. The bacterial consortium was isolated from soils in a mineral salts medium. Reactors were constructed using 160 mL sterile serum bottles with initial EDB concentrations of 200 μg/L. Each reactor was amended with propane, methane, or natural gas, then sealed and incubated. Controls were prepared without the addition of the consortium. In all active batch reactors, the ethylene dibromide concentration was reduced by >99% within 11 days of incubation. The laboratory results indicated that EDB could be biodegraded in the presence of propane, methane, and natural gas. Therefore, a technology such as cometabolic air sparging may show promising results for in situ bioremediation when applied in the field.

INTRODUCTION

Ethylene dibromide (1,2-dibromoethane [EDB]) is a persistent pollutant in groundwater due to improper disposal and spillage. EDB has been used as a pesticide and as an additive to leaded gasoline, but its use is declining due to increased regulation. EDB is extremely toxic and is a known carcinogen (Long et al., 1982). Since the discovery that EDB present in drinking water at barely detectable levels was related to increased risk of cancer, the need for effective means to treat contaminated waters has grown more urgent. The current drinking water standard for EDB is 0.02 μg/L and sites such as those at the Massachusetts Military Reserve (MMR) have EDB plumes containing over 100 μg/L. Current treatment of such plumes has been pump-and-treat, a relatively expensive option, and one that does not remediate the plume, but controls it from spreading off-site.

In a previous study, under anaerobic conditions, a methanogenic mixed culture removed >99% of initial EDB concentrations within two weeks of incubation (Bouwer et al., 1985). It has also been shown that mixed bacterial cultures, obtained from selective enrichment with EDB and vitamin-supplemented mineral salts medium in shake flasks, were capable of aerobically degrading EDB (Freitas et al., 1996). Freitas et al. (1996) also indicated that the culture completely converted EDB to bromide, H_2O, carbon dioxide, and biomass. Native soil and groundwater microbes collected from Windsor Locks, Connecticut readily degraded EDB with initial concentrations of 6 to 8 μg/L within 1 week under aerobic conditions (Pignatello, 1985).

Few studies have addressed the issue of cometabolic degradation of EDB. Cometabolic air sparging is a promising technology that has recently been proven successful at McClellan AFB, CA (Magar et al., 2001). Cometabolic air sparging has the advantage of being relatively simple to implement and inexpensive in comparison to pump-and-treat. The objective of this research was to determine if microorganisms were capable of degrading EDB under aerobic cometabolic conditions. A variety of primary substrates were tested, including methane, propane, and natural gas. For field applications, natural gas or propane were the preferred substrates since they are less expensive; however, previous studies had only shown cometabolic degradation with methane as the primary substrate (Leeson and Bouwer, 1986), so methane was also included in the testing.

MATERIALS AND METHODS

A bacterial consortium was isolated from soils collected from the Department of Defense Housing Facility in Novato, California. Mixed cultures were grown continuously in 160 mL serum bottles using 1 gram of soil and 25 mL of a mineral salts media (Foster and Davis, 1966). Each serum bottle was sealed with a Mininert™ valve (Dynatech Precision). Cultures were amended with 20 mL of propane, methane, or natural gas by injection through the Mininert™ valve using a Hamilton Gas-tight syringe fitted with a 22½-gauge needle. Making successive transfers with the mineral salts medium selectively enriched mixed cultures able to degrade propane, methane, or natural gas.

Sacrificial batch reactors were prepared in triplicate for each sampling event and consisted of 160-mL sterile serum bottles. A 10-mL aliquot of the consortium was added to 60-mL of mineral salts medium in each batch reactor. An EDB stock solution was then injected into each serum bottle to achieve initial EDB concentrations of 200 μg/L. The reactor headspace was sparged for two minutes with ultra high purity (UHP) oxygen and each reactor was sealed with a Mininert™ valve. A volume of 20-mL of the appropriate gaseous substrate was injected into each reactor using a Hamilton Gas-Tight syringe. Controls were prepared identically to the active reactors except without the addition of the consortium. The batch reactors were incubated in the dark at 28°C and 150 RPM.

Concentrations of oxygen, carbon dioxide, primary substrate (propane, methane, or natural gas), cell density, and EDB were measured in all batch reactors initially and in each sacrificed batch reactor during a sampling event. The active and control batch reactors were sampled in triplicate every 24 to 72 hours depending on percent EDB removal. Concentrations of EDB were measured using an HP 5890 Series II gas chromatograph (GC) with an electron capture detector (ECD) and split/split less injection. The detection limit of EDB was 0.029 μg/L. The samples were extracted and tested by following EPA SW846 Method 8011. Optical density was measured at each sampling point using a Milton-Roy Spectronic 21D. The optical density was correlated to cell density. Propane, methane, and natural gas concentrations also were measured at each sampling event using a Varian Star 3400 equipped with a flame ionization detector (FID) by drawing off 5 mL of the headspace of each serum bottle with a gastight syringe and directly injecting into the GC. Oxygen and carbon dioxide were tested by using an SRI 8610A gas chromatograph equipped with a thermal conductivity detector (TCD).

RESULTS AND DISCUSSION

Propane Batch Reactors. Microorganisms in active batch reactors amended with propane degraded >99% of initial EDB (to below the detection limit) after four days of incubation. Control batch reactors amended with propane showed no significant loss of EDB (Figure 1A). Simultaneous degradation of propane, oxygen utilization, and cell growth strongly suggest that the microorganisms are responsible for EDB degradation. Microorganisms in the active batch reactors

A)

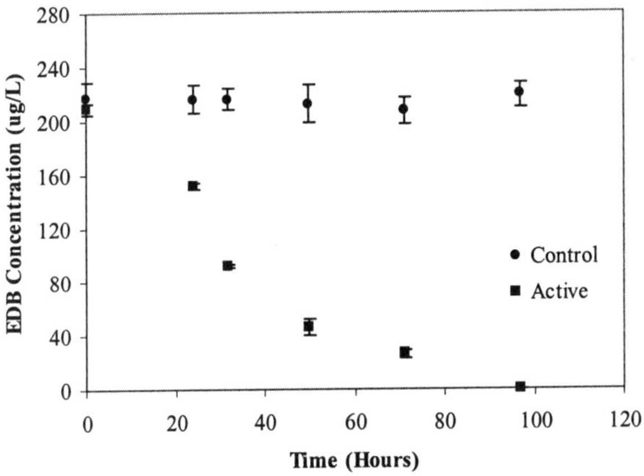

B)

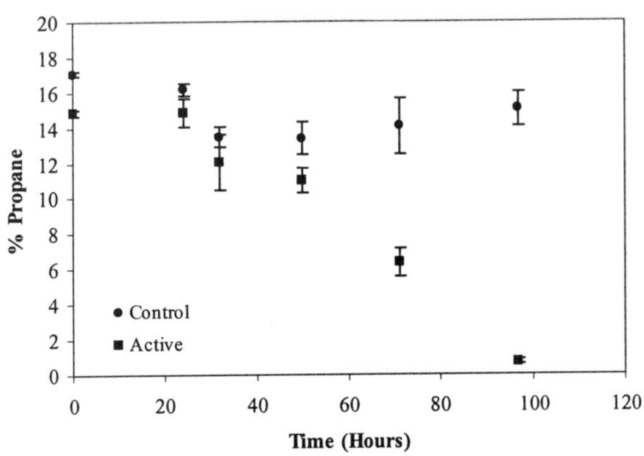

FIGURE 1. EDB (A) and propane (B) degradation in batch reactors. The average of triplicate samples with ± standard deviation is shown.

degraded >95% of initial propane within four days of incubation. In contrast, the control reactors showed a loss of <10% initial propane concentrations (Figure 1B).

Methane Batch Reactors. Microorganisms in active batch reactors amended with methane degraded >99% of initial EDB (to below reporting limit) after six days of incubation (Figure 2A). However, the control reactors showed no significant loss of EDB after 13 days of incubation. The active reactors degraded >95% of initial methane and the controls showed a 31% loss of initial methane (Figure 2B). As with the results in the propane-amended batch reactors, these results indicate that EDB is degraded by the microorganisms in the presence of oxygen and methane.

A)

B)

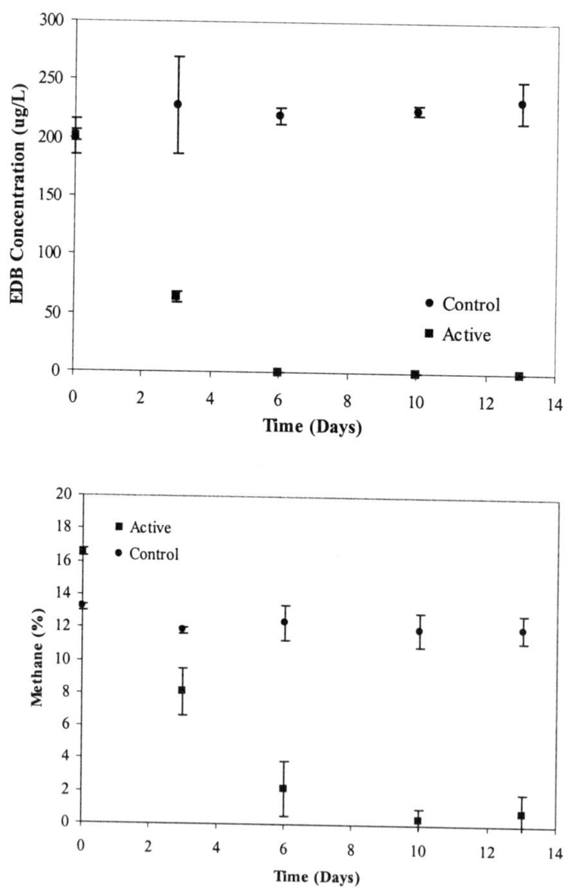

FIGURE 2. EDB (A) and methane (B) degradation in batch reactors. The average of triplicate samples with ± standard deviation is shown.

Natural Gas Batch Reactors. Microorganisms in active batch reactors amended with natural gas degraded >99% of initial EDB (to below reporting limit) after six days of incubation. Control batch reactors amended with natural gas showed no significant loss of EDB (Figure 3A). The active batch reactors degraded 100% of the initial natural gas after 10 days of incubation while the controls showed no significant loss (Figure 3B). These results confirm those from the propane- and methane-amended batch reactors, demonstrating that EDB is degraded by the microorganisms in the presence of a primary growth substrate and oxygen.

A)

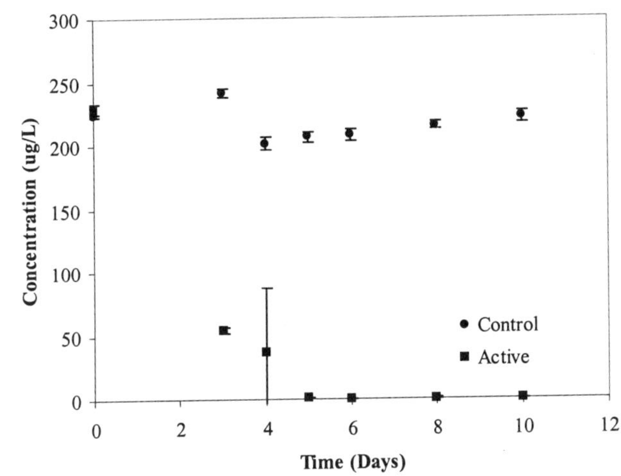

B)

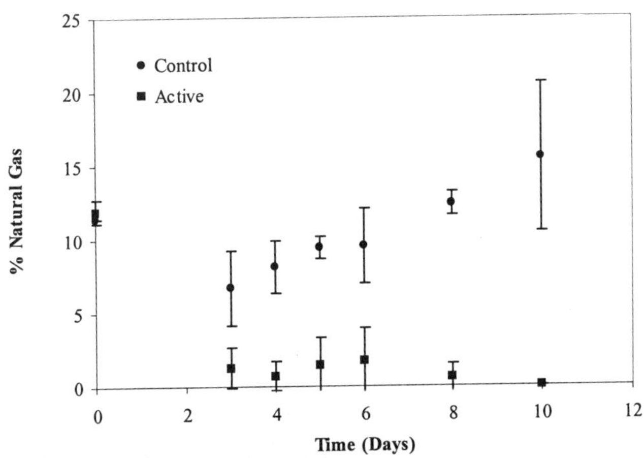

Figure 3. EDB (A) and natural gas (B) degradation in batch reactors. The average of triplicate samples with ± standard deviation is shown.

These results with natural gas are in contrast to previous studies that demonstrated that EDB could not be degraded in the presence of natural gas, possibly due to additives that inhibited growth (Leeson and Bouwer, 1986). The ability to use natural gas rather than methane or propane is desirable, since it is a cheaper and more readily available product.

CONCLUSIONS

The batch reactor experiments resulted in three mixed cultures, isolated using propane, methane, or natural gas as growth substrates, which were capable of degrading EDB. The results provide evidence of biodegradation of EDB through cometabolic mechanisms. Further studies are necessary to confirm this result.

Currently, regulations require that EDB concentrations in groundwater be reduced to 0.02 µg/L. At sites with EDB concentrations above MCLs, pump-and-treat is typically applied, followed by granular activated carbon (GAC) treatment of the extracted groundwater. This is a relatively expensive option given the length of time the systems must be operated (+30 years) and is unsatisfactory in that it does not result in destruction of EDB. The results from this research indicate that EDB can be degraded relatively quickly by microorganisms, suggesting that in situ bioremediation of EDB-contaminated groundwater is possible. In situ bioremediation of EDB could be accomplished through implementation of cometabolic air sparging, a technology that has been proven successful in the field. Cometabolic air sparging to treat EDB-contaminated groundwater has the potential to achieve contaminant concentrations below MCLs within a relatively short time frame.

REFERENCES

Bouwer, E. J., and P. L. McCarty. 1985. "Ethylene Dibromide Transformation under Methanogenic Conditions." *Appl. Environ. Microbiol.* 50(2): 527-528.

Foster, J. W., and R. H. Davis. 1966. "A Methane-Dependent Coccus, With Notes on Classification and Nomenclature of Obligate, Methane-Utilizing Bacteria." *J. Bacteriol.* 19(3): 1924-1931.

Freitas dos Santos, L. M., D. J. Leak, and A. G. Livingston. 1996. "Enrichment of Mixed Cultures Capable of Aerobic Degradation of 1,2-Dibromoethane." *Appl. Environ. Microbiol.* 62(12): 4675-4677.

Long, L. C. K., J. M. Winston, C. B. Hong, and H. Plotnick. 1982. "Carcinogenicity and toxicity of 1,2-dibromoethane in the rat". *Toxicol. Appl. Pharmacol. 63*: 155-165.

Magar, V.S., A. Leeson, L. Semprini, M. Dolan. 2001. *Use of Cometabolic Air Sparging to Remediate Chloroethene-Contaminated Groundwater Aquifers*. Final report submitted to the U.S. Air Force and ESTCP.

Pignatello, J. J. 1985. "Ethylene Dibromide Mineralization in Soils under Aerobic Conditions." *Appl. Environ. Microbiol. 51*(3): 588-592

AEROBIC COMETABOLIC DEGRADATION OF CHLOROETHENES: PILOT TEST AND FULL-SCALE DESIGN

Bernd W. Rehm, ***Thomas S. Silverman***, Jack W. Anderson, and Daniel Haak.
(RMT, Inc., Madison, Wisconsin, USA)

ABSTRACT: Engineered *in situ* aerobic cometabolic degradation was selected to remediate groundwater beneath an Indiana industrial facility contaminated with trichloroethene (TCE) and *cis*-1,2-dichloroethene (cDCE). Achieving groundwater concentrations equivalent to drinking water standards at the facility property line is required as part of the state-approved remedy for the site. Site characterization suggested a low rate of reductive dechlorination under the existing site conditions. Microcosm studies using aquifer solids and groundwater indicated that anaerobic reductive dechlorination could be stimulated by adding electron donors. However, several microcosms amended with methane and oxygen produced greater degradation rates than the anaerobic trials. An in-field pilot test was completed to evaluate the *in situ* degradation of chloroethenes and the mechanics of adding methane and oxygen to the aquifer. A two-well injection and extraction well couplet oriented transverse to the groundwater flow direction was selected for chemical injections. Numerical simulations of the pilot test area were used to establish well spacings, pumping rates, and monitoring well placement. Bromide injection was applied first to verify chemical spreading patterns within the aquifer. Methane and oxygen were introduced in two 5-day periods with subsequent groundwater monitoring. The estimated TCE degradation rate observed in the pilot test was 25 x 10^{-9} kg/L/d. Also degraded in the aquifer was cDCE; chloroethene (CE) was not detected. The pilot test results were applied to a facility-wide numerical model to design an overall remedy. A conceptual model of the remedy suggests that the engineered bioremediation solution may require as little as 15 years to achieve the remediation goals compared to over 100 years for hydraulic containment by pumping and treating at the facility property line.

INTRODUCTION

The source of the volatile organic compounds (VOCs) is a 14-acre industrial facility in Indiana. Trichloroethene (TCE) was used at the facility as a solvent for parts degreasing between 1943 and 1995. Five solvent source areas were identified through historical reviews and site investigations.

Groundwater in a 6-m– to 21-m–thick sand and gravel aquifer below the facility flows toward the south-southeast. The hydraulic conductivity of the aquifer is on the order of 2 x 10^{-2} cm/s, resulting in an average groundwater velocity on the order of 120 m/yr. Approximately 3 m of silt and clay overlie the aquifer, and clay-rich till underlies the aquifer. A southward-flowing river runs along the eastern edge of the facility property, while a clay-rich moraine borders the facility to the west.

The release of TCE at the facility has produced a 1,500-m–long plume of TCE and *cis*-1,2-dichloroethene (cDCE), with total VOC concentrations on the order of 400 µg/L. Total concentrations of TCE and its degradation products near one source area are as high as 3,700 µg/L. No dense nonaqueous-phase liquids (DNAPLs) have been observed, but the continued presence of the dissolved VOCs indicates the presence of sorbed VOCs and nonmobile residual NAPL in the aquifer pores.

The remediation goal is to preclude chloroethene migration beyond the facility property lines at concentrations above cleanup criteria and, ultimately, to return the aquifer downgradient of the facility to a potential potable water supply. Hydraulic containment options and *in situ* source reduction technologies were considered. On the basis of the evaluations described below, *in situ* enhanced bioremediation was selected as the remedy for the site.

SITE CHARACTERIZATION AND TREATABILITY STUDIES

Groundwater beneath the facility is generally more reducing, with lower dissolved oxygen (DO) and slightly higher dissolved methane, than background conditions. Petroleum hydrocarbons were identified in some groundwater samples at the site, but at relatively low concentrations. Concentrations of major nutrients required to sustain microbial growth in groundwater (organic carbon, organic nitrogen, and orthophosphate) are relatively low, suggesting that microbiological activity may be limited by nutrient availability. Assays identified moderate populations of heterotrophic organisms (about 10^4 to 10^5 colony forming units (CFU)/g or mL) and petroleum-degrading organisms (10^4 to 10^5 CFU/g or mL). The presence of methanotrophs was also indicated. TCE is the only halogenated organic compound known to have been released to soil and groundwater. The appearance of cDCE and CE suggests that reductive dechlorination of TCE occurred to a limited extent at some point below the site. However, the rate of degradation is not high enough to keep the VOCs from leaving the facility.

A two-phase laboratory treatability study was undertaken to assess the potential for enhancing bioremediation rates beneath the facility. The first phase used microcosms to assess what conditions would enhance TCE and cDCE degradation. In the second phase, column studies were run using the most promising conditions from the first phase.

A total of 16 sets of replicate, sacrificial microcosms were tested (Schanke et al., 1997). Seven anaerobic microcosms were initiated with three options for the electron donor (none, acetate, and yeast extract) and two options for the electron acceptor (none and sulfate). Nine aerobic microcosms were initiated with no additional electron donors or with three options for the electron donor (none, methane, and methane and acetate). The methane was introduced at varying initial concentrations (low, medium, and high). Oxygen was added as the electron acceptor in each aerobic microcosm. Each aerobic and anaerobic microcosm was amended with three different options for nutrients (none, ammonium phosphate, and yeast extract).

After 118 days, reductive dechlorination was observed in eight of the nine amended anaerobic treatments. In the anaerobic microcosms, the treatment with acetate as an electron donor and no electron acceptor addition resulted in the greatest degradation, with a 79 percent reduction in TCE concentrations. After an initial increase, concentrations of cDCE decreased by about 50 percent of its peak concentration by the end of the test. CE concentrations did not exhibit any increase, suggesting that CE was degraded. In the aerobic microcosms, the addition of methane stimulated methanotrophic bacteria. Methane at 5 mg/L produced the greatest overall degradation with 65 percent TCE concentration reductions, 78 percent cDCE reductions, and 100 percent CE reductions after 112 days. The methane degradation rate was rapid at the onset of the study (375 µg/L/day), but decreased to approximately 17 µg/L/day between 2 to 6 weeks. The addition of higher methane concentrations or the addition of acetate as a potential electron donor may have inhibited the degradation of TCE and cDCE.

The microcosms demonstrated both reductive dechlorination and aerobic cometabolic degradation with methane and oxygen. Creating aerobic conditions below the facility was expected to be more achievable than creating anaerobic conditions. The second phase of the study therefore proceeded with aerobic columns with methane added as the electron donor. A total of six aerobic columns were established: two replicate columns amended with DO and an abiotic control and two replicate columns amended with methane and DO, and an abiotic control. The initial DO and methane concentration of the column was 11 mg/L and 0.53 mg/L, respectively. In each of the columns, the average inlet concentrations of TCE and cDCE were between 520 and 580 µg/L and between 2,500 and 3,500 µg/L, respectively. The column studies were run for a total of 129 days at flow rates and temperatures comparable to those measured beneath the facility. The degradation of VOCs began quickly and achieved steady state within 30 days for both sets of columns. For the oxygen-amended columns, degradation rates averaged from18 µg/L/day to 97 µg/L/day for TCE and cDCE, respectively. For the methane and oxygen–amended columns, degradation rates averaged from 25 µg/L/day to 78 µg/L/day for TCE and cDCE, respectively. Although VOC degradation rates were not significantly different between the oxygen-amended and the methane and oxygen–amended columns, evaluating the results of the microcosm and column studies together indicates that the addition of both methane and oxygen would yield higher *in situ* degradation rates.

PILOT TEST DESIGN

The treatability studies confirmed that microbial communities beneath the plant have the potential to degrade VOCs. Therefore, an in-field pilot test was designed to assess the means by which the conditions beneath the plant can be changed to cause the degradation observed in the laboratory. The pilot test was designed to evaluate a two-well injection and extraction system (Figure 1) similar to that described by Kawakami et al. (1998) and McCarty et al. (1998). Groundwater will be extracted from one well, amended with methane and oxygen, and then re-injected into the aquifer in the second well. Re-injecting the groundwater spreads the amended groundwater radially, while continued pumping of the extraction well draws the amended groundwater toward the extraction well. Pumping ceases after sufficient breakthrough of amended groundwater occurs in the extraction well. The plume of introduced chemicals will then migrate with the ambient groundwater flow through the source area and the aquifer immediately downgradient of the source area. Extracting and injecting groundwater will occur periodically, rather than continuously, to maintain concentrations of amended groundwater most suitable for VOC degradation and to minimize the potential for excessive microbial growth near the injection well.

Methane and oxygen were dissolved into the extracted groundwater using two parallel gas diffusion modules. The gas diffusers consist of a semipermeable membrane (Gantzer and Cosgriff, 1995). A regulator maintains a constant gas pressure in the diffusers. The difference between the gas pressure and the water pressure outside of the diffuser controls the rate of diffusion of the gases into the water. The target concentrations of dissolved gases are established by setting a diffusion rate with respect to the rate of groundwater flowing past the diffuser.

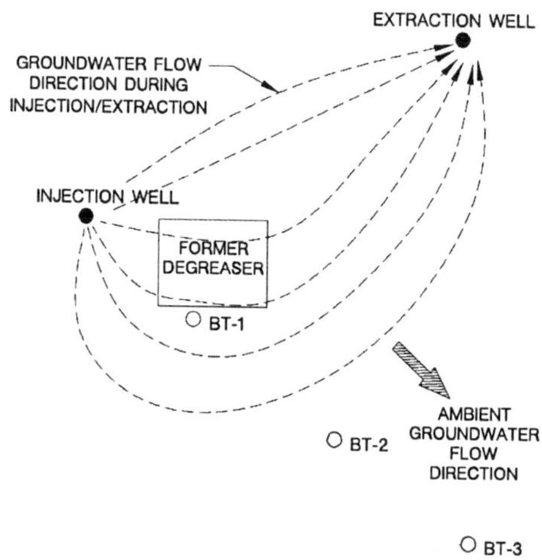

FIGURE 1. Injection/Extraction well couplet and monitoring points.

A numerical model (Visual MODFLOW, version 2.8.2, Waterloo Hydrogeologic, 2000) was used to help design the injection and extraction system. A particle tracking code (MODPATH) was used in conjunction with MODFLOW to evaluate travel time, well spacing, pumping rates, and monitoring well placement, such that the affected area of amended groundwater was optimized. Based on the modeling results, the injection and extraction wells were installed 10 m apart in a direction oriented transverse to the groundwater flow direction. Multi-depth samplers were installed at three points, BT-1, BT-2, and BT-3, at approximately 5, 10, and 16 meters, respectively, downgradient of the injection well (Figure 1).

Before proceeding with the methane injection, a tracer test using sodium bromide was completed to evaluate subsurface chemical migration. Sodium bromide was added to the extracted groundwater and subsequently re-injected until breakthrough ($C/C_0 = 0.5$) was achieved in the extraction well. Pumping then ceased, and the amended water was allowed to migrate with the ambient groundwater flow. The dissolved bromide moved through the aquifer at a rate of about 180 to 260 m/yr, which is approximately 2 to 4 times faster than estimates based on aquifer permeability test results.

RESULTS

Methane Injections and Biodegradation Assessment. The concentrations of TCE and cDCE in the pilot test area prior to methane injection were 320 ± 140 μg/L (arithmetic mean ± 1 standard deviation from the monitoring points shown on Figure 1) and 90 ± 27 μg/L, respectively. Figure 2 shows the dimensionless concentration (C/Co) of TCE and cDCE with time at the closest downgradient sampling point, BT-1. Table 1 summarizes the dimensionless concentrations of TCE and total DCE at BT-1 over time.

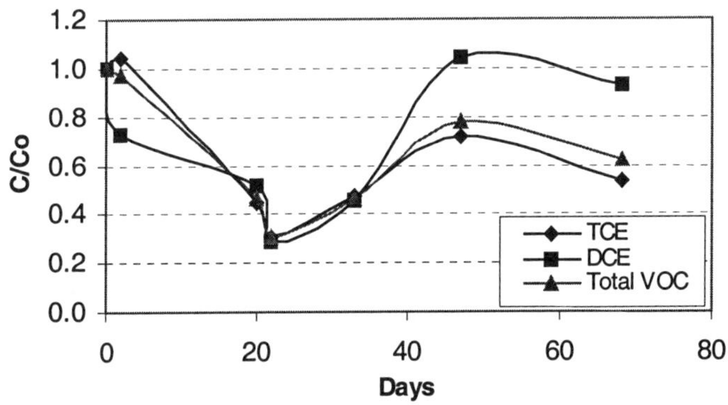

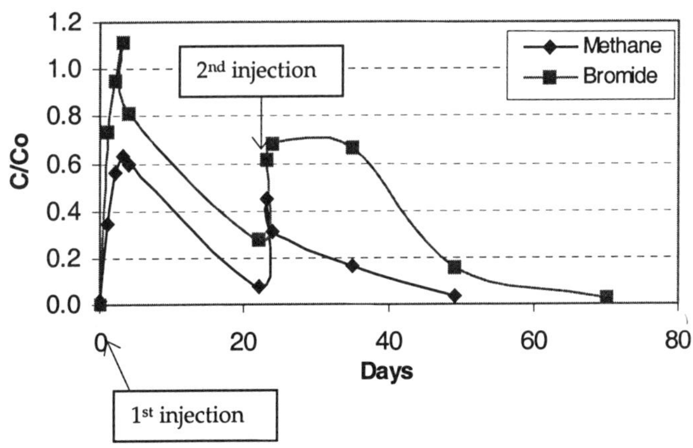

FIGURES 2 (upper) and 3 (lower). Dimensionless concentrations versus time. Symbols indicate mean values measured at BT-1.

TABLE 1. Summary of dimensionless concentrations of TCE and total DCE measured at the multi-depth samplers at BT-1.

		Days After Initial Injection					
		2	20	22	33	47	68
TCE	high	1.13	0.48	0.27	0.52	0.94	0.65
	average	1.04	0.44	0.30	0.47	0.72	0.54
	low	1.0	0.41	0.23	0.41	0.54	0.48
Total DCE	high	0.77	0.65	0.24	0.41	1.22	1.00
	average	0.73	0.52	0.29	0.45	1.04	0.93
	low	0.67	0.44	0.37	0.58	0.86	0.86
Total VOCs	average	0.97	0.46	0.30	0.47	0.78	0.62

The first chemical injection lasted for 48 hours, with initial injection concentrations of 3.1 mg/L of methane, 7.3 mg/L of DO, and 15 mg/L of bromide. Within the first 4 days of the start of the injection, the center of mass of the injected volume of water passed the BT-1 sampling point (based on the bromide peak on Figure 3). The vertical distribution of bromide in BT-1 was relatively uniform, indicating little heterogeneity that could cause significant nonuniform distribution of treatment chemicals to the aquifer. Methane concentrations at BT-1 decreased faster than the bromide concentrations (Figure 3), indicating that the methane was consumed in the aquifer rather than dispersed by groundwater flow. DO concentrations decreased to between 0.6 mg/L and 1.0 mg/L at BT-1, and are therefore not displayed graphically.

TCE and cDCE concentrations did not change significantly within the test area by day 4 (350 ± 140 μg/L and 80 ± 38 μg/L, respectively). Data collected from the sampling point closest to the injection well (BT-1) indicate that TCE concentrations did not decrease from the start of the test, but the cDCE concentrations were about 70 percent of the starting concentrations. The lack of significant concentration decreases at this time is not unexpected because the development of a methanotrophic microbial community would take some time after the first exposure to methane.

Twenty-two days after the start of the injection, TCE and cDCE concentrations at sampling point BT-1 decreased to about 40 to 50 percent of the pretreatment concentrations (Figure 2), with TCE at 150 ± 63 μg/L and cDCE at 42 ± 15 μg/L. At 22 days, the entire area should have been exposed to the injected methane based on a 14-day travel time between the injection point and sampling point BT-3 determined during the sodium bromide tracer test. Bromide concentrations at BT-1 had decreased to about 30 percent of the starting concentration, but methane concentrations were only about 5 percent of the initial concentration (Figure 3). The more rapid rate of methane decrease relative to bromide again supports methane consumption by microorganisms rather than simple dispersion and dilution of the methane over time and distance. An in-field total VOC degradation rate of 25×10^{-9} kg/L/d was calculated for the first 22 days of treatment.

A second 48-hour injection of 2.1 mg/L of methane, 8.0 mg/L of DO, and 32 mg/L of bromide was initiated on day 22. The migration of bromide and methane into the test area was evident by day 24 (Figure 3). Degradation of chloroethenes continued, with TCE concentrations at day 24 of about 110 ± 44 μg/L and cDCE concentrations of 32 ± 15 μg/L. At BT-1, the TCE and cDCE concentrations were 25 to 35 percent of the starting concentrations (Figure 2).

The next sampling event took place 35 days after the first injection and 14 days after the second injection. Bromide concentrations from the second injection did not decrease as quickly as they did following the first injection. However, 13 days after the second methane injection, methane concentrations were close to pretreatment levels of ≤ 0.1 mg/L (Figure 3). Test area concentrations did not show any further decrease in chloroethenes, with 120 ± 57 μg/L of TCE and 36 ± 11 μg/L of cDCE. At BT-1, the concentrations of TCE and cDCE were about 45 to 55 percent of the pretreatment concentrations (Figure 2). The consistencies of the day 35 concentrations with the previous two sampling rounds suggested that the rate of degradation had reached equilibrium with the rate of dissolution and/or desorption of the residual chloroethenes. If this were the case, one would expect the concentrations of chloroethenes to increase, as the methanotrophs become dormant with decreasing methane concentrations. There were no further methane injections to test this possibility.

Samples were again collected 49 days after the first injection and 27 days after the second injection. The concentrations of TCE within the test area had increased to 180 ± 63 μg/L, and the cDCE concentrations had increased to 85 ± 21 μg/L of cDCE. The plot of relative concentrations at BT-1 also reflects an increase in TCE, to about 50 to 70 percent of the pretreatment concentrations. The cDCE data at BT-1 suggest a possible analytical problem because the reported concentrations are greater than, or equal to, the starting concentrations and the variability of the results increased substantially over other sampling rounds. One would not expect concentrations to increase over starting concentrations based on the dynamics of cometabolism or desorption. Samples collected on day 71 confirmed this when the TCE concentrations were about the same as on day 49 (50 to 65 percent at BT-1). The cDCE concentrations on day 71 were about 80 percent of the pretreatment concentrations. This slow return to the pretreatment concentrations indicated that the cometabolic degradation rates were decreasing as methane concentrations dropped and the methanotrophic organisms became inactive.

TCE and cDCE concentrations over time and the methane breakthrough curve in BT-1 were used to determine a minimum concentration of dissolved methane that effectively spurs microbial consumption of VOCs. Using the decreasing concentration of cDCE after the first methane injection as a proxy for total VOC consumption, the rate of cDCE decrease between 0 and 14 days was rapid. After 14 days, the rate slowed down. The corresponding concentration of methane in BT-1 after 14 days was approximately 20 percent of the input concentration for the first methane injection, or 1 mg/L. After the second methane injection, TCE and cDCE concentrations in BT-1 decreased rapidly for 2 days and then began to rise. The corresponding methane concentration at the point where VOC concentrations began to rise again was approximately 35 percent of the input concentration for the second methane injection, or 0.7 mg/L. From these examples, it appears that a methane concentration of on the order of 1.0 mg/L is the minimum needed to stimulate the methanotrophs and produce VOC degradation.

PROPOSED FULL-SCALE DESIGN

Numerical simulations were performed to help design the full-scale remediation. The resulting proposed remediation consists of a network of injection and extraction wells, similar to the single injection/extraction couplet used in the pilot test. The network will consist of nine couplets with the injection well located sidegradient to, and between 6 m and 12 m from, the extraction well. The location of these nine couplets corresponds to the VOC source areas beneath the facility.

Using nine injection/extraction couplets targeted to source areas is more efficient than using fewer couplets spread across the facility and operating at higher pumping rates. Locating the injection/extraction wells close to the sources of chloroethenes will directly introduce methane and oxygen to the most highly contaminated areas under the facility. More couplets will also make it easier to optimize the performance of individual couplets as well as of the system as a whole.

Design Evaluation. An approximation of the length of time to achieve the cleanup goals was developed in support of the remedial design. The approximation assumes that dissolved, sorbed, and NAPL VOCs beneath the facility are lost solely through degradation. The dissolved, sorbed, and NAPL masses were estimated to be on the order of 10 to 40 kg, 20 to 760 kg, and 380 to 38,00 kg, respectively. These combine to give a range of VOC masses from 410 to 39,800 kg, with most of the uncertainty associated with the NAPL mass. NAPLs have not been observed beneath the site, but the continued presence of VOCs in the aquifer 20 to 50 years after possible solvent releases suggests that some undefined mass of NAPLs is present, but likely not at the 10,000s of kg range. For the estimate, the aquifer below the plant was treated as if it were a nonflowing system and rate-limiting desorption and NAPL dissolution were ignored. Constant degradation rates over time equal to, and at, 20 percent of the observed laboratory and field rates were used to develop the first approximations of remedial duration presented in Table 2.

TABLE 2. Estimated duration of remediation.

Degradation Rate (kg/l/d)	Total VOC Mass (kg)	Remedial Duration (yr)
5×10^{-9}	500	5
	1,500	15
	3,000	30
25×10^{-9}	500	1
	1,500	3
	3,000	6

This range of durations suggests that, if the total mass of VOCs in the aquifer is less than 3,000 kg, the remedy will achieve the groundwater cleanup goals in less than 30 years. A 15-year duration was assumed for remedial design and costing. The present worth cost of construction and 15 years of operation was estimated to be about $600,000. A containment option using pumping and treatment by air stripping with 30 years or more of operation was estimated to have a present worth cost of about $1,400,000. In addition, calculations suggest that there would be little likelihood of cleanup goals being reached in 30 years.

References:

Gantzer, C. J. and D. Cosgriff. 1995. "Membrane Oxygen Dissolution at the Libby, Montana, Superfund Site." In Hinchee, R. E. and R.M. Miller (eds.), *3rd International In Situ and Bioreclamation Symposium.* Columbus, Ohio: Battelle Press. pp. 543-549.

Kawakami, B.T., J. Christ, M. N. Goltz, and P. L. McCarty. 1998. "Design of an *In Situ* Injection/Extraction Bioremediation System." In Wickramanayake, G. B. and R. E. Hinchee (eds.) *Remediation of Chlorinated and Recalcitrant Compounds.* Columbus, Ohio: Battelle Press. Vol. 6. pp. 33-38.

McCarty, P. L., M. N. Goltz, G.D. Hopkins, M.E. Dolan, J.P. Allan, B.T. Kawakami, and T.J. Carrothers. 1998. "Full-Scale Evaluation of *In Situ* Cometabolic Degradation of Trichloroethylene in Groundwater Through Toluene Injection." *Environmental Science and Technology.* 32(1): 88-100.

Schanke, C. A., A. D. Bettermann, L. L. Graham, and B. W. Rehm. 1997. "Biological Treatability Studies for Remediation of TCE-Contaminated Groundwater." In B. C. Alleman and A. Leeson (eds.), *4th International In Situ and On-Site Bioremediation Symposium.* Columbus, Ohio: Battelle Press. Vol. 3. pp. 267-272.

SUBSTRATE FEEDING STRATEGIES TO MINIMIZE SUBSTRATE COSTS IN A COMETABOLIC TCE TREATMENT SYSTEM

Seung-Bong Lee, Stuart E. Strand, and H. David Stensel
(University of Washington, Seattle, WA. USA. 98195)

ABSTRACT: Trichloroethene (TCE) degradation was studied in a suspended growth reactor fed air containing TCE and an actinomycetes mixed culture degraded the TCE. The actinomycetes culture developed in our laboratory has demonstrated a very high transformation capacity (1.0 mg TCE/mg volatile suspended solids [VSS]). This relatively high resistance to TCE metabolite toxicity resulted in sustained TCE degradation for 95 days at a reactor TCE loading as high as 72 mg/L-d with a relatively high transformation yield of 0.30 mg TCE/mg phenol. To further reduce the cost of phenol, two different reactor operations were investigated. One was to operate the gas treatment reactor with a longer solids retention time (SRT), with less wasting of the bacteria grown on phenol. The longer SRT operation increased the reactor biomass concentration, with less substrate usage to maintain the same treatment level. The other approach was to use alternate growth substrate. The choice of the growth substrate was dependant on whether TCE itself could induce the actinomycetes oxygenase enzyme system. The use of other easily degradable substrates, instead of aromatic compounds such as phenol or toluene, potentially could reduce the substrate cost, which is a major portion of the operating cost in the cometabolic TCE degradation system. The induction potential of TCE was investigated. A minimum amount of glucose was added into the treatment reactor as a reducing energy supply and for enzyme induction and synthesis.

INTRODUCTION

A suspended growth reactor system for trichloroethene (TCE)-contaminated gas treatment has been successfully demonstrated in the laboratory with a phenol-fed culture (Lee et al., 2000a). In this work a phenol-utilizing actinomycetes enrichment was developed for TCE degradation. High TCE removal efficiencies were maintained with a transformation yield (Ty) of 0.30 mg TCE degraded/mg phenol used. A high transformation capacity (Tc) at 1.0 mg TCE/mg volatile suspended solids (VSS) was observed with minimal phenol addition in a batch test. This high Tc significantly reduces the amount of cometabolic substrate needed, and thus the cost of the bioreactor operation. However, the cost of phenol comprises the major operating cost of the reactor.

When *Burkholderia cepacia* G4 was cultivated in a fed-batch system without cell wasting the highest reported Ty value was obtained at 0.4 mg TCE/mg toluene (Mars et al., 1996). However, this system developed a mutant and couldn't maintain the strain G4 when G4 was mixed with other toluene-degraders. The actinomycetes enrichment developed in the TCE gas treatment reactor system showed more resistance toward metabolite toxicity than the strain G4 and had a higher Tc value at 1.0 mg TCE/mg VSS.

To further reduce the cost of phenol, two approaches were investigated. One was to use a longer reactor solids retention time (SRT), which results in less biomass wasting and thus less substrate usage for the same treatment level.

The second approach to reduce the substrate usage is to use an alternate growth substrate. If TCE can induce the oxygenase enzymes needed for cometabolic degradation, then the growth substrate does not have to induce TCE degradation, making it possible to consider non-aromatic alternative substrates. Advantages would be avoiding the use of expensive and toxic aromatic compound such as phenol, and eliminating the competitive inhibition between TCE and the growth substrate.

There were a few published reports of TCE degradation without the need for induction by the growth substrate. Heald and Jenkins (1994) observed that TCE induced toluene dioxygenase enzyme activity. When whole cells of *Pseudomonas putida* containing toluene dioxygenase were incubated in the presence of TCE, toluene-stimulated oxygen uptake was observed at approximately 40 % of the rate induced by toluene itself.

McClay et al. (1995) demonstrated that toluene oxidation activity in *Pseudomonas mendocina* KR1 and *Pseudomonas* sp. strain ENVPC5 was induced by TCE, and that induction was followed by the degradation of TCE. Toluene oxidation was not induced by TCE in *Burkholderia cepacia* G4, and *P. putida* F1. In the presence of an energy-generating cosubstrate such as glutamate, the level of induction of toluene oxidation activity by TCE reached 86 % of that observed in toluene-induced cells.

Leahy et al. (1996) compared several toluene-oxidizing bacteria (*P. putida* F1, *B. cepacia* G4, *P. mendocina* KR1, and *B. pickettii* PK01) to determine their abilities to degrade TCE when they were grown on lactate or on lactate plus TCE, but in the absence of toluene. TCE degradation was measured after 18 h. High levels of TCE degradation by cells grown on lactate plus TCE were observed with G4, PK01, and KR1, but not with F1. When the organisms were grown on lactate, KR1 was the only one of the four strains which exhibited significant levels of TCE degradation.

The toluene-degradative activities of the four strains were also compared by using cells grown on lactate and lactate plus TCE. The activity of cells grown in the presence of TCE were 4.6 to 11.8 times higher than the activity of cells grown in the presence of lactate alone and was 36 to 105 % of the activity observed in toluene-induced cells, suggesting that TCE or an oxidation product induced the degradation of toluene in these strains. The role of TCE as an inducer of toluene oxygenase activity in PK01 was confirmed by performing a promoter probe analysis, in which they found that TCE activates transcription from the PK01 3-monooxygenase operon promoter.

Shingleton et al. (1998) further demonstrated the induction of *tod* operon by TCE in a *tod-lux* a bioluminescent reporter strain using molecular and biochemical techniques. *P. putida* TVA8, *P. putida* F1 derivative containing a modified mini-Tn5 chromosomal insertion of a *tod-lux* fusion, showed an increasing bioluminescence response to increasing TCE and toluene concentrations. They also measured mRNA transcript levels. Compared to uninduced TVA8 cultures, *todC1* mRNA levels increased 11-fold for TCE-treated cultures and 13-fold for toluene-treated cultures.

In this work, the pure culture actinomycetes isolate (Lee et al., 2000b) was investigated for the reduction of phenol usage.

MATERIALS AND METHODS

The reactor used for TCE degradation was a glass cylinder with a 2-L liquid volume and a 0.4-m depth and with a stone diffuser and a stir bar located at the bottom to disperse the inlet gas and mix the liquid (Lee et al., 2000a). The reactor was operated under two different SRT. The hydraulic retention time was controlled by reactor wasting and nutrient liquid feeding and equaled the SRT. TCE was fed continuously to the reactor. Phenol was fed every two hour by a peristaltic pump for the 6-day SRT operation, whereas the reactor with the 60-day SRT continuously received phenol by syringe pump, keeping the same phenol loading rate at 167 mg/L-d. The other operating conditions were as shown in Table 1.

An induction test was done by adding TCE with and without 20 mg/L of chloramphenicol using tryptic soy broth (TSB)-grown actinomycetes pure culture cells. The 125 mL cell suspension was transferred into 250-mL batch serum bottles and incubated at 25°C. TCE depletion was monitored in batch tests by head space sampling and analysis by gas chromatography (GC) with flame ionization detector (SRI 8610B with 30m x 0.53 mm RTX-1 column).

TABLE 1. Reactor operating conditions.

	Condition I	Condition II
SRT, days	6	60
Influent TCE concentration, mg/L	0.688	0.900
Phenol feeding mode	every two hours	Continuous
Air flow, L/d	310	250
Phenol loading, mg/L-d	167	167
TCE loading, mg/L-d	107	111
Applied phenol/TCE ratio	1.6	1.5

RESULTS AND DISCUSSION

At the same phenol loading rate (167 mg/L-d), the reactor with the 60-day SRT showed better treatment performance. At the higher SRT, the reactor biomass concentration was 1800 mg/L compared to 457 mg/L for the 6-day SRT. The removal efficiency increased from 54 to 65 % at the higher SRT, due to the presence of more active biomass for TCE degradation. The transformation quantity (Tq), which is expressed as the mass of TCE degraded per mass of cell wasted, was increased from 0.76 to 2.43 mg TCE/mg VSS. The large increase in Tq resulted from the small increase in percent removal and large decrease in cell wasting. The Ty increased slightly from 0.35 to 0.43 mg TCE/mg phenol. Table 2 summarizes the reactor performance.

TABLE 2. Summary of reactor performance at two different SRT conditions.

	Condition I	Condition II
SRT, days	6	60
Removal efficiency, %	54	65
VSS, mg/L	457	1800
Tq, mg TCE/mg VSS	0.76	2.43
Ty, mg TCE/mg phenol	0.35	0.43

To investigate the induction effect of TCE itself, two bottles were prepared. Each bottle contained biomass and TCE. Additionally, one bottle contained 20mg/L of chloramphenicol, but the other didn't. Chloramphenicol prevents new synthesis of cell components, thus enzyme induction. The chloramphenicol added bottle didn't show any activity toward TCE, but the other bottle showed TCE degradation activity (Figure 1). This result shows that TCE itself induced TCE degradation activities.

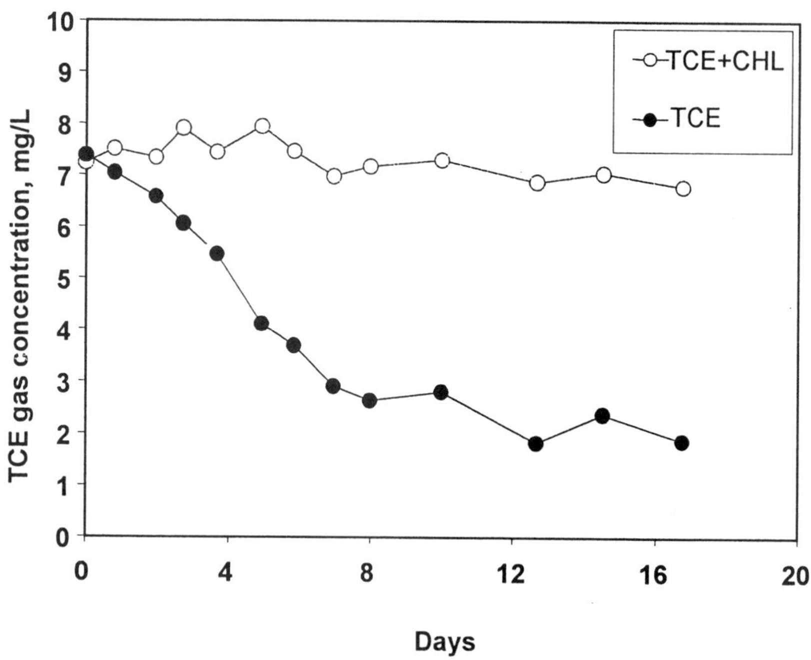

FIGURE 1. TCE degradation in batch bottle tests with actinomycetes culture with and without chloramphenicol (CHL).

Glucose addition in batch TCE degradation tests improved the TCE degradation rate and amount. The Tc for batch bottle test with prior glucose-grown cells without any external carbon addition was less than 0.1 mg TCE/mg VSS. When glucose was externally added in the batch test at 8.25 mg/L-d, the Tc value increased to 0.2 mg TCE/mg VSS. When the glucose concentration was increased further to 33 mg/L-d, the Tc for glucose-grown actinomycetes pure culture was about 1.4 mg TCE/mg VSS.

A modest amount of an external carbon source during TCE degradation improved performance. The external carbon source might have provided reducing power necessary for initial oxidation step or provided energy and carbon, which might be needed for repairing or resynthesizing damaged enzyme by toxic metabolites produced from TCE degradation.

SUMMARY

1. Reactor performance (Tq and Ty) was improved at higher operating SRT.
2. TCE induced its own degradation pathway in actinomycetes.
3. Addition of an external carbon source increased Ty value.

ACKNOWLEDGEMENTS

This research was funded by a grant from the National Institute of Environmental Health and Sciences (NIEHS), Superfund Basic Research Program (5P42ESO4696-07).

REFERENCES

Heald, S., and R. O. Jenkins. 1994. “Trichloroethylene removal and oxidation toxicity mediated by toluene dioxygenase of *Pseudomonas putida.*” *Appl. Environ. Microbiol. 60*(12):4634-4637.

Leahy, J. G., A. M. Byrne, and R. H. Olsen. 1996. “Comparison of factors influencing Trichloroethylene degradation by toluene-oxidizing bacteria.” *Appl. Environ. Microbiol. 62*(3):825-833.

Lee, S.B., S. E. Strand, H. D. Stensel. 2000a. “Sustained degradation of trichloroethylene in a suspended growth gas treatment reactor by an actinomycetes enrichment.” *Environ. Scien. Technol. 34*(15):3261-3268.

Lee, S.B., S. E. Strand, H. D. Stensel. 2000b. "Cometabolic degradation of chlorinated solvent contaminants by actinomycetes." In G. B. Wickramanayake, A. R. Gavaskar, B. C. Alleman, and V. S. Magar (Eds.), *Second International Conference on Remediation of Chlorinated and Recalcitrant Compounds: Bioremediation and Phytoremediation of Chlorinated and Recalcitrant Compounds*, pp. 455-460. May 22-25, Monterey, CA.

Mars, A. E., J. Houwing, J. Dolfing, and D. B. Janssen. 1996. “Degradation of toluene and trichloroethylene by *Burkholderia cepacia* G4 in growth-limited fed-batch culture.” *Appl. Environ. Microbiol. 62*(3):886-891.

McClay, K., S. H. Streger, and R. J. Steffan. 1995. "Induction of toluene oxidation activity in *Pseudomonas mendocina* KR1 and *Pseudomonas* sp. strain ENVPC5 by chlorinated solvents and alkanes." *Appl. Environ. Microbiol. 61* (9):3479-3481.

Shingleton, J. T., B. M. Applegate, A. C. Nagel, P. R. Bienkowski, and G. S. Sayler. 1998. "Induction of the tod operon by Trichloroethylene in *Pseudomonas putida* TVA8." *Appl. Environ. Microbiol. 64* (12):5049-5052.

THE MEASURE OF EXTRA OXYGEN DEMAND FOR THE COMETABOLIC TCE REMOVAL BY A RESPIROMETER

Chih-Jen Lu, Ching-Chia Chang, Chin-I Huang
National Chung Hsing University, Taichung, TAIWAN, ROC

ABSTRACT: This study focused on the measure of the extra oxygen demand for the cometabolic trichloroethylene (TCE) removal by phenol-degrading cells. A respirometer was used to measure the difference in oxygen consumptions between the phenol removal and both phenol and cometabolic TCE removals. The theoretical equation suggests that the oxygen consumption for phenol biodegradation is 1.65 mg-O_2/mg-phenol. The oxygen consumptions measured by the respirometer were 1.64 and 1.61 mg-O_2/mg-phenol for a duplicate test at an initial phenol concentration of 20 mg/L. The experimental values were very close to the calculated theoretical value (1.65 mg-O_2/mg-phenol). When phenol was used as the substrate to cometabolically remove TCE at initial concentrations of 500 or 1200 μg/L, TCE concentrations decreased to 10 and 403 μg/L after 27 hours of operation, respectively. With the cometabolic TCE removal, the extra oxygen consumptions were 0.2 and 0.4 mg for initial TCE concentrations at 500 and 1200 mg/L, respectively. According to the extra oxygen demand and the TCE removal, the oxygen demands for the cometabolic TCE removal were 0.41 and 0.51 mg-O_2/mg-TCE at initial TCE concentrations of 500 and 1200 μg/L, respectively. The experimental result (average 0.46) was close to the theoretical value (0.37 mg-O_2/mg-TCE) calculated from a postulated equation.

INTRODUCTION

Widespread contamination of groundwaters by volatile chlorinated solvents, such as trichloroethylene (TCE), has initiated abundant researches trying to remove these compounds from the contaminated water. Cometabolic TCE removal is one of the promising methods. Compounds that initiate the formation of oxygenase can be used as the primary substrate for cell growth to cometabolically remove TCE (Coyle, 1994; Wackett and Gibson, 1989). Therefore, methane and some aromatic compounds were used to stimulate the growth of methane- or aromatic-utilizing cell to cometabolically remove TCE. Phenol is more favorable among these compounds, because phenol-degrading cells are more effective to degrade TCE and phenol utilization has less oxygen consumption in comparison with methane (Segar and Speitel, 1995; Semprini *et al.*, 1994). Our previous study also indicated that oxygen was one of the major limiting factors for the cometabolic TCE removal (Lu *et al.*, 1998). Therefore, this study focused on the extra oxygen consumption for the TCE cometabolic removal compared to the biodegradation of phenol alone.

METHODS

The microorganisms used in this study were collected from a chemostat into which 40 mg/L of phenol was continuously fed as the substrate. The chemostat was controlled at a HRT of 6 days. After three months of operation, the microorganisms in the chemostat were ready for use in this study. A respirometer (Challenge, AER-200) was used to measure the oxygen demand for phenol biodegradation and for both phenol and TCE removals following the procedures suggested by the manufacturer. The respirometer contained eight cell reactors. One cell reactor contained only distilled water as the control. The other seven reactors contained both nutrients-containing solution and phenol-degrading microorganisms at an initial concentration of 10^7 cells/mL. Two of them were added with only TCE at 500 or 1200 μg/L without cells. Two were added with only phenol at 20 mg/L for a duplicate test. Two were added both phenol at 20 mg/L and TCE at 500 μg/L. The seventh cell reactor was added both phenol at 20 mg/L and TCE at 1200 μg/L.

The cell reactors were incubated in a water bath in dark at 25°C. Pure oxygen was continuously supplied from an oxygen tank as the dissolved oxygen was consumed from its saturated condition. Carbon dioxide produced from the biodegradation of organics was adsorbed by a KOH solution. An on-line counter connecting with a computer recorded the oxygen supply. The remaining concentrations of phenol and TCE were analyzed with a GC/FID and GC/ECD (HP-6890), respectively. The respirometer was calibrated following the procedures suggested by the manufacturer. The measuring precision was controlled by the repeatability within ± 3% Cv (in this study it was 0.8%). The measuring oxygen volume varied less than ±1% from average. The lowest measured volume of oxygen was 0.05 mg-O_2.

RESULTS AND DISCUSSION

The theoretical stoichiometry of phenol biodegradation can be expressed as equation (1):

$$C_6H_6O + 0.175\ O_2 + 0.015\ NH_3 \dashrightarrow 0.015\ C_5H_7O_2N + 0.077\ H_2O + 0.139\ CO_2 \quad (1)$$

This theoretical equation suggests that the oxygen consumption for phenol biodegradation is 1.65 mg-O_2/mg-phenol in comparison with the theoretical COD value of 2.38 mg-O_2/mg-phenol. Semprini *et al.* (1994) also suggested that the oxygen demand for phenol biodegradation was 1.6 mg-O_2 /mg-phenol. The oxygen consumptions measured by the respirometer were 1.64 and 1.61 mg-O_2/mg-phenol for a duplicate test at an initial phenol concentration of 20 mg/L with a mixture of phenol degraders. Phenol at this initial concentration was completely biodegraded after 20 hours of operation. After 20 hours of operation, the oxygen consumption for phenol biodegradation was insignificant. The experiment was ceased after 27 hours of operation. In this experiment, oxygen could be continuously supplied from the oxygen tank, the cease of oxygen consumption should not result from the oxygen availability. It should be due to the depletion of phenol. The results of phenol analysis indicated phenol concentration was less than 0.05 mg/L after 27 hours of operation. Therefore, the accumulated oxygen consumption at the end of the experiment was counted as the oxygen consumption for the biodegradation of 20 mg/L-phenol. After the deduction of the background oxygen consumption, the measured oxygen consumption was 1.59 mg-O_2/mg-phenol (average value for a duplicate test) (Table 1). The experimental result (1.59 mg-O_2/mg-phenol) was very close to the theoretical value (1.65 mg-O_2/mg-phenol) calculated from the stoichiometry equation and the value (1.6 mg-O_2/mg-phenol) suggested by Semprini *et al.* (1994). Therefore, the respirometer can be used to measure the oxygen consumption for phenol biodegradation.

The previous results indicated that phenol had the highest capability to initiate the cometabolic TCE removal compared to toluene or methane (Lu *et al.*, 1998; Semprini *et al.*, 1994). Therefore, phenol was used as the substrate to initiate the cometabolic TCE removal in this study. Our previous results also suggested that TCE removal was hindered when the remaining oxygen concentration was less than 1.5~2 mg/L (Lu *et al*, 1995). In this study, oxygen was continuously provided from a pure oxygen tank, the availability of oxygen was not the limiting factor for the phenol biodegradation. Table 1 shows the accumulated oxygen consumptions for phenol alone or both phenol and TCE measured by the respirometer. After 27 hours of operation, phenol was almost completely removed (< 0.05 mg/L). In the reactors containing both phenol and phenol-utilizing cells TCE concentrations decreased from 500 and 1200 μg/L to

TABLE 1. The oxygen demand for the cometabolic TCE removal by phenol-utilizing cells

Compounds	Initial Conc. mg/L	Final Conc. mg/L	O_2 Demand mg-O_2	Net O_2 Demd. mg-O_2 (Ave)
Phenol	20	<0.05	16.1	15.9
Phenol	20	<0.05	16.4	
TCE#	1200*	1190*	0.29	---
TCE#	500*	496*	0.30	---
Phenol	20	<0.05	16.5	16.1
& TCE	500*	9.6*	---	---
Phenol	20	<0.05	16.5	16.1
& TCE	500*	10*	---	---
Phenol	20	<0.05	16.6	16.2
& TCE	1200*	409*	---	---
Phenol	20	<0.05	16.7	16.2
& TCE**	1200*	396*	---	---
DI Water	---	---	0.39	0.39

* : TCE concentration: μg/L; #: without cells

**: results form the second run of the respirometer test

10 and 403 μg/L (average), respectively. However, TCE concentrations decreased from 1200 to 1190 μg/L in the reactor without the presence of phenol and phenol-utilizing cells. Therefore, volatilization loss of TCE was ignored in this study. With the cometabolic TCE removal, the extra oxygen consumptions were 0.2 and 0.4 mg-O_2 for the initial TCE concentrations of 500 and 1200 μg/L, respectively. According to the extra oxygen consumption and the net TCE

removal, the oxygen demand for the cometabolic TCE removal were 0.41 and 0.51 mg-O_2/mg-TCE for the initial TCE concentrations of 500 and 1200 μg/L, respectively.

The cometabolic TCE biodegradation may be assumed as equation (2):

$$C_2HCl_3 + H_2O + 1.5\ O_2 --> 2\ CO_2 + 3\ Cl^- + 3\ H^+ \qquad (2)$$

This postulated equation suggests that the oxygen demand for TCE removal is 0.37 mg-O_2/mg-TCE. It is assumed that TCE is mineralized to form CO_2 and chloride ion through the cometabolism and TCE is not utilized as the carbon source for cell growth. According to the postulated equation, the calculated oxygen demand for the cometabolic TCE biodegradation is 0.37 mg-O_2/mg-TCE. The measured value, 0.46 mg-O_2/mg-TCE (average), was close to that calculated from the postulated equation. This result suggests that the cometabolic TCE removal by phenol-utilizing cells required extra oxygen, but the total amount of the extra oxygen demand for TCE removal was relatively small, especially at low TCE concentrations.

SUMMARY

In this study, a respirometer was used to measure the extra oxygen demand for the cometabolic TCE removal by phenol-degrading cells. The measured oxygen consumption was 1.59 mg-O_2/mg-phenol. The experimental result was very close to the theoretical value (1.65 mg-O_2/mg-phenol) calculated from the stoichiometry equation. When phenol was used as the substrate to stimulate the cometabolic TCE removal, the experimental results showed that the cometabolic TCE removal consumed extra oxygen compared to the biodegradation of phenol alone. According to the postulated equation, the calculated oxygen demand for the cometabolic TCE biodegradation was 0.37 mg-O_2/mg-TCE. The experimental results showed that the oxygen demand for the cometabolic TCE removal was 0.46 mg-O_2/mg-TCE (average). The measured value was close to the theoretical one calculated from the postulated equation. This result suggested that the cometabolic TCE removal by phenol utilizers required extra oxygen, but the total amount of the extra oxygen demand for TCE removal is relatively small.

ACKNOWLEDGMENTS

This study was partially funded by the National Research Council, TAIWAN, ROC (NSC-88-2211-E005-011).

REFERENCES

Coyle, C. G. 1994. "Phenol-Induced TCE Degradation by Pure and Mixed Cultures in Batch Studies and Continuous-flow Reactors." In R. E. Hinchee, A. Leeson, L. Semprini, and S. K. Ong (Eds.), *Bioremediation of Chlorinated and Polycyclic Aromatic Hydrocarbon Compounds*, pp. 339-343. Lewis Publishers, Boca Raton, FL.

Lu, C. J., C. Y. Chang, and C. M. Lee. 1995. "Aerobic Biodegradation of Trichloroethylene by Microorganisms that Degrade Aromatic Compounds." In R. E. Hinchee, A. Leeson, and L. Semprini (Eds.), *Bioremediation of Chlorinated Solvents*, pp. 1-7. Battelle Press, Columbus. OH.

Lu, C. J., C. M. Lee, and M. S. Chung. 1998. "The Comparison of Trichloroethylene Removal Rates by Methane- and Aromatic-Utilizing Microorganisms." *Wat. Sci. Tech. 38*(7): 19-24.

Segar, R. L, Jr. and G. E. Speitel, Jr. 1995. "Cometabolism of Chloroethene Mixtures by Biofilms Grown on Phenol." In R. E. Hinchee, A. Leeson, and L. Semprini (Eds.), *Bioremediation of Chlorinated Solvents*, pp. 245-253. Battelle Press, Columbus. OH.

Semprini, L., G. D. Hopkins, and P. L. McCarty. 1994. "A Field and Modeling Comparison of In Situ Transformation of Trichloroethylene by Methane Utilizers and Phenol Utilizers." In R. E. Hinchee, A. Leeson, L. Semprini, and S. K. Ong (Eds.), *Bioremediation of Chlorinated and Polycyclic Aromatic Hydrocarbon Compounds*, pp. 248-254. Lewis Publishers, Boca Raton, FL.

Wackett, L. P. and D. T. Gibson. 1989. "Degradation of Trichloroethylene by Toluene Dioxygenase in Whole-cell Studies with Pseudomonas Putida F1." *Appl. Environ. Microbiol. 54*(7): 1703-1708.

FULL-SCALE REMEDIATION OF CHLORINATED HYDROCARBONS IN MULTIPLE AQUIFERS

***James J. Kang*, Ph.D., P.E.** (URS Corporation, Santa Ana, California)
Taras B. Kruk, RG, CHG (URS Corporation, Santa Ana, California)

ABSTRACT: A full-scale in situ soil/groundwater remediation system was designed and constructed to remediate the soil and groundwater impacted with chlorinated hydrocarbons, primarily trichloroethylene (TCE), beneath a 14-acre former aerospace parts manufacturing facility located in Anaheim, California. Results of subsurface investigation indicated that multiple sources of soil contamination were found beneath and around the manufacturing buildings, and that two aquifers beneath the site (A and B Zones) have been impacted by TCE with concentrations as high as 23,000 μg/L in the A Zone and 7,400 μg/L in the B Zone. The impacted groundwater plume extends off site and is approximately 2,500 feet (758 m) long by 700 feet (212 m) wide. The facility was developed for commercial/residential use. URS worked closely with the new property owner's architect and engineers to design and construct a full-scale remediation system, consisting of a methane/air sparging system, a soil vapor extraction (SVE) and treatment system, and a 300 gpm groundwater extraction and treatment system. To expedite the in situ remediation process, methane was periodically injected to stimulate the aerobic biodegradation of TCE via cometabolism. The extracted groundwater is first treated with a 300 gpm low-profile air stripper followed by a particulate filter and GAC treatment system prior to being reinjected upgradient into the B Zone aquifer to conserve the groundwater resources and speed up the remediation. The results show that the geometric mean concentrations of indicator wells at the site show a general decline in TCE concentration over time due to pumping, air/methane sparging, SVE, and in situ aerobic biodegradation.

INTRODUCTION

The project site, located in Anaheim, California, has a roughly triangular shape and occupies approximately 10 acres of land. An additional four acres along the southwest border of the site were recently assimilated into an expansion of the Santa Ana Freeway (I-5). The site was developed in the late 1950s by Chrysler Corporation as a regional display, presentation, and research facility. In the 1960s the site was purchased by the Brunswick Corporation for their Circle Seal Controls operation. The site was then modified for manufacturing stainless steel and brass precision valves for the aerospace industry. Manufacturing operations included cuttings, washing, and degreasing. The operation ended in 1994. In 1997, the property was sold to Sa-Rang Church Group. They converted the former manufacturing facility into a 3,000-people community church. The site is surrounded by residential housing to the north and west of the site and the Santa Ana Freeway (I-5) to the south of the site. Commercial development is located to the east of the site.

The lithology of the subsurface soil generally consists of fine sand from the surface to 30 feet below ground surface (bgs); interbedded layers of silt, clay, or silty clay from 30 to 55 feet bgs; and fine to medium sand from 55 to approximately 85 feet bgs (Figure 1). A localized perched aquifer (A Zone) is located approximately seven feet above a semi-confined regional aquifer (B Zone). The A Zone aquiclude beneath the perched aquifer is an average of 4 feet thick beneath the northern portion of the site and increases to approximately 9 feet thick beneath the southern portion of the site. The average depth of the A Zone aquiclude is 55 feet bgs.

The B Zone aquifer is a semi-confined aquifer in the vicinity of the site and an unconfined aquifer in areas where a low permeability confining layer is not present. The B Zone aquifer consists of poorly graded fine to medium sands with limited amounts of silt. The lower confining layer (B Zone aquitard) of clay and silty clay at approximately 100 feet bgs defines the lower limits of the aquifer. The B Zone aquitard is approximately 20 feet thick.

A deeper aquifer (C Zone) begins at 110 feet bgs and is estimated to be at least 40 feet thick. Regional groundwater data suggest that the C Zone might be part of the upper portion of a drinking aquifer. Depth to the piezometric groundwater surface is on the order of 50 feet bgs in the A Zone, 60 feet bgs in the B Zone, and 80 feet bgs in the C Zone.

FIGURE 1. Site Lithology.

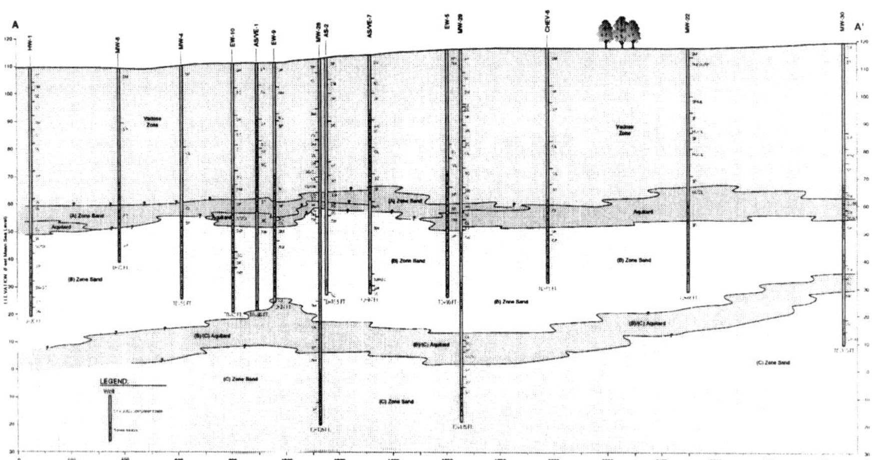

NATURE AND EXTENT OF THE CONTAMINATION

The results of the site investigations indicate the presence of chlorinated hydrocarbons (CHCs), primarily trichloroethylene (TCE), with some 1,1-dichloroethylene (1,1-DCE) and perchloroethylene (PCE) in groundwater beneath the site. The impacted groundwater plume extends off site to the north and southeast. The investigation results show detectable TCE concentrations ranging from 9.2 to 23,000 µg/L in the A Zone, and from 6.1 to 7,400 µg/L in the B Zone. In addition, some trace amounts of CHCs were detected in the soils. The impacted groundwater plume extends off site to the southeast and is approximately 2,500 feet (758 m) long by 700 feet (212 m) wide. There is no significant groundwater impact in the C Zone. Subsequently, groundwater remediation was focused on the A and B Zones with continued monitoring of the C Zone.

CLOSURE CRITERIA

The California Maximum Contaminant Level (MCL) for TCE in drinking water is 5 µg/L. As the lead agency, the California Regional Water Quality Control Board (CRWQCB) recognized that the current best available groundwater remedial technologies might not be able to reduce the contaminant concentrations to the prescribed MCLs within a reasonable time frame. Therefore, we are proposing closure criteria for the A Zone aquifer to be 250 µg/L for TCE based on the risk-based cleanup levels (RBCLs) calculated for the Commercial Land Use scenario. As for the B Zone aquifer, recognizing the potential of impacting the underlying C Zone groundwater quality, its closure criteria would be to remediate the B Zone aquifer to reduce the contaminant concentration to reach an asymptotic level at which no significant increase in net CHC mass removal rates can be achieved even with pulsed mode or intermittent pumping operations. In addition, continued groundwater monitoring demonstrates that the residual CHC plume poses no threat to existing or planned drinking water receptors. After a point of diminishing returns has been reached, a site closure will be granted by the CRWQCB.

REMEDIAL APPROACH

It is known that short-chain aliphatic CHCs, such as TCE, 1,1-DCE, etc., can be aerobically biodegraded to form carbon dioxide, water, and chloride in the presence of a primary substrate such as methane (Henry et al. 1986; Semprini et al. 1990; Kang 1993; Kang 1994). Air/methane sparging introduces molecular oxygen and methane gas into A and B Zone aquifers to increase the dissolved oxygen and methane concentrations in the groundwater, thereby stimulating the in situ aerobic biodegradation of TCE, 1,1-DCE, etc. In addition, air sparging enhances in situ air stripping of residual CHCs from groundwater into the vadose zone soil where the vapors can then be captured by a soil vapor extraction (SVE) system. To maintain a healthy growth of methanotroph, methane injection was implemented in a periodic or pulsed mode.

Due to the presence of elevated levels of CHCs, our initial remediation strategy is to extract the groundwater from both A and B Zone aquifers in an effort to dewater the A Zone aquifer and lower the B Zone aquifer table. This will allow the SVE system to effectively remediate the impacted capillary fringe zone soils, the A Zone aquifer sediment materials, as well as the A Zone aquitard soils. The air sparging system will then be initiated to enhance the volatilization of the residual CHCs to the vadose zone and then captured by the SVE system.

WELL FIELD DESIGN CONSIDERATIONS

To design an effective groundwater extraction well field to extract impacted groundwater from both A and B Zone aquifers, aquifer testing and subsequent groundwater flow and transport modeling were performed using MODFLOW, MODPATH, and MT3D. The results indicate that vertical hydraulic communication exists between the A and B Zones, and little or no communication exists between the A, B, and C Zones.

Aquifer test results indicate that A Zone is moderately permeable with a hydraulic conductivity (k) of 2.2 x 10^{-3} cm/sec, while B Zone is quite permeable with a hydraulic conductivity (k) of 3.9 x 10^{-2} cm/sec with an observed radius of pumping influence that exceeded 233 feet during the pumping test. Therefore, pumping groundwater from extraction wells that are screened through both the A and B Zone aquifers should dewater the A Zone while extracting impacted groundwater from the B Zone. Results of the tests also indicate that the B Zone is responding in a semi-confined to confined manner. On the basis of this observation, we found that the B Zone extraction rate should be controlled to minimize the dewatering of B Zone to avoid compressing the skeletal matrix of the aquifer, which would result in reduction in hydraulic conductivity and the zone of capture. Therefore, the A and B Zone groundwater extraction rates were limited to 25 to 30 gpm per well. The modeling results show that, at these extraction rates, a total of 10 A and B Zone groundwater extraction wells and four reinjection wells would be adequate to provide hydraulic containment of the dissolved phased plume. To control the drawdown, each groundwater extraction well would be equipped with an electric submersible pump and level sensor.

To address the residual CHCs in the soil and aquifer materials, 10 methane/air sparging wells and 26 SVE wells were installed to expedite the soil/groundwater remediation process.

FULL-SCALE REMEDIATION SYSTEM DESIGN AND CONSTRUCTION

Based on the selected remedial alternative, a full-scale soil and groundwater remediation system was designed and constructed in May 1999. We worked with the church's architect and engineers to incorporate our remediation design into their plan to minimize future disruption to their activities. To this end, the above-mentioned well field was designed to accommodate their floor plan. The system consisted of the following components: 10 6-inch A and B Zone groundwater extraction wells, 26 4-inch SVE wells, 10 2-inch air/methane sparging wells, 4 10-inch B Zone reinjection wells, a 300 gpm groundwater treatment and reinjection system, a 600 SCFM vapor extraction and granular activated carbon (GAC) treatment system, and a 90 cfm air/methane injection system. The 300 gpm groundwater treatment system consisted of an influent holding tank (3,000 gal. capacity), dual 10-micron influent in-line particulate filters, a 300 gpm 6-tray low-profile air stripper equipped with an electric preheater, four vapor phase GAC adsorbers (2,000 lbs. each), one aqueous phase GAC polishing adsorber (5,000 lbs.), an effluent holding tank (3,000 gal. capacity), dual 10-micron effluent in-line particulate filters, and a reinjection system.

RESULTS AND DISCUSSIONS

The remediation system was started up in June 1999. To date, approximately 200 lbs. of CHCs have been removed via groundwater and vapor extraction (Figure 2). In addition, an unknown amount of CHCs was biodegraded to form carbon dioxide, water, and chloride. Figures 3 and 4 demonstrate the effectiveness of the groundwater remediation system. The geometric mean concentrations of indicator wells at the site show a general decline in TCE concentration over time due to pumping, air/methane sparging, SVE, and in situ aerobic biodegradation. We are currently monitoring the natural attenuation parameters such as dissolved oxygen, redox potential, nitrate, nitrite, chloride, sulfate, sulfides, etc. in indicator wells to document the biological attenuation of the residual CHCs.

CONCLUSION

This project demonstrates that enhanced remediation of CHC-impacted aquifers is possible at an active site, while minimizing the disruption to the aboveground activities. The geometric mean levels of CHCs in the groundwater have been declining since the implementation of the enhanced remediation process. This information, in conjunction with natural attenuation documentation, will be used to negotiate with the CRWQCB to obtain site closure approval.

The authors would like to thank James A. Carney of Brunswick Corporation for his full support on this project. We also would like to thank Devon Hass and Gary Hann of URS – Santa Ana, California for their valuable contribution to this project.

FIGURE 2. Cumulative Mass Removal of CHCs.

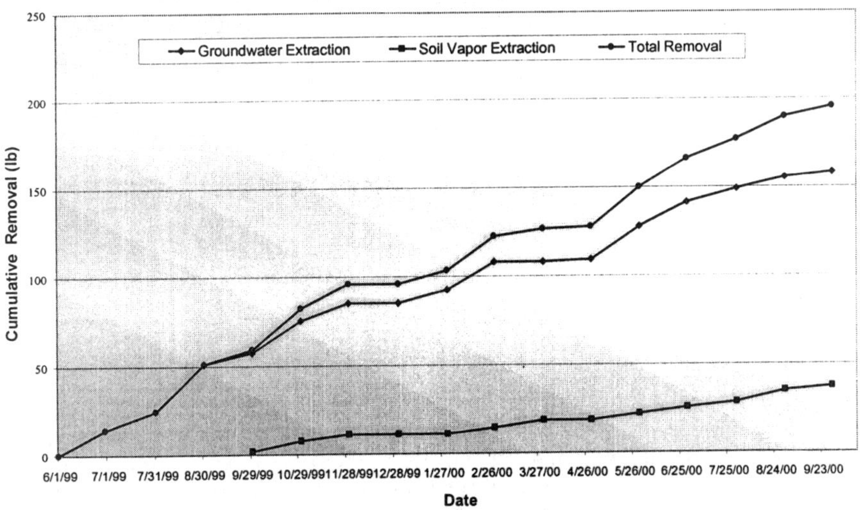

FIGURE 3. Historical TCE Concentrations A Zone Monitoring Wells. (existing wells)

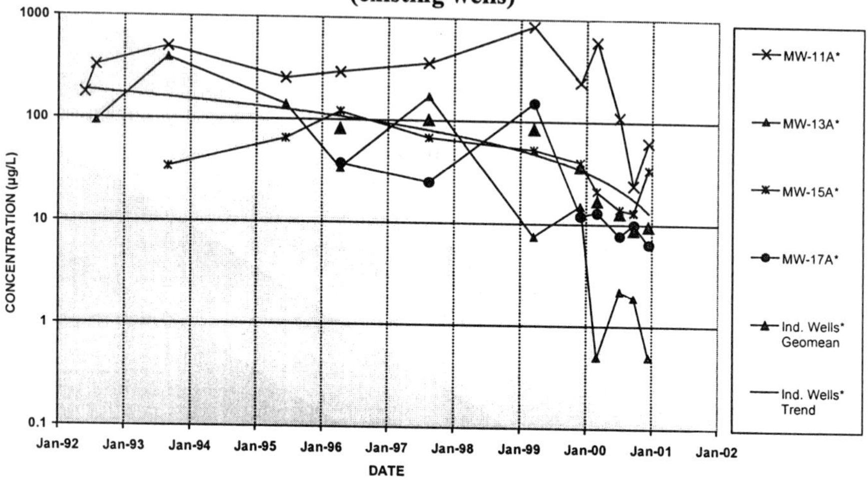

FIGURE 4. Historical TCE Concentrations B Zone Monitoring Wells. (existing wells)

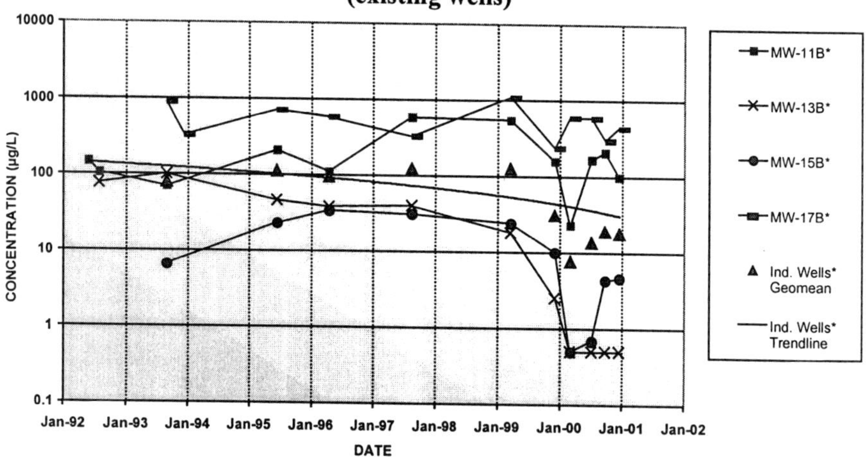

REFERENCES

Kang, J. J. 1993. "Assessment of In Situ Air/Methane Sparging/Bioventing Potential in Remediating a TCE Contaminated Site," presented at the 4th Annual West Coast Conference on Hydrocarbon Contaminated Soils and Groundwater, Long Beach, California, March 4.

Kang, J. J. 1994. "Evaluation of the Feasibility of Implementing In Situ Air/Methane Sparging in an Alluvial/Fractured Bedrock Aquifer," Presented at the 5th Annual West Coast Conference on Hydrocarbon Contaminated Soils and Groundwater, Long Beach, California, March 30.

Henry, S. M. and D. Grbic-Galic. 1986. "Aerobic Degradation of Trichloroethylene by Methanotrophic Isolated from a Contaminated Aquifer," Proceedings of Annual Meeting, Am. Soc. Microbiol, p. 294.

Semprini, L., P.V. Roberts, G. D. Hopkins, and P. L. MacCarthy, 1990. "A Field Evaluation of In Situ Biodegradation of Chlorinated Ethenes: Part 2, Results of Biostimualtion and Biotransformation Experiments," Ground Water 28: 715-727.

IN SITU AEROBIC BIOREMEDIATION OF METHYLENE CHLORIDE IN LOW-PERMEABILITY BEDROCK

Theresa Santangelo-Dreiling (Colorado Dept. of Transportation; Denver, CO);
Patricia Boyd-Kaygi, Brad Toth, Paul Weaverling, Jon Wilken,
Don Smallbeck (Harding ESE; Denver, CO and Novato, CA); and
Brian LaFlamme (Engineering Management Support, Inc.; Denver, CO)

ABSTRACT: Enhanced biodegradation was evaluated as an alternative to groundwater pump and treat for remediation of contaminated groundwater at a Colorado Department of Transportation (CDOT) facility in Denver, Colorado (Site). A pilot-scale test for enhanced *in situ* aerobic bioremediation of methylene chloride (MeCl) and higher chlorinated volatile organic compounds (VOCs) was performed at the Site over a 12-month period. The pilot-scale test followed a successful bench-scale study that indicated aerobic biodegradation of MeCl and higher chlorinated VOCs was possible if the naturally occurring conditions for biodegradation in the formation could be enhanced and maintained. Concentrations of MeCl in groundwater that exceed 2,000,000 micrograms per liter (µg/L), low-permeability bedrock, and complex hydrogeology presented unique challenges for the system design and implementation. The pilot-scale test included low-flow injection of nutrients and oxygen into low permeability bedrock sandstones. Methane was added to the injection stream to stimulate the potential cometabolic degradation of low concentrations of tetrachloroethene (PCE), trichloroethene (TCE), and 1,1-dichloroethene (1,1-DCE). The concentrations of MeCl and the higher chlorinated VOCs in the source area have decreased significantly over the 12-month test period.

INTRODUCTION

This paper presents results from a pilot-scale test of an enhanced bioremediation system constructed to stimulate biodegradation of methylene chloride (MeCl) and higher chlorinated volatile organic compounds (VOCs) within two low-permeability sandstones of the Denver Formation. Summarized herein are the pilot-scale test objectives, site description, design basis and operating parameters, pilot-test results and discussion, and conclusions reached from the pilot test.

PILOT-SCALE TEST OBJECTIVES

There were three objectives for the pilot-scale test: 1) design a low-flow injection system to deliver nutrients and oxygen into a low-permeability formation, 2) propagate the nutrient-enriched injection water through the low-permeability formation, and 3) enhance the naturally occurring microbial population for the active biodegradation of MeCl and higher chlorinated VOCs. To assess these objectives, water-level data, field-parameter measurements, and groundwater samples were collected during the pilot-scale test for evaluation.

SITE DESCRIPTION

Site History. Historical activities at the Site included asphalt testing in which the asphalt cement was extracted using MeCl. The used solvent was disposed down a Dry Well located next to the Site materials laboratory over an approximate 10-year span from 1966 to 1976. The Colorado Department of Transportation (CDOT) initiated an Interim Measures (IM) investigation in November 1998 to address contamination in the Dry Well area as part of a Compliance Order for the Site. The IM consisted of abandoning the Dry Well, performing aquifer tests for two zones, biodegradation bench-scale testing, and implementing an *in situ* pilot-scale test.

Geology and Hydrogeology. The Site stratigraphy consists of unconsolidated, unsaturated loess overlying bedrock of the Denver Formation. The top of bedrock occurs between 3.9 to 5.5 meters [m] below ground surface (bgs) and consists of interbedded claystones, siltstones, and sandstones (Figure 1). Two sandstone stratigraphic units separated by a laterally continuous claystone were identified beneath the Dry Well. The sandstones in the uppermost unit are locally continuous, are very fine- to fine-grained, and are generally silty to clayey. The lower sandstone stratigraphic unit is relatively thick and more laterally continuous than the sandstones in the upper unit. This unit grades from a coarse-grained sandstone at the base to a fine-grained sandstone with interbedded siltstone or claystone stringers near the top.

Beneath the Dry Well area, the bedrock is moderately to highly weathered to a depth of approximately 14.6 m bgs. The sandstones in the weathered zone are generally loose to poorly consolidated and more permeable than the interbedded claystones. The sandstones have been interpreted to be the primary pathways of groundwater flow and contaminant transport.

Two groundwater flow systems are interpreted to be present beneath the Dry Well: an unconfined upper zone (base of the upper sandstone stratigraphic unit to the top of the water table) that extends to approximately 9.8 m bgs, and a semi-confined/confined lower zone (lower sandstone stratigraphic unit) that extends from approximately 10.4 to 16.5 m bgs (Figure 1). The hydraulic conductivities of the upper and lower flow zones, estimated from aquifer pumping tests and core samples, are approximately 3.15×10^{-5} and 8.65×10^{-5} centimeters per second (cm/sec), respectively. The respective estimated groundwater flow velocities are 2.7 to 5.5 meters per year [m/yr] and 12 m/yr.

Contaminant Distribution. The primary contaminant identified in groundwater beneath the Dry Well is MeCl, exceeding 2,500,000 micrograms per liter (µg/L) in the upper zone at the source. The area

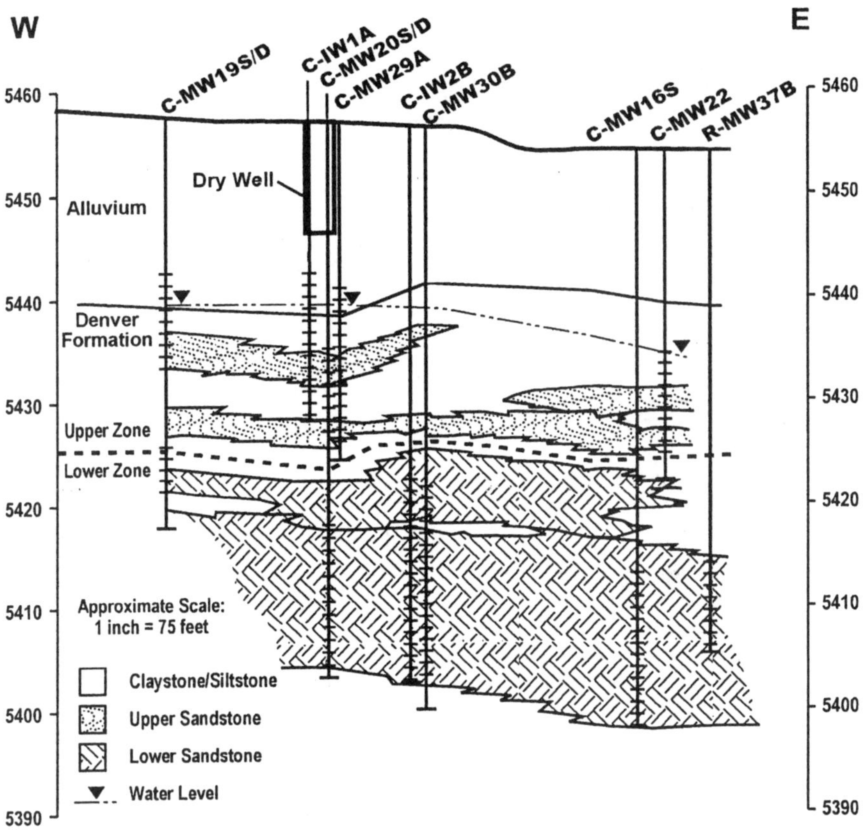

FIGURE 1: Geologic cross-section through the Dry Well

of elevated MeCl concentrations in the upper zone extends from well C-MW20S at the source area downgradient to monitoring well C-MW22 (Figure 2). Other chlorinated VOCs, including tetrachloroethene (PCE) and trichloroethene (TCE), were also detected in this area but at concentrations four orders of magnitude lower than MeCl. The historical concentrations of PCE and TCE in groundwater range between 200 μg/L and 300 μg/L and between 300 μg/L to 400 μg/L, respectively.

In the lower zone, MeCl concentrations were lower with only a slightly different extent of contamination than the upper zone (Figure 3). MeCl was detected at 1,800 μg/L in monitoring well C-MW30B, located 11.5 m downgradient of the Dry Well. Very low concentrations of other chlorinated VOCs were also present. The highest contaminant concentrations in the lower zone were present in monitoring well C-MW16S, located 33.5 m downgradient of the Dry Well. The pre-test MeCl concentration in this well was as high as 480,000 μg/L.

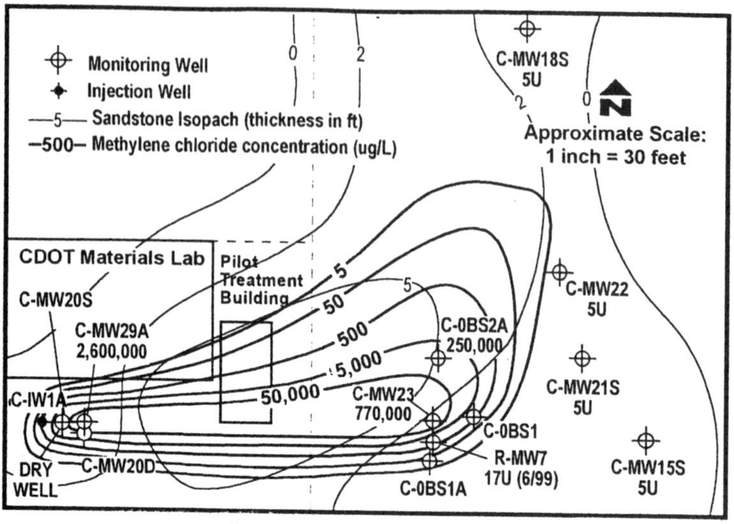

FIGURE 2: Extent of Methylene Chloride in groundwater, Upper Zone, Fall 1999

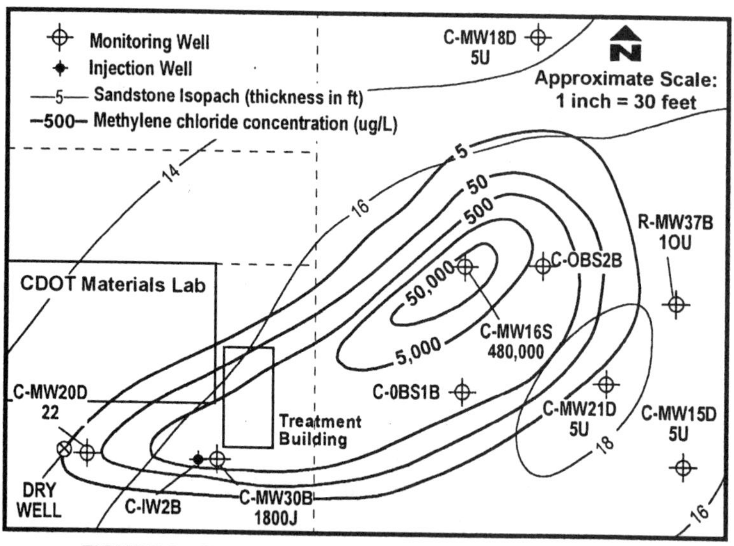

FIGURE 3: Exent of Methylene Chloride in groundwater, Lower Zone, Fall 1999

DESIGN BASIS AND OPERATING PARAMETERS

Bench-scale Test Results. A treatability study was performed on soil and groundwater samples from the Site to evaluate the potential effectiveness of enhanced aerobic biodegradation for reducing concentrations of MeCl and other chlorinated VOCs in groundwater. The bench-scale test results indicate that the addition of nutrients and oxygen enhanced the growth of the microbial population, resulting in nearly complete MeCl degradation during a 58-day incubation period.

Pilot-scale Design and Operation. The pilot-scale design included the installation of two injection wells, two monitoring wells, a temporary treatment building, electrical service, a water supply system, and a chemical mixing system.

A chemical mixing system constructed in the treatment building consisted of two 757 L mixing tanks, chemical metering pumps with a range of 0.05 to 0.5 0.19 to 1.9 liters per minute, and flow meters. The mixing tanks provided capacity for an injection period of approximately two days. Ammonium nitrate and potassium phosphate were added to chlorine-free potable water at concentrations of approximately 50 mg/L and 5 mg/L, respectively. Concentrations of hydrogen peroxide in the injection stream were slowly increased from 100 mg/L at startup to as high as 800 mg/L. In order to enhance potential cometabolism, methane gas was added to the injection stream at a rate of approximately 1.14 mg/L. A static mixer was located in the treatment system immediately downstream of the methane injection port to thoroughly mix the solution.

Operational Monitoring Parameters. To monitor the performance of the system, water-levels and field parameters were collected on a weekly basis for the first 15 weeks, semi-monthly through the remainder of the first six months of operation, and then quarterly for the last six-month period. Field parameters collected during each event included phosphate, carbon dioxide, pH, electrical conductivity, redox potential, dissolved oxygen, and temperature.

Groundwater samples were collected twice per month during the first six months of operation and quarterly thereafter. Groundwater samples were collected using micro-purge technology and submitted for analysis of VOCs, light hydrocarbons, anions, total and dissolved organic carbon, chemical oxygen demand, general chemistry parameters, and heterotrophic plate count.

RESULTS AND DISCUSSION

Hydrogeologic System Response. Water-level data and phosphate travel time were used to evaluate the groundwater hydraulic gradient and the rate of the treatment front breakthrough at downgradient monitoring points.

Water levels near the injection well in the upper zone increased approximately 0.3 m within one week after the pilot test began. Water levels in the lower zone monitoring wells near the injection well increased by approximately 0.6 m within one week after the pilot test began. These results indicate that the injected water was migrating into each zone.

To further assess the hydrogeologic system response, the baseline groundwater velocities were compared to the observed travel time of phosphate for each of the zones. In the upper zone, the estimated groundwater velocity prior to injection was 1.2 m/yr. After injection began, the hydraulic gradient in the area of the injection well increased from 0.0208 m/m to approximately 0.12 m/m, increasing the estimated groundwater velocity to approximately 7 m/yr.

Groundwater velocities estimated using the phosphate travel time to wells C-MW20S and C-MW29A range from 15 to 86 m/yr. In the lower zone, the estimated groundwater velocity prior to injection was 2.6 m/yr. After injection began, the hydraulic gradient in the area of the injection well increased from 0.0105 m/mto approximately 0.0438 m/m, increasing the estimated groundwater velocity to approximately 11 m per year. Groundwater velocities estimated using the phosphate travel time to wells C-MW20D and C-MW16S, range from 48 to 75 m/yr. These increases in estimated velocities are a result of the increased head and pressure at the injection wells, and provide an additional indicator that there are hydraulic responses to injection in both zones.

Degradation Response. Selected groundwater analytical results for upper zone monitoring well C-MW29A and lower zone monitoring well C-MW30B are summarized in Table 1. The corresponding dissolved oxygen (DO) concentrations measured during sample collection are also included.

TABLE 1. Summary of analytical results and dissolved oxygen concentrations for Wells C-MW29A and C-MW30B.

	Well C-MW29A					Well C-MW30B				
Day of Operation	**Baseline**	**84**	**177**	**289**	**377**	**Baseline**	**84**	**177**	**289**	**377**
Methylene chloride	2,600,000	2,100,000	30,000	21	<5	1,800	1	<5	<5	<5
Trichloroethene	1,400	880	13	12	13	11	<5	<5	<5	<5
Tetrachloroethene	1,000	660	33	28	22	5.4	<5	<5	<5	<5
1,1-Dichloroethene	670	450	3	3	5	3	<5	<5	<5	<5
cis-1,2-Dichloroethene	22	53	<5	<5	<5	<5	<5	<5	<5	<5
trans-1,2-Dichloroethene	540	320	3	3	3	2	<5	<5	<5	<5
1,1,1-Trichloroethane	86	75	<5	<5	<5	<5	<5	<5	<5	<5
Vinyl chloride	4	12	<2	<2	<2	<10	<10	<2	<2	<2
Chloroethane	<10	11	<10	<10	<10	<10	<10	<10	<10	<10
Methane (mg/L)	2.16	2.44	0.317	<0.001	0.503	0.922	130	14.3	<0.001	0.774
Sulfate (mg/L)	150	140	56	54	61	600	43	69	44	79
Dissolved Oxygen (mg/L)	2.46	0.29	>20	>20	>20	1.04	>20	>20	>20	>20

Upper Zone Degradation Response. The baseline concentrations for MeCl in monitoring wells C-MW20S and C-MW29A were 280,000 µg/L and 2,600,000 µg/L, respectively. Lesser concentrations of PCE and TCE were detected in the baseline groundwater samples. These wells exhibited anaerobic to slightly aerobic conditions during baseline sampling. Through the third month of operation, the MeCl concentrations at the two monitoring wells remained above 200,000 µg/L and 2,000,000 µg/L. A slight decrease in concentrations of TCE and PCE over the same period was accompanied by an increase in reductive dechlorination products such as vinyl chloride, chloroethane, and chloride. During this time, DO and redox potential measurements remained anaerobic and were apparently unaffected by the injection system. Subsequently, monitoring well C-MW20S was converted to an injection well and aerobic conditions were established in C-MW29A as indicated by the increased DO and redox potential. As shown in Figure 4, through the first six-month period of the test, concentrations of MeCl in well C-MW29A decreased by two orders of magnitude from 2,600,000 to 30,000 µg/L. After remaining stable for three consecutive bimonthly sampling events as the system equilibrated, the second six-month period of the test resulted in a five order of magnitude decrease in the MeCl concentration. The most recent results indicate no detected concentrations of MeCl in well C-MW29A. The analytical data and field measurements collected during the pilot test suggest that as an aerobic environment is attained and maintained, aerobic respiration will continue to degrade VOCs.

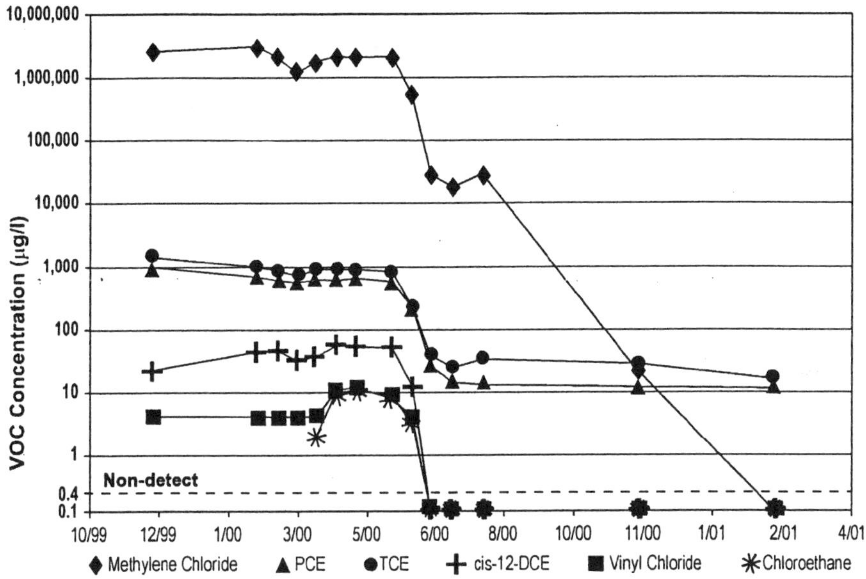

FIGURE 4: VOC concentrations versus time, Well C-MW29A - Upper Zone

Lower Zone Degradation Response. The baseline concentration for MeCl at well C-MW30B was 1,800 µg/L and field parameters were indicative of mildly anaerobic conditions. The first performance monitoring samples from C-MW30B indicated a significant increase in DO and redox potential and a corresponding decrease in VOC concentrations. These groundwater chemical changes were concurrent with the observed hydraulic affects in the lower zone. Given the relatively low baseline VOC concentrations at well C-MW30B, the induced aerobic conditions resulted in rapid VOC degradation. Sample results since the fifth month of operation indicate no detected concentrations of MeCl or other chlorinated VOCs in well C-MW30B.

Injection effects were evident in well C-MW20D located 11.5 m upgradient; DO and redox potential increased in the well throughout the test. The concentration of MeCl gradually decreased from 52 µg/L to nondetect. Effects of injection water were less evident at downgradient monitoring wells C-MW9 and C-MW16S. The presence of lesser-chlorinated VOCs such as vinyl chloride and chloroethane and the absence of an oxygenated environment suggest no impact from the injection well. It appears that reductive dehalogenation remains the predominant microbial process controlling VOC degradation outside of the radius of influence of the lower zone injection well.

Dilution Response. To evaluate the potential for concentration decreases resulting from dilution, a mass balance was performed using sulfate and PCE; two parameters assumed to be relatively stable under the conditions established in the test area. Using the baseline and measured sulfate concentrations in C-MW29A and the sulfate concentration in the injection water, the volume of water affected by injection at C-MW20S was estimated for each sampling event. These volumes were then used to estimate the expected concentrations of MeCl that might be attributable solely to dilution. The results for sulfate indicate that injection water has the potential to dilute MeCl concentrations up to six times (6X) whereas the analytical data show a six order of magnitude decrease in concentration for MeCl. A similar analysis was performed using PCE as a semi-conservative parameter to assess potential dilution affects. The key assumption here is that the observed decrease in PCE concentrations reflects the percent decrease of each analyte concentration due solely to dilution. During the pilot-scale test, a 96 percent decrease was observed in PCE concentrations in C-MW29A. When a similar decrease is applied to the starting MeCl concentration, the expected concentration would be 85,000 µg/l if simple dilution were the sole controlling factor. However, the observed endpoint concentration for MeCl in C-MW29A is less than 5 µg/l. Based on a mass balance approach, dilution does not account for the observed concentration decreases of MeCl in the test area.

CONCLUSIONS

Based on the data collected during a 12-month pilot-scale test implemented at the Dry Well and the assessment of that data within the hydrogeologic and geochemistry framework, the conclusions for the pilot-scale test are as follows:

- Despite the complex stratigraphy, water and nutrients can be injected at a sufficient rate to affect groundwater gradients, flow rates, and chemistry in both the upper and lower zones;
- Aerobic conditions can be induced in both bedrock zones and natural microbial populations can be stimulated to degrade MeCl;
- Although direct evidence for cometabolism of higher chlorinated solvents is not readily observed, VOC concentrations of the higher chlorinated solvents in the upper and lower zone in the Dry Well area are decreasing, suggesting that this process is occurring;
- Mass balance of conservative compounds in aerobic conditions indicate dilution effects do not account for the extent of observed decreases in VOC concentrations; and
- Enhanced *in situ* biodegradation of MeCl and more highly chlorinated VOCs is a feasible technology for source removal in the immediate Dry Well area and downgradient locations.

ACKNOWLEDGEMENT

The authors thank the CDOT Region 6 for their contributions to the project and for their cooperation in the installation of the pilot-scale test system.

HYBRID BIOTREATMENT SYSTEM: DEGRADATION OF CHLORINATED SOLVENT IN SOIL

Yoji Ishikawa, Kensuke Fujii, Hirokazu Tsuji and Kazuki Ide
(Obayashi Corporation, Tokyo, Japan)

ABSTRACT: A combined treatment system of physical and biological processes has been developed as the "Hybrid Biotreatment System" to degrade chlorinated solvents such as trichloroethylene (TCE) and dichloroehylene (DCE) in soil using the phenol-induced microbe which we screened as a TCE-degrading bacterium. The microbe was found to have no toxicity, and when it is applied to soil phenol is removed from the microbe in advance in this system. TCE in contaminated soil from an actual site (Soil-A) reached the cleanup standard level (0.03 mg/L) within two hours; the decrease of cis-1,2-DCE in the soil was even faster. The importance of controlling moisture content in the soil was stressed. TCE in modeled soil (Soil-B) did not reach 0.03 mg/L only with turn-over and aeration; the addition of the bacterial cells lowered the concentration well below 0.03 mg/L. Noticeable influence of the added microbe on the soil fauna was not found, and the number of the added microbes decreased after treatment.

INTRODUCTION

It is required in some cases that soil contaminated with chlorinated solvents, in particular trichloroethylene (TCE), be excavated and actively cleaned *ex-situ*. Bioremediation is one of the promising methods of soil treatment because it requires little energy and can leave few by-products after treatment. Since there has been no report on microorganisms growing on TCE as a sole carbon source, the aerobic biodegradation of TCE must occur through the process of co-metabolism induced by other chemical compounds such as methane, toluene, or phenol (Oldenhuis et al., 1989; Nelson et al., 1987; Kimura et al., 1997).

In order to utilize the advantages of both biological and physical processes, we developed the "Hybrid Biotreatment System", which is shown schematically in Figure 1. In the first step, the vaporization and removal process is enhanced through turning-over of the soil combined with air flow, resulting in a lower concentration of TCE in the soil. The volatalized TCE is trapped in gas treatment facility. The second step is biodegradation of the remaining TCE in the soil by mixing the soil with a phenol-degrading microorganism that mometabolically degrading TCE.

We applied the "Hybrid Biotreatment System" to a soil contaminated with chlorinated solvents from actual site, and a modeled soil prepared in the Laboratory.

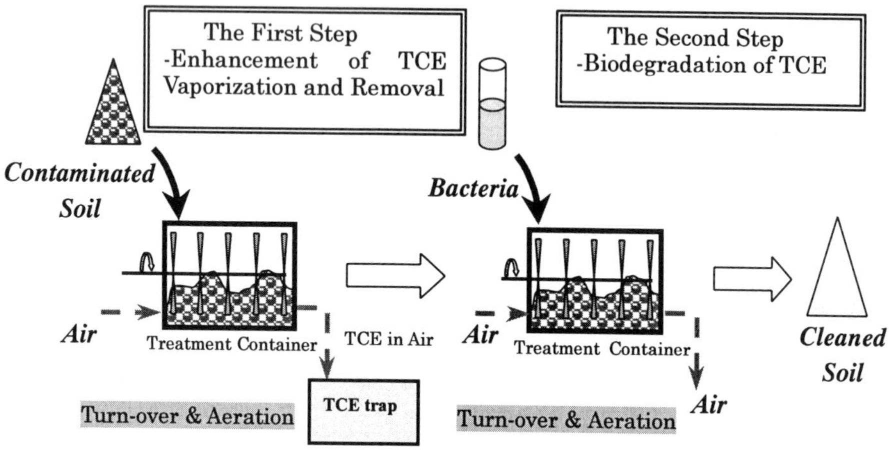

FIGURE 1. Schematic Diagram of "Hybrid Biotreatment System"

The purposes of this study were: (1) to determine overall efficiency of the Hybrid Biotreatment System, (2) to evaluate the efficiency of the biological process on contaminated soil, and (3) to examine the influence of the added microbes on the soil fauna and on the fate of TCE/DCE in the soils.

MATERIALS AND METHODS

Experimental Apparatus. The Hybrid Biotreatment System was examined in a small scale sealed (55 L) housing two-axis rotating blades. The rotation cycle was variable, from 0 to 60 rpm. Compressed air was introduced into soil in the container through the blade tips, and was eventually exhausted to an activated carbon filter where vaporized TCE was trapped. The air flow rate could be varied up to 600 L/min. This container was used for both physical and biological process experiments.

Bacterial Strain. The phenol-degrading bacterial strain was isolated from activated sludge. It was used to degrade TCE through cometabolic process. Its bacterial characteristics and its gene sequence suggest that it does not fall into any known species. The strain grows well at 20 to 30 degree Celsius, and also at neutral pH. This strain can degrade 50 mg/L of TCE in one day when 10^9 cfu/mL of the strain is applied. The safety of the strain was examined in the manner described in the guideline of the Ministry of Agriculture, Forestry, and Fishery. The results indicated the strain has no toxicity to either humans or the environment. We used the resting cells of the strain to degrade TCE. 500 ppm of phenol, as an inducer, had been added to activate the strain. Then, the strain was separated from phenol-rich solution though a centrifugal force, and washed twice with buffer solution. In this way, inducer-free strain can be added to the soil.

Soil. As TCE is easily vaporized and removed from sand just by turning-over or aeration, we focused on soils including silt and clay in this study. The soils we report here were; (1) Soil-A, contaminated soil from actual site, and (2) Soil-B, modeled contaminated soil. The characteristics of the Soil-A were: fine grain type soil containing 44% of silt and clay, moisture content; 43.8%, pH; 6.0, ignition loss; 4.6%. Soil-B: silt type sand containing 47% of silt and clay, moisture content; 25.7%, pH; 6.5, ignition loss; 4.0%. The Soil-A had been contaminated with 1.8 mg/L of TCE (trichloroethylene) and 2.0 mg/L of *cis*-1,2-DCE (dichloroethylene), measured as eluted concentrations (see below). The Soil-B was air-dried and then mixed with TCE, thus artificially contaminated at arbitrary TCE concentration (approximately 1 mg/L TCE).

Analysis. Eluted concentrations of TCE and *cis*-1,2-DCE were measured in the method described in the Japanese Industrial Standard, which means the eluted portion in H_2O from the soil sample was analyzed by GC. The soil cleanup standards in Japan are 0.03 mg/L for TCE and 0.04 mg/L for *cis*-1,2-DCE, using this analytical method. On the other hand, the contents of TCE and *cis*-1,2-DCE in soil were measured by either GC-MS for eluted portion in solvent or PID-GC for head space gas.

The fauna after the treatment was analyzed by standard colony counting method.

RESULTS AND DISCUSSION

Cleanup Efficiency. The Soil-A was expected to be relatively easily remediated because its soil quality lets TCE elute quickly from the soil. The soil was contaminated with more than one kind of contaminant, and preliminary tests were conducted to assure the cometabolic biodegradability of the contaminants. Laboratory tests in ideal condition; using slurry reactions indicated that the soil contained no chemical compounds to inhibit the activity of the added bacteria, and that 3 x 10^9 cfu/g-dry soil of the bacterial cells were necessary for TCE degradation.

TCE biodegradation results in Soil A are shown in Figure 2. The eluted concentration of TCE from the sample reached the cleanup standard level (0.03 mg/L) within 2 hours from the start of the treatment. Decrease of *cis*-1,2-DCE, which are not shown here, was usually quicker than TCE for Soil-A, and in this case it reached cleanup level (0.04 mg/L) within 1 hour. We found that the higher the moisture content was, the better the efficiency of the biodegradation, presumably because of better contact between bacterial cells and contaminants and more abundant supply of dissolved oxygen for microbes. On the other hand, the first step – physical turn-over and aeration – was found to function better with lower moisture contents, and thus the actual application requires the control of the moisture contents between each step.

Soil-B results are shown in Figure 3. The TCE concentration did not reach 0.03 mg/L with turn-over and aeration alone. However, the addition of the

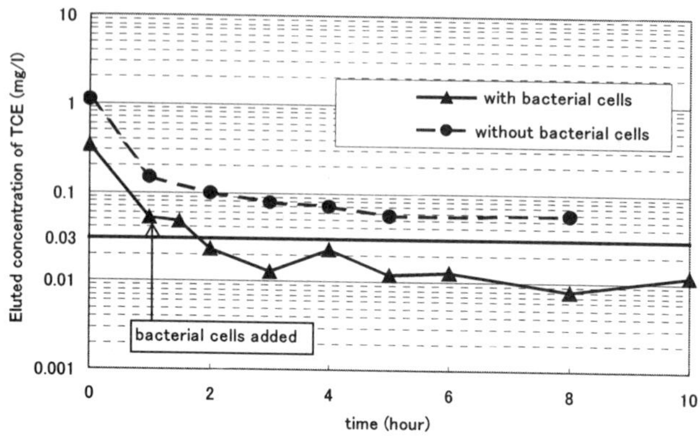

FIGURE 2. Biodegradation of TCE by Soil-A with/without innoculation with phenol-degrading bacteria (addition of 6 x 10^9 cfu/g-dry soil, continuous turning-over, 30 L/min of air flow)

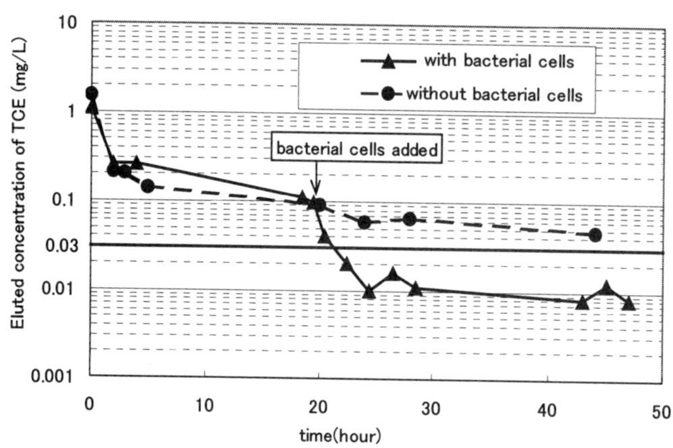

FIGURE 3. Biodegradation of TCE by Soil-B with/without innoculation with phenol-degrading bacteria (addition of 3 x 10^{10} cfu/g-dry soil, continuous turning-over, 10 L/min of air flow)

bacterial cells lowered the concentration well below the 0.03 mg/L level. The phenol-degrading bacteria clearly enhanced TCE degradation in the "Hybrid Biotreatment System".

In all the cases, dichloroacetic acids (DCAA) were found during biotreatment. As the DCAA is known to be by-product of TCE biodegradation (Nakajima, et al., 1992), it is another proof that the bacteria degraded TCE. The final concentrations of DCAA were below the water quality standard (0.02 mg/L) in all the cases.

Safety Regarding the Addition of the Bacterial Cells. As mentioned above, non-toxicity of the bacteria had been already confirmed; the Hybrid Biotreatment System requires no introduction of a toxic inducer – phenol in this case – into soil. The remaining concern is possibility of the potential alternation of the natural ecology – fauna – in the soils, and the long-term influence of residual bacteria added to the soil. The fauna in soils was followed by counting the number of the added strain after the biotreatment was finished. The number of *Escherichia coli* and Fungi remained around 10^3 to 10^4 MPN/g-dry soil before and after the treatment, and did not indicate a noticeable influence of the added microbe. The number of the added bacteria, which was around 10^{10} cfu/g-dry soil just after treatment, decreased to 10^6 or 10^8 cfu/g-dry soil within 2 to 8 weeks in the four examined cases.

ACKNOWLEDGEMENTS

This project was supported in part by the fiscal year 1999 Fund for Cleanup of contaminated underground water and soil from Environment Agency, Government of Japan.

REFERENCES

Kimura, T., Y. Oda, M. Yamashita, Y. Okamura, and K. Numata. 1997. *"Aerobic* Degradation of Chlorinated Solvents by Aromatics-Using Bacteria" In Situ and On-Site Bioremediation, *3*: 133-138

Nakajima, T., H. Uchiyama, O. Yagi, and T. Nakahara. 1992. "Purification and Properties of a Soluble Methane Monooxygenase from Methylocystit sp. M." *Biosci. Siotech. Biochem.* 56: 736-740.

Nelson, M. J. K., S. O. Montgomery, W. R. Mahaffey, and P. H. Pitchard. 1987. "Biodegradation of Trichloroethylene and Involvement on an Aromatic Biodegradative Pathway" *Appl. Environ. Microbiol. 53*(5): 949-954.

Oldenhuis, R., R. L. J. M. Vink, D. B. Janssen, and B. Witholt. 1989. "Degradation of chlorinated aliphatic hydrocarbons by Methylosinus Trichosporium OB3b Expressing Soluble Methane Monooxygenase" *Appl. Environ. Microbiol.* 55(11): 2819-2826.

IMPROVED BIOSLURPER OPERATIONS WITH THE USE OF IN-WELL OIL-WATER SEPARATION

Sarah McCall, Matt Place, Travis Williamson, Abe Chen
(Battelle, Columbus, Ohio, USA)
Ron Hoeppel (Naval Facilities Engineering Service Center, Port Hueneme, California, USA)

ABSTRACT: Bioslurping combines vacuum-enhanced LNAPL recovery with bioventing and soil vapor extraction to simultaneously recover free product and treat the unsaturated zone. Bioslurping with the use of an in-well oil-water separation technique has been demonstrated to be effective at reducing oil/water emulsion formation, and petroleum hydrocarbon concentrations in aqueous and vapor discharge streams. This technique has been tested at six sites under different geologic and contaminant conditions. The results show a near complete elimination of the emulsion formation and an average reduction of petroleum hydrocarbon concentrations of 99% and 33% in the aqueous and vapor discharge streams, respectively.

INTRODUCTION

Bioslurping has proven to be an effective method for removing light, non-aqueous phase liquid (LNAPL) from the subsurface. Bioslurping can reduce the time required to remediate LNAPL-contaminated sites, because it enhances LNAPL recovery in comparison to conventional skimming and pump-drawdown technologies (Battelle, 1997; 1998). Bioslurping combines vacuum-enhanced LNAPL recovery with bioventing and soil vapor extraction (SVE) to simultaneously recover free product and treat the vadose zone and capillary fringe. Mathematical modeling of drawdown pumping and bioslurping (Parker, 1996) predicts that bioslurping removes free product three times faster than drawdown pumping while withdrawing seven times less groundwater.

Conventional bioslurper systems use a liquid ring pump to create vacuum in the subsurface to extract LNAPL, groundwater, and soil gas in the same process stream. The simultaneous extraction of the LNAPL, groundwater, and soil gas often results in the mixing of these components (either in the liquid ring pump or in the extraction well manifold). As a consequence, floating solids and oil/water emulsions are produced which may accumulate in downstream treatment devices, such as an oil/water separator (OWS), and give the aqueous discharge stream a milky white appearance. The formation of the floating solids and oil/water emulsions may cause operational problems and require costly downstream treatment.

Many modifications to conventional bioslurper systems have been attempted to reduce the problems associated with the production of the floating solids and emulsions. The modification that demonstrates the most promise is the use of a dual drop-tube configuration that extracts groundwater, soil gas, and LNAPL from the well in two separate streams. The main advantage of this configuration is the reduction of petroleum hydrocarbon concentrations in the water and vapor discharge streams from the bioslurper system and the near complete elimination of floating solids and emulsions.

This paper reports the results of a multiple-site study that evaluates the effectiveness of the in-well oil-water separation configuration at reducing petroleum hydrocarbon concentrations in the aqueous and vapor discharge streams from bioslurper systems. Demonstrations were carried out at multiple sites under a variety of geologic and contaminant conditions. To date, demonstrations have been completed at six sites, and the aqueous and vapor discharge samples from those sites have been analyzed.

DEMONSTRATION SITE SELECTION

Several criteria were taken into account during the site selection. The presence of LNAPL contamination at the candidate sites was the most important factor. The geologic conditions at the sites had to have permeable soils to allow groundwater and LNAPL flow into extraction wells. The geologic conditions must also present a soil environment, which would produce a vacuum to start the movement of groundwater and LNAPL into extraction wells. Six sites were chosen for the short-term demonstrations. Table 1 lists the location of each site, along with type of LNAPL present, soil type, and depth to groundwater.

DEMONSTRATION METHODS AND ANALYSES

Six short-term demonstrations (lasting two to three weeks each) were performed to evaluate the use of the dual-drop configuration. Similar to the conventional bioslurper configuration, the dual-drop tube configuration uses a vacuum to draw LNAPL, groundwater, and soil gas to extraction wells. The dual-drop tube configuration uses one drop tube to remove LNAPL and the other drop tube, similar to the one used for the conventional bioslurper system, to remove groundwater and soil gas. A shield guards the drop-tube that extracts groundwater and soil gas. The use of a shield allows groundwater to be drawn through the bottom of the shield and soil gas through the top of the shield while LNAPL is prevented from entering the shield. The drop-tube that extracts LNAPL is placed outside the shield, with the opening of the tube positioned approximately 0.25 inches above the oil/water interface. After extraction, the recovered groundwater and soil gas enter the liquid ring pump. The groundwater then exits the pump to the OWS, and the soil gas exits the pump to a stack. The recovered LNAPL, drawn to the surface by the vacuum created by the same liquid ring pump, is captured in a separate tank (under vacuum) for temporary storage. The conventional bioslurper configuration and the in-well separation bioslurper configuration are presented in Figure 1 and Figure 2, respectively.

The performance of the dual-drop tube configuration was evaluated by comparing the analytical results of aqueous and vapor discharge samples in the dual-drop tube configuration with those of similar samples collected from a conventional configuration. The quality of the aqueous process stream was evaluated during the demonstrations to determine the effectiveness of the dual drop tube system at reducing the formation of floating solids. In addition, the operation and maintenance requirements for operating the system in the different configurations were compared.

TABLE 1. Summary of Selected Sites.

Site Location	Type of LNAPL	Soil Type	Depth to Groundwater (bgs)
Naval Air Station (NAS) Fallon, Nevada, USA	JP-5	Sands and Silts	10 ft
Naval Air Weapons Station (NAWS) China Lake, California, USA	JP-5	Sands and Silts	45 ft
Bolling Air Force Base (AFB), District of Columbia, USA	No. 2 Fuel Oil	Silt and Clay	8 ft
Tyndall AFB, Florida, USA	JP-4	Sands	5 ft
Naval Fuel Depot (NFD) Point Molate, California, USA	Bunker and JP-5	Sands and Clay	15 ft
Marine Corp Air Station (MCAS) Cherry Point, North Carolina, USA	No. 2 Fuel Oil	Sands	5 ft

Demonstration Schedule. Short-term demonstrations at each of the sites were performed using a single extraction well to allow for a side-by-side comparison of the in-well separation configuration to conventional bioslurper operation.

The demonstration schedule included a series of tests starting with the conventional drop-tube configuration, followed by the dual-drop tube configuration. At most sites, each of these tests lasted approximately four days. At some of the sites, low LNAPL recovery reduced the length of the tests to 48 or 72 hours. A conventional single drop tube test was repeated at the end to help establish a baseline.

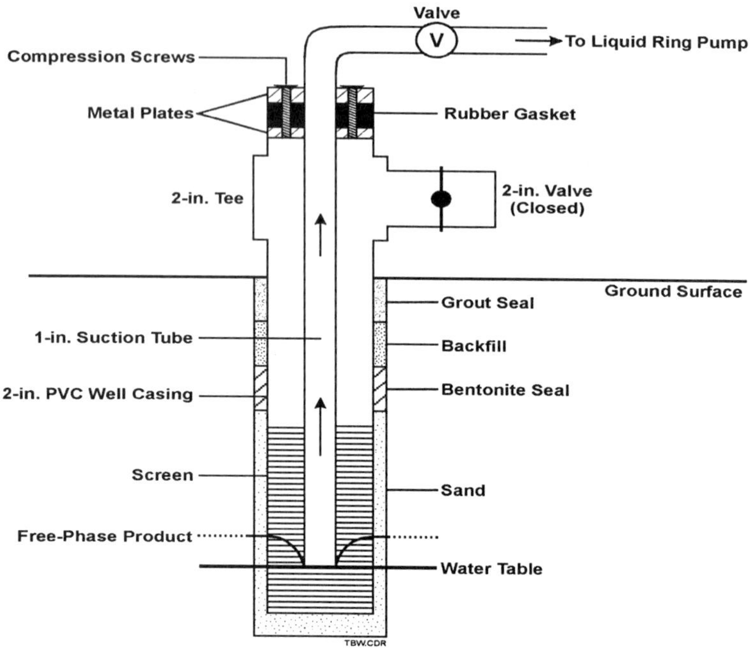

FIGURE 1. Conventional bioslurper configuration.

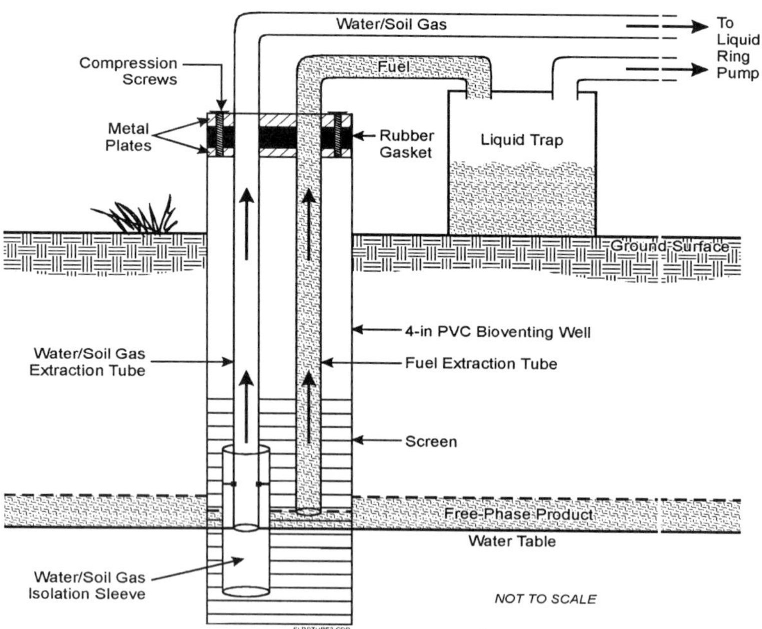

FIGURE 2. In-well separation bioslurper configuration.

Sampling and Analyses. Water and off-gas samples were collected downstream of the liquid ring pump during operation in each of the configurations to allow for direct comparison of the analytical results. Table 2 provides the sampling schedule used for each of the demonstrations.

Prior to initiating the demonstration, two groundwater samples were collected and analyzed for total petroleum hydrocarbons (TPH) using EPA method SW846-8015 to provide a baseline for the degree of emulsification of the LNAPL in the process water that occurs during the bioslurper operation. In addition, during the conventional configuration testing, two samples of the floating solids within the OWS were collected.

TABLE 2. Sampling schedule for each demonstration.

Sample Type	Sampling Location	Sampling Frequency
Groundwater	Extraction well through drop-tube	Prior to beginning testing
Water in seal water tank	Seal water tank for the liquid ring pump	Daily
OWS effluent water	OWS effluent sampling port	Daily
Floating solids	OWS	Daily – if formed
Stack gas (with a handheld meter)	Sampling port in off-gas stack	Daily
Stack gas (with Summa a canister)	Sampling port in off-gas stack	Twice during each configuration

Samples of the bioslurper process water were collected periodically during each test to determine the TPH concentrations in the process stream of the bioslurper. At most of the demonstration sites, samples were collected from both the seal water tank of the liquid ring pump and the OWS effluent port. Typically, the process water samples were collected 24, 48, and 72 hours after the testing had begun in each configuration, and were analyzed for TPH extractable using method EPA SW846-8015.

TPH concentrations in the off-gas were quantified using a calibrated, handheld meter (Gastech GT 201) at 2, 4, 8, 12, and 24 hours after initiating each test. Thereafter, off-gas TPH concentrations were monitored at each 24-hour interval. The measurements were done by directly connecting the handheld meter to the stack until the TPH concentration stabilized to within ±10 ppmv.

To validate the off-gas measurements obtained from the handheld meter, two samples of the off-gas were collected during each test of the demonstration using 1-L, evacuated, polished stainless steel, Summa™ air-sampling canisters. Generally these samples were collected at 24 and 72 hours after the initiation of the test. These samples were analyzed using method EPA TO-13 for TPH.

At every system monitoring event, the volumes of floating solids and recovered LNAPL were recorded using a graduated cylinder or drum. Samples of the floating solids were collected for qualitative analysis of the appearance and consistency of the material. A qualitative assessment of the process water also was made to evaluate the degree of emulsification produced during the bioslurping process.

RESULTS AND DISCUSSION

The six sites at which the demonstrations were performed had different geologic conditions, contaminant levels (LNAPL thickness), and LNAPL types, which resulted in a variety of different average LNAPL and groundwater recovery rates. Table 3 presents the average LNAPL recovery in gal/day of pure product for each test of the six demonstrations. Table 4 present the average groundwater recovery rates for each test of the seven demonstration sites.

TABLE 3. LNAPL recovery rates (gal/day).

Site Location	1st Conventional Configuration	Dual-Drop Tube Configuration	2nd Conventional Configuration
NAS Fallon, NV	246	123	65
NAWS China Lake, CA	47	35	34
Bolling AFB, DC	2.8	3.8	2.8
Tyndall AFB, FL	0.52	4.9	0
NFD Point Molate, CA	0.36	0.12	0
MCAS Cherry Point, NC	0	0.34	0

TABLE 4. Groundwater recovery rates (gal/min).

Site Location	1st Conventional Configuration	Dual-Drop Tube Configuration	2nd Conventional Configuration
NAS Fallon, NV	4.4	4.3	4.0
NAWS China Lake, CA	2.9	1.6	1.1
Bolling AFB, DC	1.2	1.1	1.0
Tyndall AFB, FL	0.99	1.0	0.95
NFD Point Molate, CA[1]	-	-	-
MCAS Cherry Point, NC	4.5	4.4	4.6

The results from the demonstrations indicate that the dual-drop tube configuration did not affect the recovery of the LNAPL relative to operation in the conventional configuration. In general, the LNAPL recovery rates decreased throughout the demonstration, but did not significantly decrease when operating in the dual-drop tube configuration. In addition, the dual-drop tube configuration did not appear to alter the groundwater recovery rate.

The results of in-well separation testing demonstrated near complete elimination of the floating solids production and an average reduction in the TPH concentration of the process water of 99% compared to a conventional bioslurper

[1] NFD Point Molate groundwater recovery rate was not measured due to the metering in of water because of existing well design.

without the use of in-well dual-drop tube configuration. Table 5 presents the results from the analysis of the samples from the seal water tank.

TABLE 5. Analytical results of liquid ring pump effluent water.

	Average TPH Conc. (mg/L)		
Site Location	**Conventional Configuration**	**In-Well Configuration**	**Percent Reduction**
NAS Fallon, NV	4,944	32.5	99%
NAWS China Lake, CA	4,400	7.2	99%
Bolling AFB, DC	588	1.2	99%
Tyndall AFB, FL	278	ND	99%
NFD Point Molate, CA	42	2.6	98%
MCAS Cherry Point, NC	30	0.48	98%

In addition, the use of the dual-drop tube configuration resulted in an average reduction of 33% for off-gas TPH concentrations. Table 6 presents the results from the analyses of the off-gas. Sites that were contaminated with No. 2 Fuel Oil had an average reduction of 42% (Bolling AFB and MCAS Cherry Point), the sites contaminated with JP-4 had a reduction of 52% (Tyndall AFB) and the sites contaminated with JP-5 or a mixture of JP-5 and Bunker had an average reduction of 21% (NAS Fallon, NAWS China Lake, and NFD Point Molate).

The results have shown that the in-well dual-drop tube configuration is very effective at reducing the TPH concentrations in the aqueous and vapor effluent. It has also shown almost complete elimination of floating solids.

TABLE 6. Results of off-gas analyses.

	Average TPH Conc. (ppmv)		
Site Location	**Conventional Configuration**	**In-Well Configuration**	**Percent Reduction**
NAS Fallon, NV	2,940	2,350	20%
NAWS China Lake, CA	2,820	2,250	20%
Bolling AFB, DC	160	100	37%
Tyndall AFB, FL	7,950	3,850	52%
NFD Point Molate, CA	130	100	23%
MCAS Cherry Point, NC	140	74	47%

A long-term demonstration was performed at NAS Fallon to do a cost comparison between the conventional and the in-well separation configurations. This demonstration lasted for approximately four months and was performed on multiple wells. The analytical results for the pump effluent TPH concentrations show a 99% reduction in TPH when the in-well separation configuration is in use. Using the in-well separation configuration, the labor did not increase and provided a considerable cost savings in comparison to the conventional configuration.

In conclusion, the in-well separation configuration enhanced the bioslurper system performance. There was a 98-99% reduction in the TPH in the pump effluent water at all of the sites. There was an average reduction in the TPH in the off-gas of approximately 33%. There was near complete elimination of the floating solids and emulsion in the pump effluent water. The in-well separation technology works at a variety of sites that include tidal influence, varied geologic conditions, and varied LNAPL types and thickness. The in-well separation technology also provides a considerable cost savings when compared to the conventional configuration.

ACKNOWLEDGEMENTS

The authors would like to thank Cathy Vogel (Environmental Security Technology Certification Program), John Spivey (Air Force Research Laboratory), Art Fisher (NAS Fallon), Agnes Peters (Bolling AFB), Jim McDonald (NAWS China Lake), Joe McLernan (Tyndall AFB), Michelle Gallice Sondrup (Navy Southwest Division), and John Myers (MCAS Cherry Point). This demonstration was funded by Environmental Security Technology Certification Program.

REFERENCES

Battelle. 1997. *Engineering Evaluation and Cost Analysis for Bioslurping Initiative.* Prepared for U.S. Air Force Center for Environmental Excellence, Technology Transfer Division, Brooks AFB, TX.

Battelle. 1998. *Application Guide for Bioslurping.* Prepared for Naval Facilities Engineering Service Center, Port Hueneme, CA.

Parker, J.C. 1996. "Evaluating the Effectiveness of Product Recovery, Bioventing, and Bioslurping Systems." *Environmental Systems and Technology, Inc. News, 4* (Winter): 4-6. ES&T, Blacksburg, VA.

REMOVAL OF A LARGE LNAPL SPILL BY MULTIPHASE VACUUM EXTRACTION

Haimo Tonnaer, Charles G.J.M. Pijls, Berto J.B. Collet, Pedro J.P. Aarnink, Jentsje de Jong (Tauw BV, P.O.Box 133, 7400 AC-NL, The Netherlands)

ABSTRACT: At a large industrial site in central Poland, a multiphase vacuum extraction system (MPVE) has been installed for the removal of a large floating layer of aromatic solvents originating from past spills. The removal of the free phase light nonaqueous phase liquid (LNAPL) is probably the first in situ remediation in Poland. Prior to the design and installation of the MPVE system, pilot tests were carried out to determine the most suitable technique for the removal of the free product. The recovery techniques compared were: passive skimming, soil vapor extraction, and MPVE. The results of the tests were used for the design of the full-scale system. The removal system has been in operation since April 2000 and works effectively. In the first 8 months, about 41 m^3 of pure product has been removed.

INTRODUCTION

Since the early 1990s, the use of in situ techniques to remove light nonaqueous phase liquid (LNAPL) spills from soil has increased. The more conventional removal techniques are mainly passive and gravity-enhanced skimming. Through the years, more intensive techniques have been developed and applied. Examples are: soil vapor extraction, multiphase vacuum extraction (MPVE), also referred to as bioslurping, and steam injection. Bioslurping is still maturing, but has started to become a proven technology and is used more and more in full-scale applications (Lessard et al., 1999; U.S. Army, 1999). In this paper we present the results of a full-scale LNAPL-removal project with MPVE, including the results of the pilot tests, and the design and operation of the full-scale system.

SITE DESCRIPTION

The site in question is a paint production facility located in Poland. Due to spills at loading stations, underground storage tanks, and production plants, a large floating layer has developed. The major part of contamination consists of aromatic solvents (e.g. xylenes, ethylbenzene, trimethylbenzene) and light aliphatic compounds ($< C_{12}$).

At the start-up of the remediation project, the extent of the floating layer was not assessed in detail. It was decided to delineate the extent of the free product area in combination with the installation of a remediation system. The delineation during remediation found that the floating layer covers a total surface area of approximately 10,000 m^2. Based on the current knowledge, the amount of recoverable product is estimated at 100 m^3.

The site is situated in a complex hydrogeological system near a river. The soil in the eastern and southern sections of the site consists of quartinary sands in a layer until about 20 m below ground level (bgl). In this area, piezometric heads between 53 and 54 m above Baltic Sea Level (BSL) are found (3 to 4 m bgl). The more central and northern part of the site show a clay type isolation, interpreted as an upconing tertiary formation. This formation stretches to the river. The piezometric heads found in this area are 49 and 50 m +BSL.

The groundwater level contours show the presence of a "hydrological pit" at the central/northern part of the site (Figure 1). This phenomenon is caused by the presence of a ditch north of the site. Water from the ditch is infiltrating and causes a local increase of the water level over a distance of approximately 30 m. From the "pit" the contaminated groundwater is migrating in northern direction via the deeper groundwater under the ditch towards the river. Most of the pure product has ended up in this "pit".

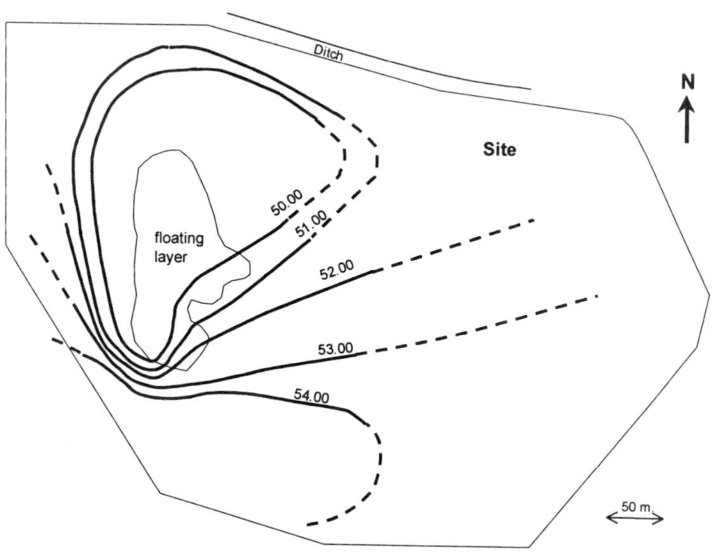

FIGURE 1. Groundwater level contours (m +BSL) and location of floating layer.

The aim of the remediation is to remove the LNAPL to an extent at which natural attenuation is capable of removing the remaining contamination in the groundwater. More specifically, it should prevent migration of contaminants from the site border. The performed site assessment proved that the natural redox conditions on site are aerobic, indicating a high potential for natural biodegradation in the groundwater.

The full-scale MPVE system will remove the major part of the floating layer, but capillary-trapped product will remain. If the remaining product is not removed, continuous leaching of contaminants will occur. Therefore, additional measures to remove the remaining product have been installed (air sparging).

BAIL-DOWN TESTS

Bail-down tests were used to estimate the actual LNAPL thickness in the soil as well as to determine product removal efficiency in case of passive skimming. The usefulness of bail-down tests for these purposes are described and theoretically founded by several authors (Aral and Liao, 2000; Zhu et al., 1993).

The recovery process is illustrated in Figures 2 and 3. The LNAPL floats as free product above the water capillary zone (a). At the start of the bail-down test, the free product is removed by pumping, causing the groundwater table to rise (b). Initially the LNAPL will flow into the well with a continuous rate until it reaches the top level of the water capillary zone (c). When that level is reached, the flow will slow down (d).

The maximum recovery rate that can be reached with passive skimming corresponds with the initial flow of LNAPL reached in the test (c). The actual thickness of the floating layer in the soil is assessed graphically, as the LNAPL thickness in the monitoring well between the end of phase (c) and the LNAPL at equilibrium end of phase (d).

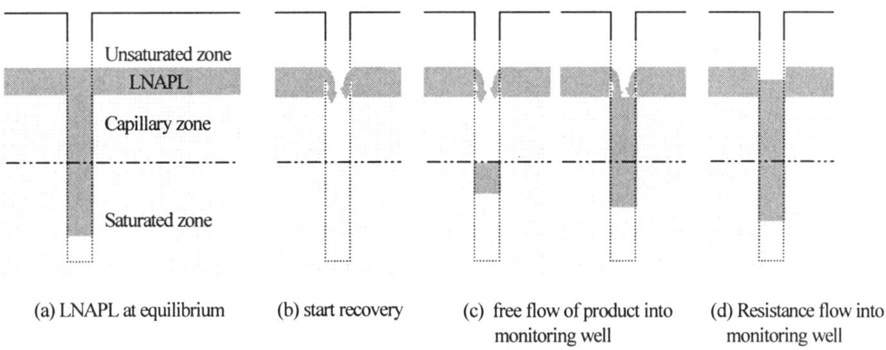

FIGURE 2. Bail-down and recovery of LNAPL in a monitoring well.

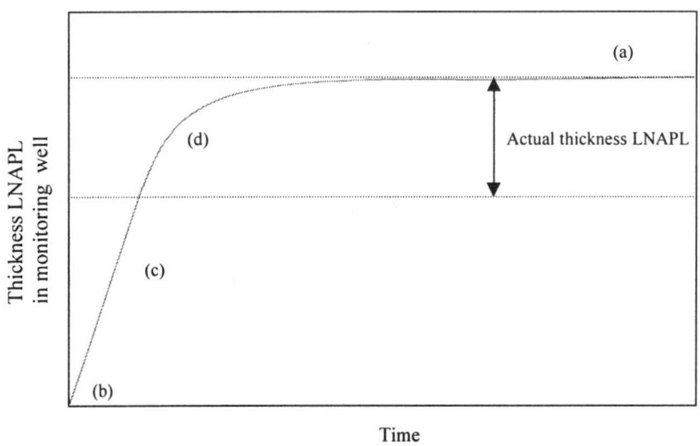

FIGURE 3. Graphical illustration of recovery process.

Figure 4 shows the results of a bail-down test performed on site. The well is located in coarse, permeable sand. During the start of the recovery, the LNAPL is removed (b) and the groundwater table rises. After LNAPL removal, the initial LNAPL flow in the monitoring well (c) is 0.35 cm/min. The diameter of the monitoring well is 9 cm, which results in a maximum LNAPL recovery rate of 1.05 kg/hr. The actual thickness of the LNAPL is estimated at 13 to 14 cm.

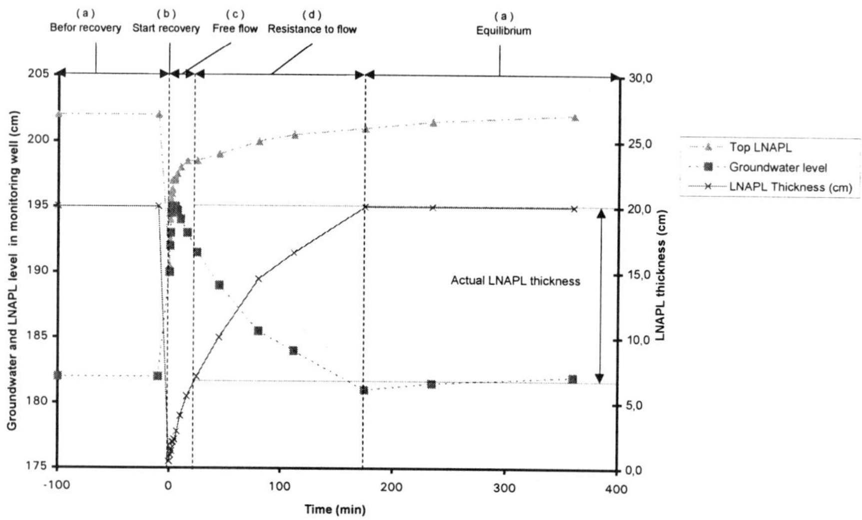

FIGURE 4. Bail-down test in a monitoring well.

PILOT TRIAL

Prior to the design of the full-scale removal system, pilot tests were carried out to select the most suitable technique for the cost-effective removal of the product and to obtain the relevant design parameters for a full-scale removal system. Three in situ techniques were compared: passive-skimming, soil vapor extraction (SVE), and MPVE. For determining the removal efficiency of passive-skimming the bail-down test as described in the previous section was used. The tests were carried out over a period of several days. The results of the pilot tests are summarized in Table 1.

TABLE 1. Removal rates of different techniques during pilot trial (per well).

	Removal rates (kg/hr)		
	Skimming	**MPVE**	**SVE**
Removal with soil vapor	-	0.31	0.23
Removal as pure product	1.05	3.00	-
Total production rates	**1.05**	**3.31**	**0.23**

The results of the pilot trial showed that with MPVE the product recovery rate is approximately 3 times higher than the recovery rate with skimming. The radius of influence was determined by under pressure measurements and estimated on 5 meters, which is much larger compared to skimming (1 to 2 meters). Based on the results it was decided to use MPVE as the preferred floating layer removal technique.

FULL SCALE SYSTEM

Based on the extent of the floating layer the remediation system consists of about 150 MPVE wells for the removal of pure product. Delineation is established simultaneously with the installation of the MPVE wells. The same amount of air sparging wells is installed for release and removal of trapped capillary product and dissolved contaminants. After the major part of the free product is removed, the MPVE system will be transformed to a soil vapor extraction system. The soil vapor extraction system will remove the evaporated product from the capillary zone as well as some residual product in the unsaturated zone.

The MPVE system is operated in sections. Each section, existing of 8 to 10 extraction wells, is in operation for about 2 weeks. After all sections have been treated the cycle is repeated, until the recovery rate has dropped to a low level. At this point the air sparging system is started up near the section in question, in order to mobilize the remaining product. The MPVE-system stays in operation as long as product is mobilized.

For the multiphase vacuum extraction a liquid ring pump is used with a capacity of 280 Nm^3/h. The extracted product is collected in a settling tank with a volume of 11 m^3. The settling tank is situated between the wells and the pump. Thus, there is no contact between the pump and the recovered product. The extracted air which contains high concentrations of contaminants (10 to 30 g/m^3) is treated in a biofilter, after buffering in an activated carbon filter. The recovered product is reused by an external company.

A considerable amount of the product is extracted as an emulsion. The ratio between product, emulsion, and groundwater is about 10:20:70. In the settling tank, the emulsion settles into layers of pure product, residual emulsion, and water. The residual emulsion is treated with hydrochloric acid, which results in an almost complete separation of product and water and some remaining biomass and ferruginous sludge.

The full-scale system has been in operation since the end of April 2000. During the first eight months, about 41 m^3 (33 tons) of pure product (density 0.81 kg/dm^3) was extracted. A significant part of product is removed as vapor. The results are shown in Figure 5.

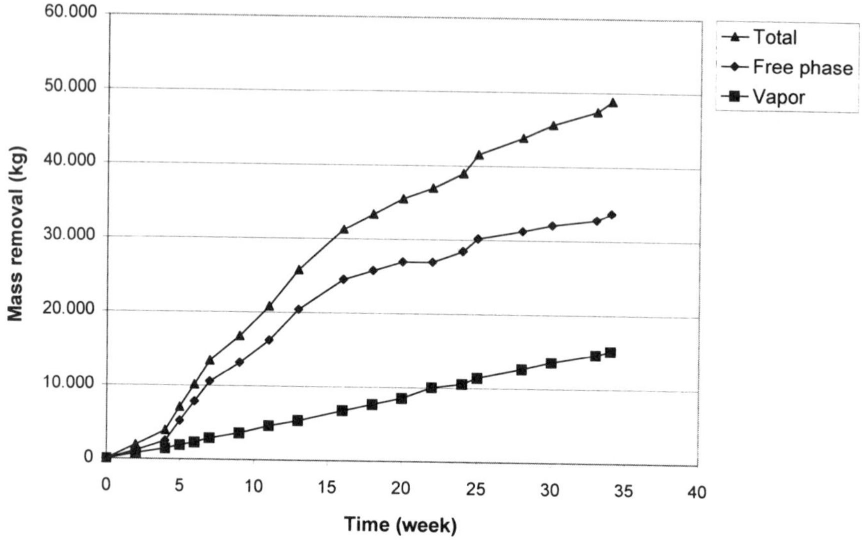

FIGURE 5. Mass removal MPVE system.

CONCLUSIONS

After 8 months of operation it is clear that the MPVE system is operating properly, in line with the expectations from the pilot tests. It is expected that the estimated amount of 100 m^3 of free product will be recovered in this year. The results of full-scale removal system prove that MPVE is very well suited for the removal of light LNAPLs and should be considered as a mature technology.

REFERENCES

Aral, M.M. and B. Liao. 2000. "LNAPL Thickness Interpretation Based on Bail-Down Tests". *Ground Water. 38*(5): 696-701.

Lessard, D. F., D.L. Rizutto, and D. Bonaldo. 1999. "Optimizing LNAPL Recoveries Using Combined Extraction Technologies." In B.C. Alleman and A. Leeson (Eds.). *In Situ Bioremediation of Petroleum Hydrocarbon and Other Organic Compounds*, pp. 163-168. Battelle Press: Columbus, Ohio.

U.S. Army Corps of Engineers. 1999. "Multi-Phase Extraction." Prepared for the Department of the Army U.S. Army Corps of Engineers, Washington D.C. 20314-1000.

Zhu, J.L, J.C. Parker, D.A. Lundy, and L.M. Zimmerman. 1993. "Estimation of Soil Properties and Free Product Volume From Baildown Tests". *Ground Water Management 17*(1): 99-111.

DESIGNING HIGH- EFFICIENCY RECOVERY WELLS FOR MULTI-PHASE EXTRACTION SYSTEMS

Matthew P. Peramaki, P.E. and Brad A. Granley, E.I.T.
(Leggette, Brashears, & Graham, Inc.
St. Paul, Minnesota)

ABSTRACT: Dewatering of soils containing residual non-aqueous phase liquids and rapid removal of these contaminants via volatilization are critical to the success of a multi-phase extraction (MPE) system. Researchers have proposed that the primary operational metrics of an MPE system should be maximum applied (well casing) vacuum to promote the flow of soil vapor, and maximum drawdown/ground-water flow rates in extraction wells to dewater the surrounding aquifer (Peargin, 2000). If these objectives are not met, then the MPE system will be inefficient or ineffective.

Meeting the primary operational goals for an MPE system require that engineers maximize the efficiency of the system by minimizing the overall pressure losses. This is critical because vacuum pumps have a practical vacuum ceiling of ~25 inches of mercury. Higher horsepower vacuum pumps are not capable of generating additional vacuum.

Previous research has shown that the most significant pressure losses in an MPE piping network occur within the vertical drop tube (Peramaki et al. May 2000). In many cases, the pressure losses in a drop tube can exceed the sum of the pressure losses from the rest of the piping network. In fact, vertical-piping pressure loss has been measured to be in excess of 20 times that in horizontal piping for a given set of flow conditions. (Peramaki et al. May 2000). Therefore, designing an efficient extraction well and associated drop tube is critical to effective MPE system design.

In order to better understand the significance of a drop tube on remedial effectiveness, field tests were conducted at an active MPE remediation site. Results of these tests show that a properly sized drop tube can nearly double ground-water and soil-vapor flow rates. In addition, the impact of installing "slots" in drop tubes was investigated.

VERTICAL TWO-PHASE FLOW REVIEW

Vertical Two-Phase Flow Patterns. MPE systems extract both soil vapor and ground water simultaneously in one pipe. The mixture of these two fluids results in reproducible flow "patterns" that are a function of the volumetric ratio of the gas and liquid in the pipe. Vertical two-phase flow patterns include bubble flow, slug flow, churn flow, ripple flow and annular flow. Churn, ripple and annular flow patterns are common in the typical MPE system. Visual representations of typical vertical flow patterns are presented in Figure 1.

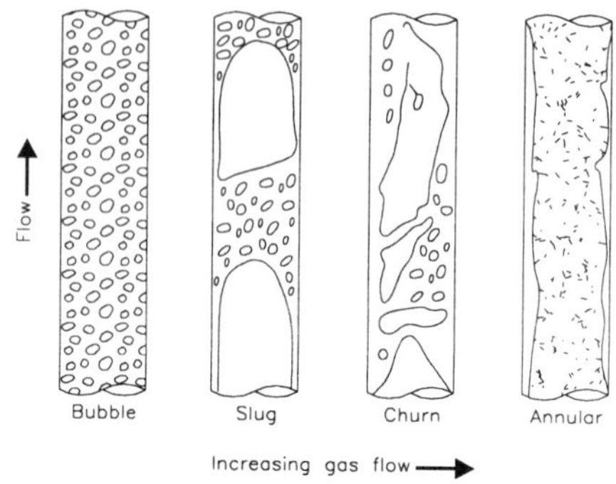

Figure 1 – Flow Patterns in Vertical Piping

Source Figure 1: Taitel, Y., et al. Reproduced with permission of the American Institute of Chemical Engineers.

Flow Pattern Prediction. Two-phase flow patterns can be predicted utilizing a chart prepared by Govier et al. (1957) (Figure 2). By calculating the gas to liquid volume ratio of the two-phase flow in the pipe and the superficial velocity of the liquid flowing in the pipe, designers can use Figure 2 to predict vertical flow patterns in small diameter pipes (0.5 to 2 inches in diameter).

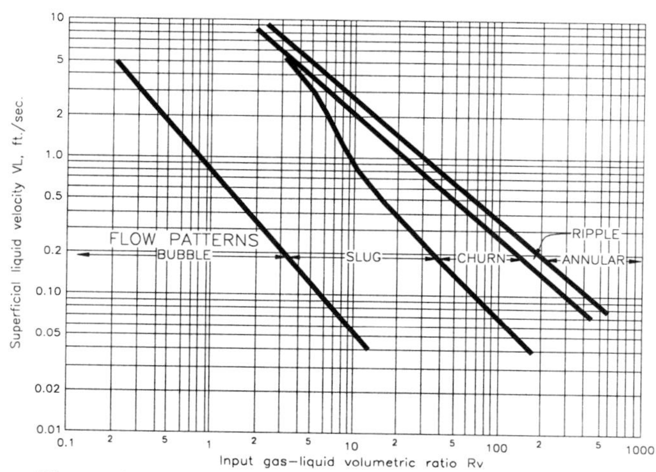

Figure 2 – Two-Phase Flow Pattern Prediction Chart

Source: Govier, G.W., B.A. Radford and J.S.C. Dunn, (1957). "The Upwards Vertical Flow of Air Water Mixtures: I. Effect of Air and Water-Rates on Flow Pattern, Holdup and Pressure Drop," *Can J. Chem. Eng.*, *35*, 66.

NOTE: This chart was developed using a 1.025-inch diameter smooth pipe with air and water at 70 degrees F at 36 pounds per square inch actual.

Significance of Two-Phase Flow Patterns. Figure 3 presents a vertical two-phase flow pressure drop curve. The total vertical pressure loss has two components, the losses due to friction and the losses due to hydrostatic lift requirements. When these two components are summed, the resulting curve has three inflection points, two of which have significance in the design of MPE systems. The important pressure drop inflection points exist in early churn flow (pressure drop maximum) and within ripple flow (pressure drop minimum). Figure 3 displays the total pressure drop over various air to water ratios.

Churn flow is the most chaotic two-phase flow pattern. The maximum inflection point shown on Figure 3 during early churn flow is due to friction losses resulting from the chaotic interaction between the gas and liquid phases. Therefore, vertical piping, including extraction well drop tubes, should be designed to avoid churn flow and to operate in the ripple flow pattern when possible. Churn flow can also produce significant "water hammer" that has the potential to damage plastic piping. A more detailed discussion of vertical flow patterns is presented by Peramaki et al. (March 2000).

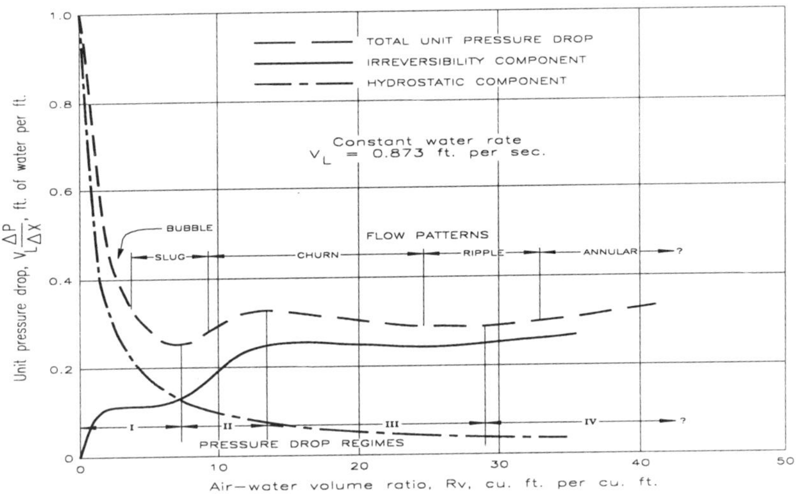

Figure 3. Vertical Two-Phase Flow Pressure Drop Curve

Source: Govier, G.W., B.A. Radford and J.S.C. Dunn, (1957). "The Upwards Vertical Flow of Air Water Mixtures: I. Effect of Air and Water-Rates on Flow Pattern, Holdup and Pressure Drop," *Can J. Chem. Eng.*, *35:* 66.

Vacuum-Lift Limits for MPE Systems - Deep Extraction. Traditional vacuum-lift pumping techniques are limited to extracting ground water from a depth less than 28 feet below grade level. This is because vacuum pumps are limited to producing approximately 25 inches of mercury (~28 feet of water) vacuum. It takes 1 foot of water vacuum (0.88 inches of mercury) to lift a full column of water 1 foot in elevation (Figure 4). However, if a gas (such as soil vapor) is added to the water column, the hydrostatic component of the total pressure drop is greatly reduced due to the decrease in the average fluid density within the column.

MPE systems extract both soil vapor and ground water simultaneously through a single pipe. Therefore, the addition of soil vapor into a ground-water column within a drop tube will greatly reduce the pressure losses compared to traditional vacuum-lift pumping techniques (Figure 4). Therefore, if an aquifer/MPE well produces sufficient soil vapor (or air is mechanically introduced to the well) when vacuum is initiated, ground water can be extracted from far in excess of traditional vacuum-lift limits. Theoretically, MPE may be able to extract ground water from depths exceeding 100 feet (30.5 m) below grade level if equipment and piping are sized appropriately.

Figure 4. Two-Phase Flow Pressure Drop Curve – Including Traditional Vacuum-Lift Dewatering and MPE Operational Ranges

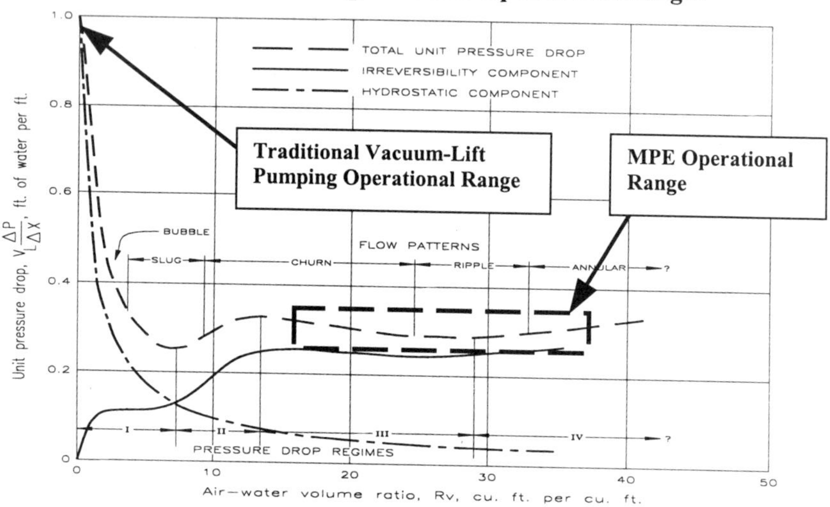

Source: Govier, G.W., B.A. Radford and J.S.C. Dunn, (1957) "The Upwards Vertical Flow of Air Water Mixtures: I. Effect of Air and Water-Rates on Flow Pattern, Holdup and Pressure Drop," *Can J. Chem. Eng.*, 35: 66.

Two-Phase Pressure Loss Prediction. Pressure loss calculations are more complex for two-phase flow than for single-phase flow due to interaction between the vapor and liquid phases. Two-phase flow pressure losses in an MPE system can be as much as ten times the sum of the superficial gas and liquid pressure drops. Minimizing overall pressure losses in an MPE system is important because vacuum pumps can only generate ~25 inches of mercury. Specifying a larger horsepower vacuum pump cannot make up for pressure losses not accounted for by a system designer. A more comprehensive discussion of two-phase pressure drop is presented in Peramaki et al. (May 2000).

EFFICIENT MPE WELL DESIGN

MPE wells are similar to ground-water production wells; both are designed to produce maximum fluid flow rates. For maximum efficiency in ground-water wells, the gravel pack and well screen are among the most the critical design parameters. For MPE wells however, the diameter of the MPE well casing and the drop tube within the well may be the most critical design parameters.

Drop Tube Significance. In deep MPE systems (i.e. those installed at sites having a water table greater than 20 feet below grade level), this relatively short length of pipe can account for 50 percent (or more) of the total pressure losses in the entire piping network (Peramaki et al, May 2000). Pressure losses associated with two-phase vertical flow can easily be 20 times that of the horizontal flow, even in pipe of the same diameter (Peramaki et al. May 2000).

Field tests were conducted at an active MPE remediation site to show the effects that drop tube diameter and slotting have on ground-water and soil-vapor recovery rates. Results of these tests indicate that using the proper drop tube design is critical to overall system performance.

FIELD TESTING AND DATA COLLECTION

The data utilized to investigate the significance of drop tube diameter and slotting was collected from tests conducted on an active MPE remediation system. Pressure loss in the drop tube was precisely quantified by measuring the vacuum directly from within the MPE well casing and at the top of the drop tube. The difference in the vacuum measurements is therefore equal to the pressure loss through the known length of the drop tube. Soil-vapor flow rates (measured after the vacuum pump) were converted to standard conditions, and ground-water flow rates (measured by recording the changes in water level in the vapor/liquid separator over time) were recorded along with observed flow patterns.

Measurements were recorded over a range of soil-vapor and ground-water flow rates through three drop tube diameters, each having three slot patterns. All testing was conducted in a single 4-inch diameter MPE well. The depth to water was approximately 35 feet (10.7 m) below grade level and the base of the well is approximately 40 feet (12.2 m) below grade level.

Drop tubes diameters of 1-inch, 1.5-inches and 2-inches were investigated. The effect of installing slots in drop tubes was conducted by comparing data collected using solid drop tubes with no slots (designated Pattern A), drop tubes having 20 slots over the lower 4 feet (Pattern B) and drop tubes having 34 slots over bottom 7 feet (Pattern C). Each of these slot patterns was cut into drop tubes for all three diameters, resulting in a total of nine separate tests. NOTE: A 'slot' consists of a cut on both sides of a drop tube with a hack saw. The two cuts were staggered slightly to preserve strength in the drop tube. For reasons of practicality, the duration of each test was limited to approximately 30 minutes and steady-state conditions were not achieved. The collection of transient data may account for outlying data points.

RESULTS

Tabulated Results. Results of the nine field tests are tabulated in Table 1.

Table 1 - Summary of Field Data

Drop Tube Dia.	Pattern A (no slots)					Pattern B (20 slots)					Pattern C (34 slots)				
	Pdt	Pwc	Pd	Qw	Qa	Pdt	Pwc	Pd	Qw	Qa	Pdt	Pwc	Pd	Qw	Qa
	"Hg	"Hg	"Hg	gpm	scfm	"Hg	"Hg	"Hg	gpm	scfm	"Hg	"Hg	"Hg	gpm	scfm
1"	16.4	4.5	11.9	2.2	18.2	14.2	3.9	10.3	1.7	20.0	15.6	6.1	9.5	1.6	15.6
1.5"	11.5	5.7	5.8	3.5	22.3	13.1	7.1	6.0	3.0	24.6	14.0	8.2	5.8	2.6	14.0
2"	12.3	5.2	7.1	3.6	25.7	12.8	7.2	5.6	2.9	- - -	14.0	8.5	5.4	3.0	27.0

Pdt = vacuum at top of drop tube
Pwc = vacuum in well casing
Pd = pressure loss through drop tube
Qw = ground-water flow rate
Qa = soil-vaopr flow rate
"Hg = inches of mercury
gpm = gallons per minute
scfm = standard cubic feet per minute

Significance of Drop Tube Diameter. The impacts of varying drop tube diameter are presented in Figures 5 and 6. Figure 5 shows that with an increasing diameter, the pressure losses through the drop tube decreases. This results in an increase in the well casing vacuum as shown in Figure 6.

Figure 5 - Effect of Diameter on Pressure Loss Through Drop Tube

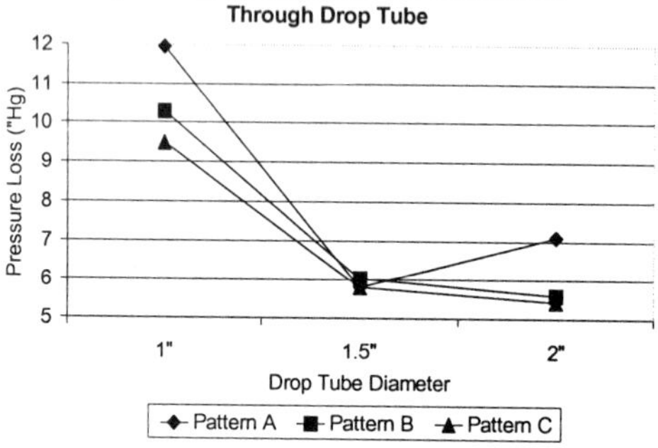

Figure 6. Effect of Drop Tube Diameter on Applied (Well Casing) Vacuum

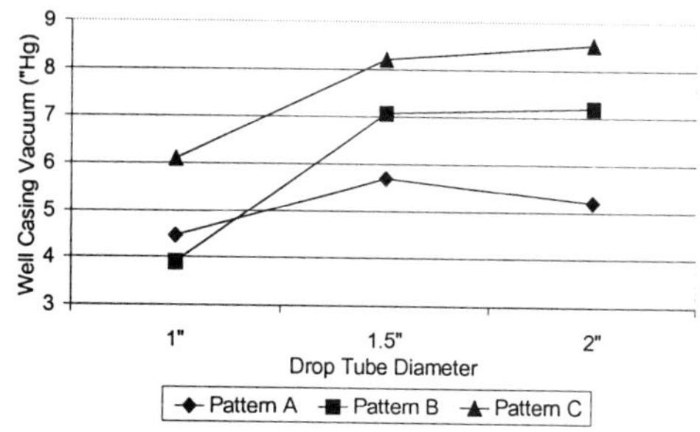

Figures 7 and 8 indicate that the increase in the applied (well casing) vacuum results in significant increases in ground-water and soil-vapor flow rates.

Figure 7 - Ground-Water Flow Rate vs Drop Tube Diameter

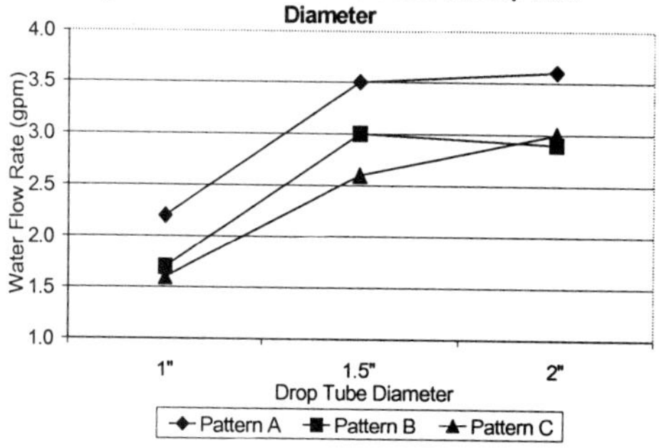

Figure 8 - Soil-Vapor Flow Rate vs Drop Tube Diameter

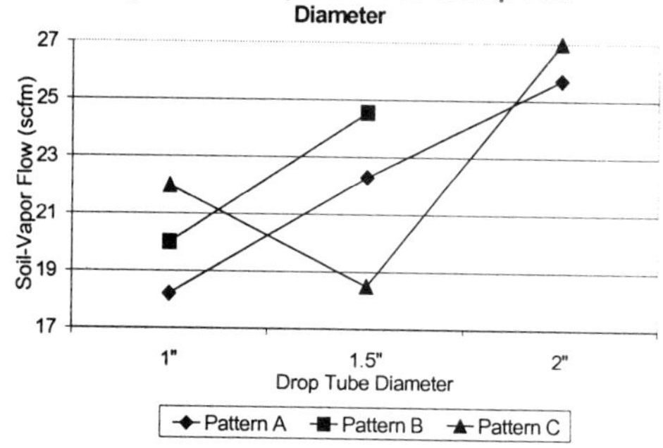

Figure 9 indicates that increasing the numbers of slots in a given drop tube diameter results in a decrease in ground-water production and an increase in soil-vapor flow rates. Figure 10 shows that additional drop tube slots result in a higher applied vacuum in the well casing.

Figure 9 - Ground-Water and Soil-Vapor Flow vs Number of Slots

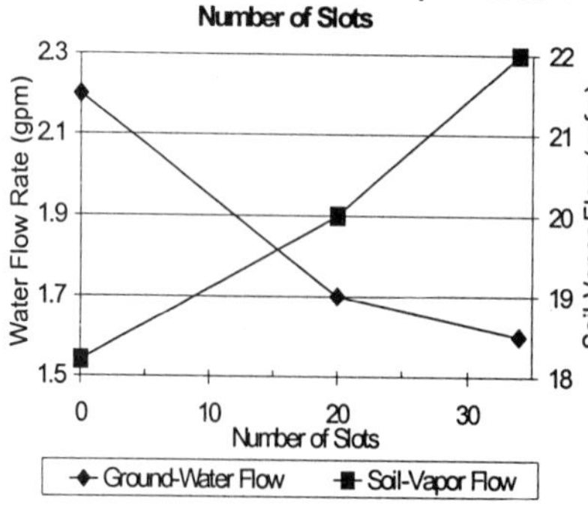

Figure 10 - Well Casing Vacuum vs Number of Slots

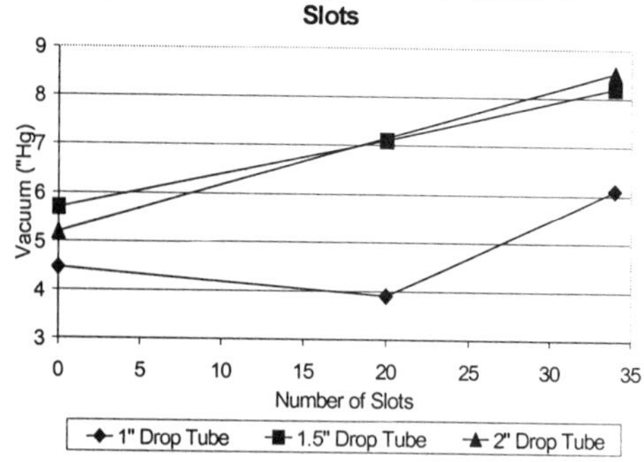

Figures 11 and 12 present data collected from the three tests on drop tubes with no slotting.

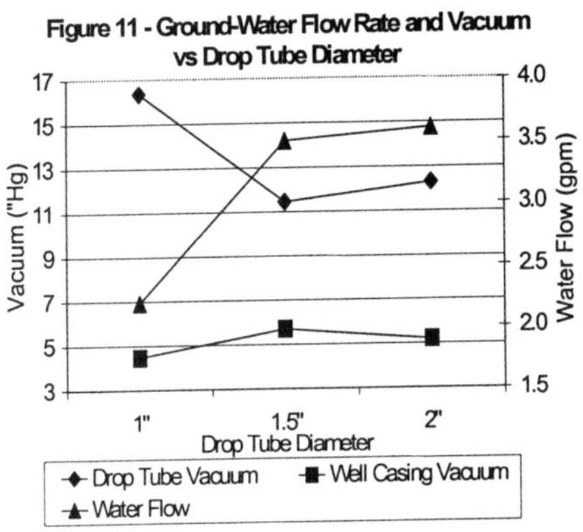

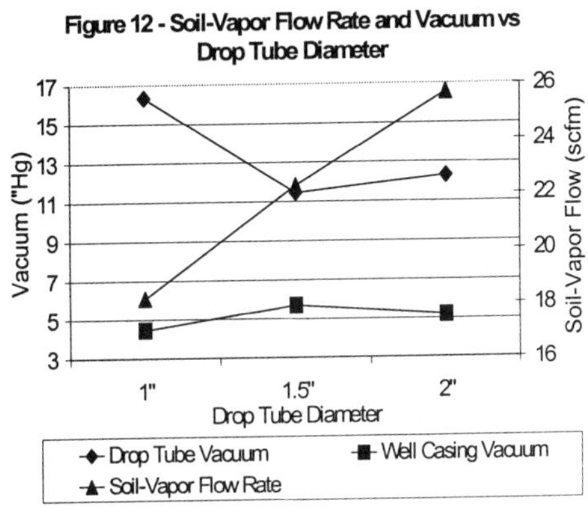

Figures 11 and 12 graphically display that a 1-inch diameter drop tube produced excessive pressure losses, which resulted in decreased ground-water and soil-vapor production. The 1.5-inch diameter tube has significantly lower pressure losses and a resulting increase in water and vapor flow rates.

Figure 11 also indicates that the 2-inch drop tube may have been sized sufficiently to allow the extraction of ground water near the yield rate of the aquifer. Increasing the drop tube diameter beyond 2-inches may not have resulted in a significant increase in ground-water production.

Figure 12 shows that the soil-vapor flow rate continued to increase with the larger diameter drop tube.

Table 2 summarizes the remedial benefits of a properly designed MPE well drop tube. The changes in operational metrics presented in the table are the difference between the least efficient diameter and slot pattern and the most efficient diameter and slot pattern.

Table 2 - Changes in Key MPE Operational Metrics Due to Proper Drop Tube Sizing

Benefit	Pattern A (no slots)	Pattern B (20 slots)	Pattern C (34 slots)
Ground-Water Flow Rate Increase (%)	64%	76%	88%
Soil-Vapor Flow Rate Increase (%)	41%	23%	93%
Drop Tube Pressure Loss Reduction ("Hg)	6.1	4.7	4.1
Drop Tube Pressure Loss Reduction (%)	51%	46%	43%

DISCUSSION

Designing Efficient MPE Wells. In order to design an efficient MPE well, engineers must begin with the drop tube. Designers can utilize expected maximum soil-vapor and ground-water flow rates as determined from pilot testing to calculate pressure losses through a range of pipe diameters using the methods outlined in Peramaki et al. (May 2000).

One additional design consideration is size of the outer casing of an MPE well. Once the drop tube has been properly designed, the outer well casing should be sized to ensure sufficient annular space between the outer well screen and the inner drop tube for the expected soil-vapor flow rates. If the annular space between the outer well casing and the drop tube are insufficient, then excessive pressure drops will result prior to any fluid entering the drop tube.

Drop Tube Slotting. Slotting an MPE well drop tube may be advantageous at a site when the aquifer to be remediated can be effectively dewatered at relatively low ground-water flow rates. In this case the higher soil-vapor flow rate is advantageous. However, engineers must be confident that the design dewatering rates are trustworthy. Otherwise, the resulting drop in ground-water flow rates will be detrimental to remedial effectiveness.

REFERENCES

Govier, G.W., B.A. Radford and J.S.C. Dunn, (1957). "The Upwards Vertical Flow of Air Water Mixtures: I. Effect of Air and Water-Rates on Flow Pattern, Holdup and Pressure Drop," *Can J. Chem. Eng.*, *35:* 66.

Peargin, Tom R., 2000. "Engineering and Hydrogeologic Factors Affecting Successful Application of Multi-Phase Extraction to LNAPL Source Zone Remediation", *2000 Contaminated Site Remediation Conference.* Melbourne, Australia.

Peramaki, M., Nelson, M., Granley, B., Carlson, R., March 2000. "The Significance of Two-Phase Flow Regimes in Designing Multi-Phase Extraction Systems," *American Institute of Chemical Engineers Conference: Remedial Process Optimization*, Atlanta, GA.

Peramaki, M., Granley, B., Carlson, R., and Nelson, M. May 2000. "Designing Piping Networks for Multi-Phase Extraction Systems", Physical and Thermal Technologies, pp. 101-112. Battelle Press, Columbus, Ohio.

Taitel, Y., D. Bornea and A.E. Dukler, (1980) "Modeling Flow Pattern Transitions for Steady Upward Gas-Liquid Flow in Vertical Tubes," *AIChE J., 26*, 345.

SOURCE REDUCTION FOR NATURAL ATTENUATION USING TWO-PHASE EXTRACTION

Mehmet Pehlivan and Gregory Buchanan
(Tait Environmental Management, Inc., Santa Ana, California)

ABSTRACT: This paper presents the results of a case study using Bubblex[SM1] two-phase extraction (BTPE) method for removal of hydrocarbons in preparation of the site for development by risk based closure or natural attenuation. The site selected for this study is located in Orange County, California, where a release from an underground storage tank (UST) contaminated the soil and shallow groundwater. Fourteen (14) groundwater wells were installed on-site for periodic monitoring. A 90-day BTPE pilot test began in mid January of 2000. Initially two wells (EW-1, EW-2) were connected to the extraction system. The BTPE involved the use of vacuum developed in the extraction system to draw groundwater and soil vapor through an extraction pipe. The pipe was open to the screened portion of the well, which allowed the BTPE to create turbulent flow, stripping volatile organic compounds (VOC's) from the water. These vapors were subsequently destroyed in the catalytic oxidizer. The extracted water was polished by two carbon vessels and discharged to the storm drain under National Pollution Discharge Elimination System (NPDES) permit guidelines. During the first two months of operation, approximately 72,000 gallons of groundwater were extracted and treated. After two months of operation, three additional groundwater wells and two vapor wells were connected to increase the recovery rate and zone of influence of the treatment cell. By the end of the third month, all wells showed a significant reduction in benzene concentrations in groundwater. Benzene concentrations in two of the wells (MW-1 and MW-9) dropped from 6,000 parts per billion (ppb) to less than 1 ppb, and 260 ppb to less than 1 ppb, respectively. All but two wells (MW-5 and MW-6) showed a marked decrease in total petroleum hydrocarbons as gasoline (TPH-G) concentrations in groundwater. After completion of first 90-day test, BTPE test was continued for additional 90 days to extend the observed significant reductions in hydrocarbon concentrations in groundwater. After operating the BTPE system for six months, a rebound test was performed. The results of the rebound test indicated that static vapor and water concentration in wells after 15 days of non-operation were low enough to consider risk based closure and/or natural attenuation for the final phase of cleanup. Currently, site closure monitoring is being performed to obtain final closure from the regulatory agency.

INTRODUCTION

The purpose of this paper is to provide a case study of Bubblex[SM] two-phase extraction-stripping (BTPE) method used for source reduction. The uniqueness and interest to the scientific community is that a new method of two-phase extraction was used with some enhancements, which allowed us to use a low vacuum system for extraction and stripping of volatile hydrocarbons (VOCs).

---

[1] Bubblex is a service mark of Tait Environmental Management, Bubbling Extraction Method, a patented method (Patent No. 5,906,204) for extraction and in well stripping of volatile organic compounds.

SITE SETTING

The site is located in the Orange County, California, was used as a car dealership prior to 1992. A 2000 gallon unleaded gasoline underground storage tank was used to fuel cars. During the removal of the tank in 1988, contaminated soil and shallow groundwater was encountered. A light non-aqueous phase liquid (LNAPL) layer was observed up to 4 feet thick in some wells. Passive LNAPL recovery was performed for over 7 years until the LNAPL was completely removed from the wells. Shallow perched groundwater rests in the silty sand and sand zone at 10 feet below ground surface (bgs). A clay layer, over 5 feet thick with permeability less than $1.0X10^{-7}$ cm/sec underlies the perched shallow groundwater zone at 20 feet bgs. The natural groundwater flow is to the west with a mild gradient of 0.006.

REMEDIATION PROCESS

BTPE method was used to cleanup both soil and groundwater at this site. The BTPE method allows a limited amount of air/vapor from the unsaturated zone and a limited amount of water from the saturated zone to enter an extraction pipe through a patented, modified screen (Pehlivan et al, 1999). Both vapor and water flow in a combined stream within the pipe until it reaches the separation tank. The new design of the extraction pipe screen facilitates lifting water from depths greater than 33 feet (10.06 meters), extracting vapor from the vadose zone, and stripping VOCs from the extracted water. Up to 99% of VOCs stripping efficiencies were observed in the extraction pipe during previous studies (Pehlivan et al, 2000a and 2000b). The system that was operated at this site consisted of a thermal/catalytic oxidizer with a positive displacement blower that is capable of applying 150 inches of water column (WC) vacuum and producing 200 cubic feet per minute (CFM) vapor flow. Applied vacuum during the pilot test ranged between 80 and 120 inches of WC. These vacuum ranges were enough to lift the water from 10 ft bgs to 5 ft above ground surface and produce up to 5 gallons per minute of water flow from five extraction wells. The extracted water and vapor were fed into a separation tank. The vapor is sent to a thermal/catalytic oxidizer, and the water is discharged through a granulated activated carbon (GAC) polishing unit to a local storm drain under a National Pollution Discharge Elimination System (NPDES) permit. The extraction piping was modified to allow water to drain under gravity flow into the air/water separation tank after extraction from the wells. This modification reduced the friction loss in the horizontal pipes and greatly increased the recovery of groundwater.

FINDINGS

Observed vacuum readings showed a declining trend as the water level was drawn down because a larger vadose zone was available for vapor flow (Figure2). An estimated total of 1375 lbs of hydrocarbon was extracted through the vapor stream. The hydrocarbon removed in the vapor stream included the hydrocarbon that stripped off from the water during extraction (Figure 3). Approximately 6 lbs of hydrocarbon was adsorbed by the granular activated carbon. Due to stripping in the extraction pipe, the carbon use was minimal.

The total petroleum hydrocarbon as gasoline (TPH-G) and benzene concentrations in groundwater declined as the system continued to operate. When the system was started, only two extraction wells (EW-1 and EW-2) were operating. In March 2000, three monitoring wells (MW-5, MW-6, and MW-9) were also connected to the system to enhance the water and vapor removal. The system was shut down for two weeks to see the rebound concentrations in August 2000. The TPH-G concentrations in two wells (MW-1 and EW-1) increased after two weeks of non-operation. TPH-G and benzene concentrations in most wells remained below 1000 and 10 micrograms per liter (ug/l) respectively after the rebound test. The average benzene concentrations remained below 10 ug/l in groundwater as observed in quarterly groundwater samples collected on December 7, 2000, three months after the system shut down (Figure 4).

CONCLUSIONS

Our conclusions are summarized below:

- The two-phase extraction was effectively applied at this site without using a liquid ring pump or high vacuum.
- Elevating the extraction pipes and having water flow into the knockout tank under gravity drainage increased the recovery of water and reduced the friction loss in the extraction pipe.
- The carbon use was minimal due to stripping in the extraction pipe.
- The two-phase extraction method was able to reduce elevated concentrations of benzene and TPH-G in groundwater significantly to consider site closure.

ACKNOWLEDGMENTS

We are grateful to Dr. Ken Tait and Mr. Tom Tait for supporting the project and research associated with the patent. We are thankful to Tait Environmental Management Inc. personnel for their help during the fieldwork and other related activities.

REFERENCES

Pehlivan, Mehmet, Neal Beaty, Tom Dixon, Scott Ek, Richard Coffman, 2000a, "MTBE and BTEX Stripping During Bubblex^{SM} Two-Phase Extraction", G. B. Wickramanayake, A.R. Gavaskar, J.T. Gibbs, and J.L. Means (Eds.) *Case Studies in the remediation of Chlorinated and Recalcitrant Compounds (C2-7), pp 41-48*, Battelle Press, Columbus, Ohio

Pehlivan, Mehmet, J. Neal Beaty, 2000b, "Remediation of Petroleum Hydrocarbons Using Bubblex^{SM} Two-Phase Extraction Method – A Case History", *Proceedings of ETCE/OMAE 2000, February 14-17, 2000, New Orleans, Louisiana*, Paper No. ETCE2000-10153, A CD-ROM Published by ASME, New York, NY.

Pehlivan, Mehmet, James R. Miller, 1999, "Method for Bubbling Extraction of Groundwater", *U.S. Patent Office*, Patent No. 5,906,241, Issued on May 25, 1999.

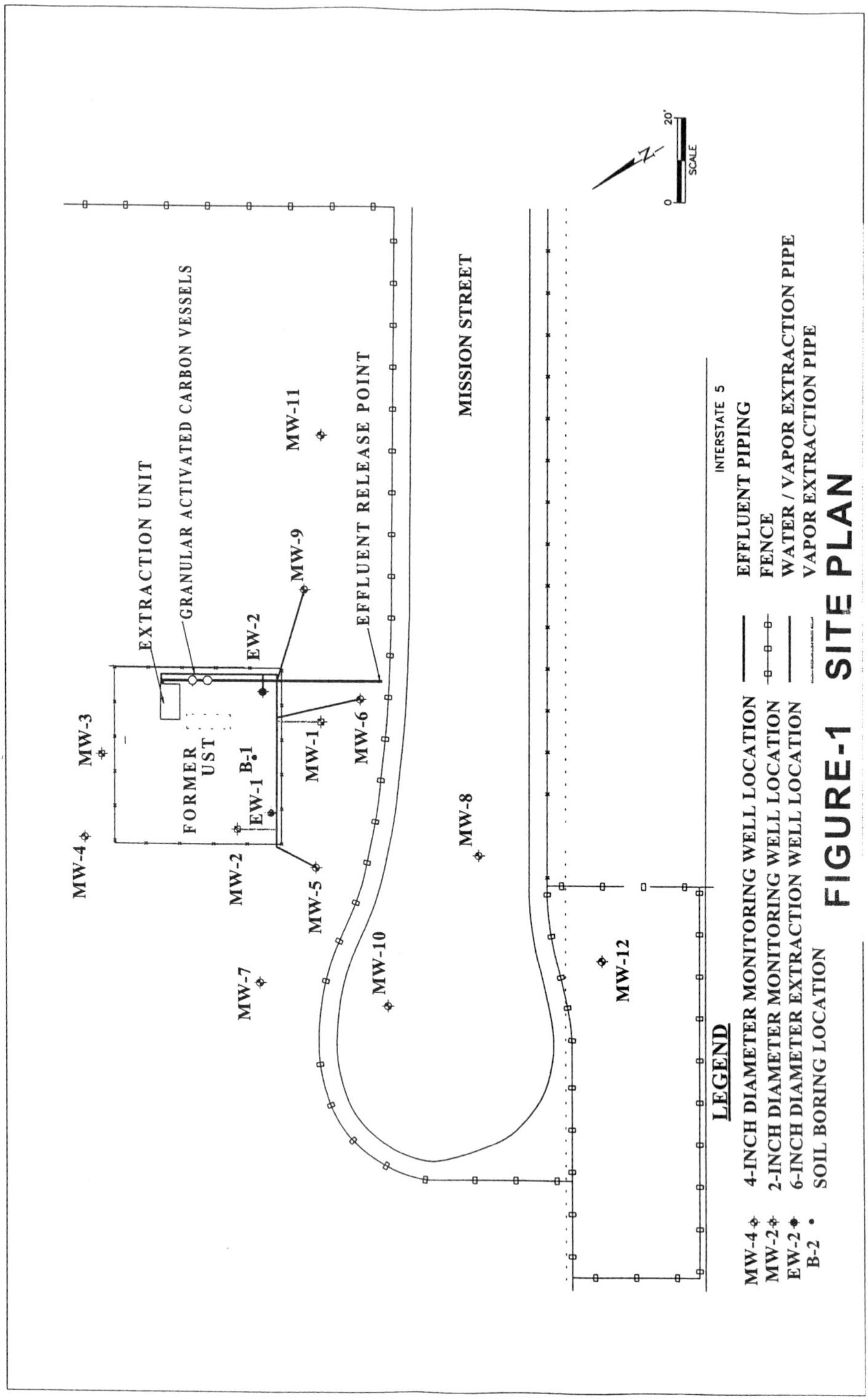

FIGURE-1 SITE PLAN

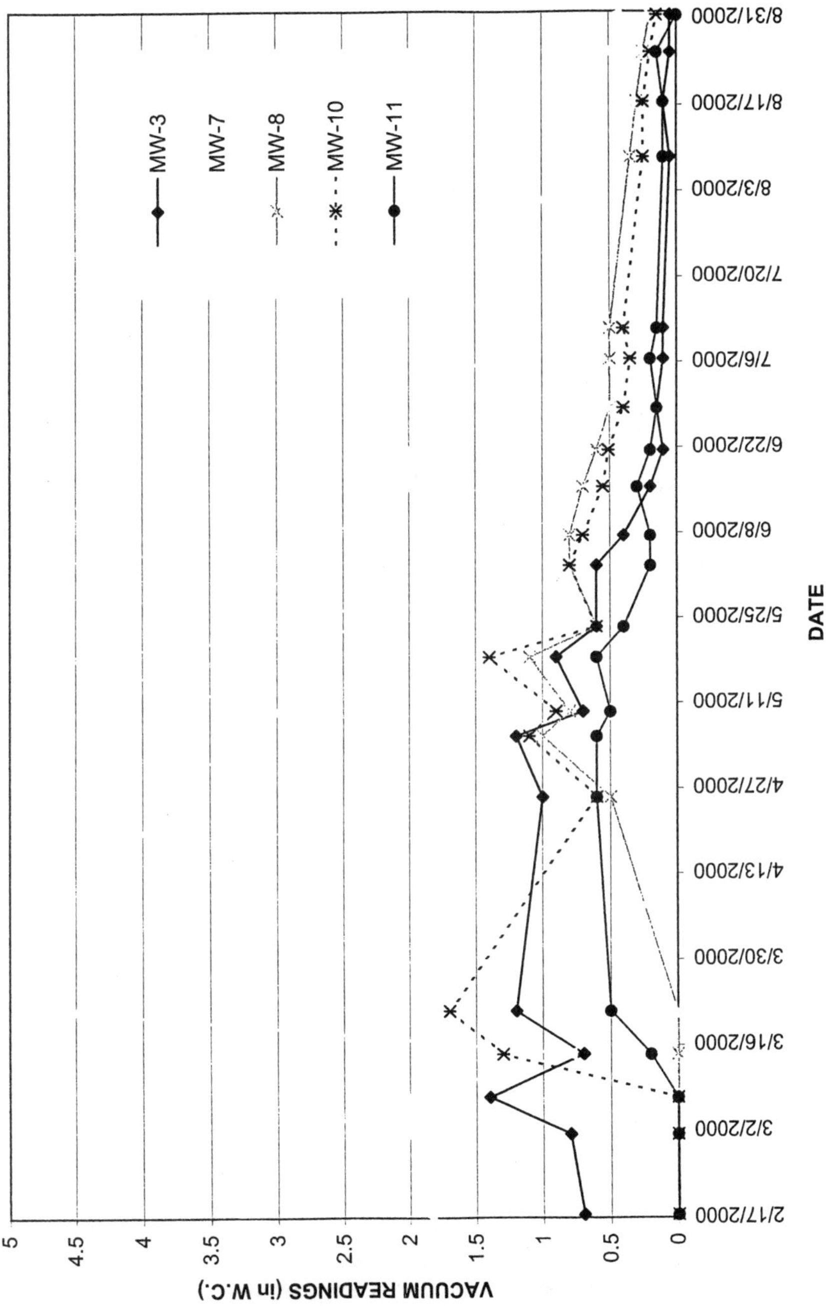

FIGURE 2 - VACUUM READINGS IN OBSERVATION WELLS

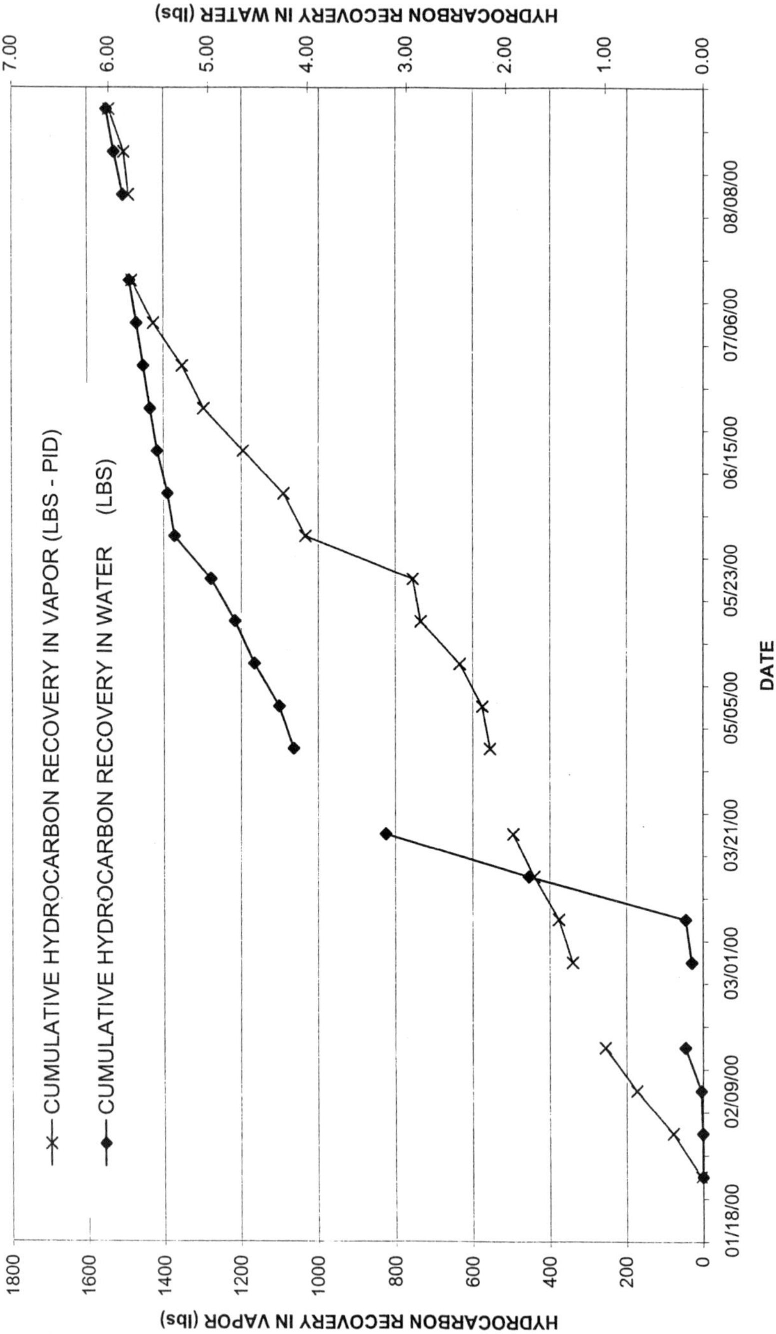

FIGURE 3 - CUMULATIVE HYDROCARBON RECOVERY

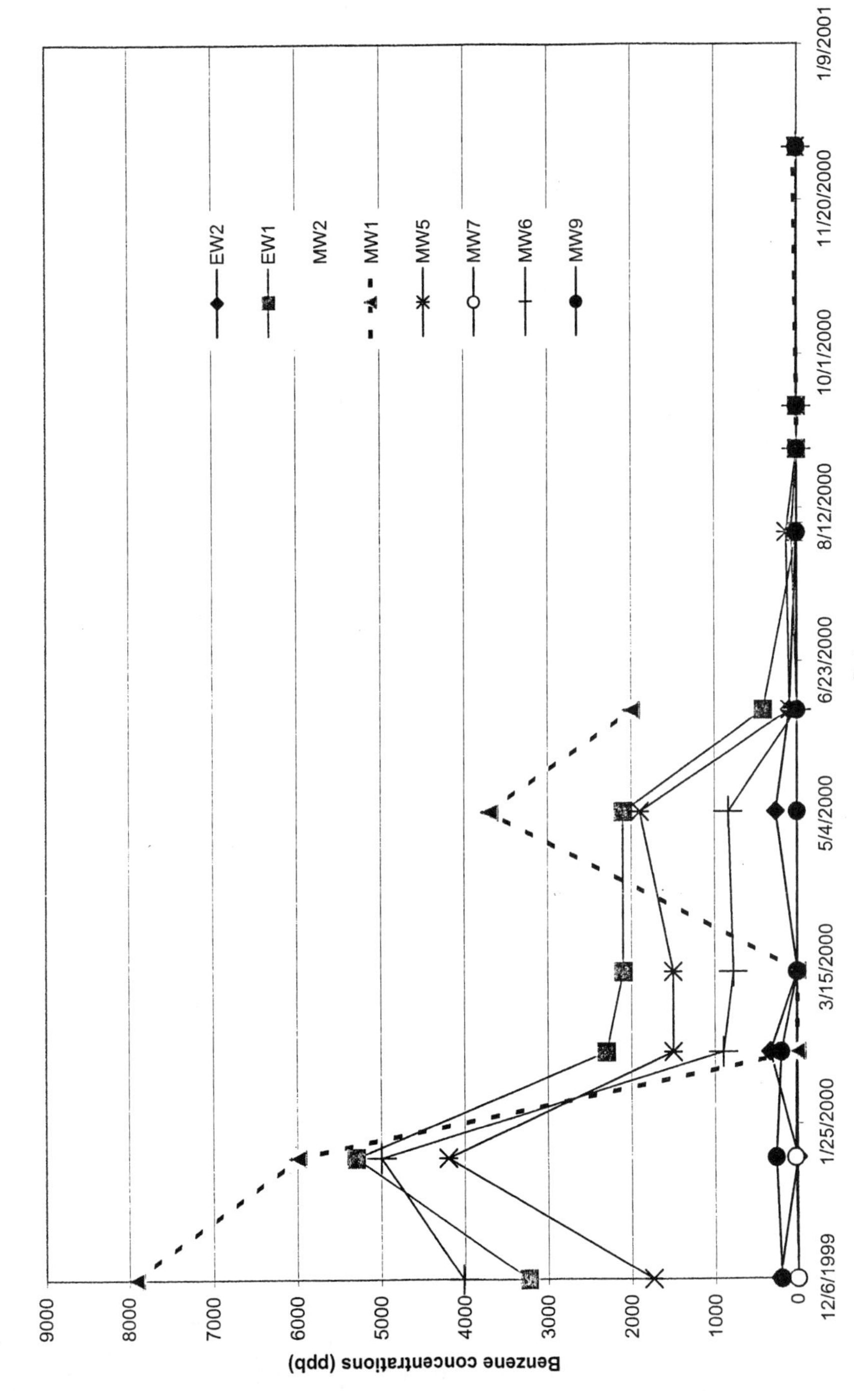

FIGURE 4 - BENZENE CONCENTRATIONS IN GROUNDWATER

TWO-PHASE EXTRACTION DESIGNING, PERFORMANCE EVALUATIONS AND VOCs STRIPPING

Mehmet Pehlivan (Tait Environmental Management, Inc., Santa Ana, California)

ABSTRACT: This paper presents the results and lesson learned from several two-phase extraction tests and system operations data from the last three years. The paper provides an outline for the proprietary and non-proprietary two-phase extraction methods, designing and implementing effective pilot tests, performance monitoring, and design considerations for volatile hydrocarbons stripping and minimizing the friction losses. Two-phase extraction is an effective method for the remediation of soil and groundwater contaminated with volatile hydrocarbons. Two-phase extraction removes both groundwater and soil vapor in a common stream while enhancing biodegradation of the hydrocarbons in the vadose and the saturated zone. The method should be used after serious consideration for the water lifting mechanism of the vacuum, and the friction losses due to handling large volume of water. In most applications, a conventional vacuum blower system with a carefully designed knockout and piping system can perform as well as a liquid ring pump system. The success of two-phase extraction depends on the ability of the piping design to deliver maximum vacuum to the wellhead with minimal friction loss in the extraction pipe. One major challenge is to lift the water in the well from depths greater than 33 feet. There are several proprietary and non-proprietary methods available presenting their own challenges and benefits. Another major challenge is to measure the vapor flow at the individual wellhead when extracting from more than one well. A flow splitter was developed to overcome the difficulty of measuring flow of water and vapor separately. Two-phase extraction strips volatile compounds from the groundwater with stripping efficiency up to 99% for BTEX and/or MTBE observed during two full-scale system operations and over seven pilot tests.

INTRODUCTION

The purpose of this paper is to provide a brief overview of two-phase extraction methods for remediation of soil and groundwater contaminated with volatile organic compounds (VOCs).

Two-phase extraction is a method of vapor and water removal in a combined flow stream using a vacuum and a drop tube. Two-phase extraction is widely used for remediation of VOCs and can be applied as a good source reduction method to prepare any site for natural attenuation and risk-based site closure. Several records related to the use of two-phase extraction, multiphase extraction and dual phase extraction methods in the environmental cleanup area, can be found in literature going back to over a decade. This paper will briefly introduce few of these methods and applications that had significant impact on the use of two-phase and multiphase extraction methods.

One of the challenges of performing a two-phase or multiphase extraction is to overcome the limitations of the vacuum when extracting water from depths greater than 33 feet (10.06 m) below ground surface (BGS). The maximum available vacuum lift for water is 33.9 feet (10.33 m; 29.92-inches [76 cm] of mercury), which is the displacement of the Earth's atmospheric pressure. In order to obtain greater lift, the extraction pipe should have a mechanism to allow air inflow with the water at the same time. This can be accomplished by injecting air during extraction or allowing air to be extracted with the water by placing pipe intake at the air/water interface (water table). The applications of two-phase extraction can be summarized in two major categories: (1) non-proprietary methods (non-patented technologies) and (2) proprietary methods (patented technologies).

Non-Proprietary Methods:

Bioslurping. This method uses a drop tube set at the air/water interface to allow both air and water enter simultaneously into the drop tube when vacuum is applied to the distal end of the drop tube (Baker, 1996 and Trowbridge et al, 1991). This method is mostly used at the sites where LNAPL is present on top of the water table (multiphase extraction).

Dual Phase Extraction. This method uses two separate flow streams for vapor and water. Water is removed from the extraction wells using a submersible pump while extracting vapor from the vadose zone. The terms dual phase extraction and two-phase extraction have been used interchangeably in some literature. However, the dual phase extraction method is significantly different from the conventional two-phase extraction and may involve two completely separate above ground equipment and extensive piping (electrical or pneumatic conduit for each extraction well, separate extraction lines for vapor and water, etc).

Proprietary Methods:

The Morrow Method. A patented method that involves placing an air injection line that is open to atmosphere inside the drop tube (Morrow, 1991). This method is also known as *priming method* and *air assist* method. Air can be injected actively using a compressor, or passively as a response to applied vacuum to the drop tube to start two-phase flow. The drop tube draws vapor from the vadose zone, when the water level in the extraction well drops below the bottom intake of the drop tube.

The Xerox Method. A patented method that involves air injection into the drop tube using an air injection line(s) placed outside the drop tube (Mancini et al, 1994). The air injection line(s) connected to the drop tube via a manifold(s) on the side of the drop tube and to atmosphere through an injection pump (compressor or blower). The air can be injected at several points from the side and/or the bottom of the drop tube either actively, using a compressor, or passively as a response to the applied vacuum. When the water level in the extraction well drops below the bottom intake of the drop tube the vapor from the vadose zone is drawn into the drop tube.

The Bubblex[1] Method. A patented method that uses a drop tube with orifices to draw limited amount of air and vapor from the vadose zone with the water when

---

[1] Bubblex is a service mark of Tait Environmental Management, Inc., Santa Ana, California

vacuum is applied to the distal end of the drop tube (Pehlivan et al, 1999). This method draws vapor/air from the vadose zone at all times and uses this vapor/air to lift the water from depth greater than 33 feet as well as to strip VOCs from the extracted water.

Both proprietary and non-proprietary methods of two-phase extraction tried to overcome the limitation of vacuum extraction in lifting water from depth greater than 33 feet. In most cases the extraction and treatment unit used on the surface were identical with the exception of the drop tube used for extraction.

PILOT TEST AND SYSTEM DESIGN

Before using two-phase extraction, it is recommended that a pilot test be performed. A two-phase extraction test requires using a vapor extraction unit equipped with a high vacuum (liquid ring pump) blower, a water/air separating knockout tank, a carbon polishing unit for the water (or a storage tank), depending on the method selected, an air injection/priming system (a compressor or an air blower) and a drop tube system (with or without injection setup). Naturally, an effective pilot test should start with reviewing all the available assessment records, soil test results (physical and chemical properties) and any other feasibility testing results (aquifer test, vapor extraction test etc.). Prior to start of a pilot test, depth to groundwater in monitoring and extraction wells should be measured. Distances between all observation and extraction wells should be recorded. If more than one extraction well will be used for the pilot test, at least one of the wells should be placed near the source area and another one downgradient of the plume near the area where hydraulic control is desired. The extraction equipment, especially the air/water separation tank should be located at a low spot at the site to allow water flow under gravity from the wells to reduce the friction loss. Straight piping between the extraction wells and the system is preferred over a coiled hose. An effective pilot test should have the following monitoring/data collection program for the design of the system:

- The zone of influence of the vacuum and groundwater drawdown.
- Vapor and water flow rates.
- Recovery of VOCs from water and the vadose zone.
- Frequent measurements of vacuum at the system, at the extraction wellhead (drop tube) and at the well annulus (in the casing outside the extraction pipe).
- The analysis of extraction time versus water drawdown in observation wells (three wells minimum).
- The analysis of extraction time versus vacuum response in observation wells (three wells minimum).
- The analysis of extraction time versus vapor concentrations at the main extraction well or at the system.
- The analysis of extraction time versus water concentrations in the extraction stream (minimum three samples should be collected from the extracted water).
- Pre-test and post-test vapor readings from the extraction and select observation wells, including measurement of permanent gases (CO_2, CH_4, and O_2).

- Pre-test and post-test groundwater sample results from extraction and select observation wells. Collection of mid point groundwater samples from the extraction wells are also recommended for a test lasting over 4 hours.

PERFORMANCE MONITORING

The monitoring parameters listed under the Pilot Test section above should also be part of a performance evaluation monitoring when a two-phase system is operated. A performance evaluation monitoring should also include the cumulative removal data from the vapor phase, and aquatic phase with a comparative evaluation of the cost per pound removal to determine when the system is not cost effective to operate as configured. The removal rates of VOCs from individual wells should be monitored frequently to maximize the system operation. The well recoveries and biological activities in the extraction wells should be monitored frequently enough to recognize any bio-fouling in the wells.

CASE STUDIES – LESSONS LEARNED

This chapter is prepared to share experiences and lesson learned from over ten pilot tests, several full-scale two-phase and dual phase installations and operations. In each of these pilot tests and system operations I have learned valuable lessons in optimizing the system operation, maximizing the extraction of water and vapor and performance monitoring of the system operation. Following are the brief narratives and lesson learned from my personal experiences:

Vacuum Difference Between Extraction Pipe and Well Annulus. The difference between the vacuum readings in the extraction pipe at the wellhead and the vacuum readings in the well casing is related to the permeability of the vadose zone. We (my co-workers at Tait Environmental Management, Inc. and I) observed that the vacuum differences between the well casing and the drop tube varied in differing vadose zone soil types (change in permeability). Later we were able to test this concept in a display model as well. Our display model had 4-inch wells with a solid casing connected to a tank from the bottom openings below the water level of the tank. In order to simulate vadose zone permeability, we placed two extra openings to the well cap that was holding a clear PVC drop tube. A valve was placed on one of the extra opening, and a vacuum gauge on the other one. We applied 36 inches of water column (W.C.) vacuum to the drop tube and observed the vacuum in the well casing and the flow in the drop tube. When the valve setting was changed, the vacuum in the annulus was responding as a response to the increased or decreased air entry (connection to the atmosphere simulated the increase and decrease in vadose zone permeability). Since all the other parameters remained constant, we can reasonably predict that the vacuum differences between the extraction pipe and the well casing are related to changes in vadose zone permeability. As the vadose zone permeability (opening the valve) increased the vacuum difference between the drop tube and the well casing also increased. We have clearly observed this phenomenon in the field during all our pilot tests and system operations. For example in a sandy soil with approximately

5 feet of open(vadose zone) screen, 2 gallons per minute (gpm) of water flow, 40-50 CFM of vapor flow we have observed the extraction pipe (EP) vacuum to well annulus (WA) vacuum ratio to be 81/18 (~4.5/1) inches of W.C.. At the same site similarly constructed well in clayey sand and silt zone yielded 0.1 gpm water and 15-20 CFM vapor flow with an EP/WA vacuum ratio of 156/100 (~1.5/1).

Vapor and Water Flow Measurements. One of the most difficult measurements in two-phase extraction is the flow measurement. In order to measure both vapor and water flow accurately the liquid and vapor need to be separated. A flow splitter was developed to measure both vapor and water flow separately at each wellhead. The flow splitter is a looped piping that allows water flow to the lower section of the loop while routing the vapor flow to the upper section. With this apparatus we were able to establish air to water ratio for the VOCs stripping rates that we have observed in our pilot tests and performance monitoring data from the operating two-phase extraction systems (Pehlivan et al, 2000a).

Friction Loss Related to Moving Water With Vapor: When we were performing a ninety day test at a site in Southern California using a Baker Unit without the liquid ring pump, we observed that the water recovery (~ <1 gpm) from the wells were a lot less than the recoveries we observed during well development (~ 2gpm). We did some vacuum readings along the extraction pipe between the system and wellheads. We observed that the vacuum loss between the wellhead and knockout tank were over 60%. The majority of the vacuum loss was on the sloped pipe where it was rising to the knockout tank (approximately five feet of rise from the ground surface). We were using above ground piping at this site. We raised the piping, placed 5 foot of casing extension on each extraction well and provided a gentle slope in the extraction pipe toward the knockout tank. This modification nearly tripled amount of water recovery and increased the vacuum delivered to the wellhead.

Comparing the Pre-test and Post-test VOC Concentrations. During one test in Studio City, California, we have observed that the groundwater concentrations during extraction as collected from the extraction well declined in the well near the source zone and increased in the well on the perimeter of the plume. Similar observations in other tests were also noted. Although we can not conclusively predict that the change in pre-test and post test groundwater concentrations is always related to the proximity of the source, our observations in known cases made us to believe that there is a connection. Additionally, at a site where we were performing bi-weekly two-phase extraction in five wells, we observed that post test VOC concentrations in one of the well were consistently higher. We placed another well near this well observed that the VOC concentrations in groundwater was five times higher than the well that we observed increasing post-test VOC concentrations in our tests.

Consider Intermittent Operations. In one site we observed that after eight hours of extraction the VOC concentrations in the groundwater did not reached back to the near starting levels in over 5 days. In order to explain this observation, we re-evaluated the soil analytical results and the capillary pressure tests performed by a laboratory on the soil samples collected from the various soil zones at this site. We determined that the VOC contaminations were mostly in a zone with irreducible water saturation over 80%. We realized that our extraction efforts were removing the water from the 20% of the pore volume. The VOCs that was diffused into the 80% of the pore volume (irreducible water zone) would need to be diffused out of this space by replenishing the water in 20% of the pore space (effective pore space) in optimum frequencies. If we installed a system and operated continuously at this site, we would be flushing out a lot of clean water from the 20% of the pore space without impacting the irreducible water zone significantly. This would result in treatment of clean water, and unnecessary system operation costs. We realized that by operating weekly or bi-weekly we could accomplish the similar treatment efficiency with a system operating continuously.

Two-phase extraction strips VOCs. We have observed that up to 99% of the VOCs stripping is possible during two-phase extraction with a careful selection of design and operating parameters. The stripping efficiency is increased when the VOC concentrations in the vadose zone are lower, vapor flow is higher, water flow is lower, and depth to water is higher (Pehlivan et al, 2000a and 2000b).

SUMMARY

Two-phase extraction requires a good understanding of the subsurface geological and hydrogeological conditions while providing valuable information for such conditions. The observations and suggestions presented in this paper should not be accepted without first validating at a given site by a pilot test or operation and performance monitoring data. All the observations presented in the Case Study-Lessons Learned section were made by using the Bubblex Method. Using a different method may result in a different finding. I strongly suggest testing of our observations by other scientists working in this area. Additional studies related to the modeling of two-phase flow in the subsurface and stripping efficiency are recommended. Additionally, it is strongly suggested that before performing a two-phase extraction, a record search and a review of patents should be performed to avoid any legal ramifications.

ACKNOWLEDGMENTS

Author appreciates very much the comments received by the editors and technical reviewers. This paper would not be complete without their input and comments. Author also thankful to Dr. Ken Tait, Tom Tait, J.D., MBA, and Tait Environmental, Inc. personnel for supporting the project and research associated with the patent and the fieldwork.

REFERENCES

Baker, Ralph S., Ph. D., 1996, "Contamination", *Pollution Engineering*, March 1996, pp. 39-40.

Mancini, Alfonso R., Ronald E. Hess, Richard A. Williams, Douglas J. Montgomery, Heinrich J. Jurzysta, 1994, "Process and Apparatus for High Vacuum Groundwater Extraction", *U.S. Patent Office* Patent No. 5,358,357, Issued on October 25, 1994.

Morrow, Steven R., 1991, "Priming Methods for Vacuum Extraction Wells", *U.S. Patent Office* Patent No. 5,076,360, Issued on December 31, 1991.

Mehmet Pehlivan, Neal Beaty, Tom Dixon, Scott Ek, Richard Coffman, PhD, 2000a, "MTBE and BTEX Stripping During Bubblex[SM] Two-Phase Extraction", G. B. Wickramanayake, A.R. Gavaskar, J.T. Gibbs, and J.L. Means (Eds.) *Case Studies in the remediation of Chlorinated and Recalcitrant Compounds (C2-7), pp 41-48*, Battelle Press, Columbus, Ohio

Pehlivan, Mehmet, J. Neal Beaty, 2000b, "Remediation of Petroleum Hydrocarbons Using Bubblex[SM] Two-Phase Extraction Method – A Case History", *Proceedings of ETCE/OMAE 2000, February 14-17, 2000, New Orleans, Louisiana*, Paper No. ETCE2000-10153, A CD-ROM Published by ASME, New York, NY.

Pehlivan, Mehmet, James R. Miller, 1999, "Method for Bubbling Extraction of Groundwater", *U.S. Patent Office*, Patent No. 5,906,241, Issued on May 25, 1999.

Trowbridge, Bretton E., James J. Malot, 1991, "Soil Remediation and Free Product Removal Using In-Situ Vacuum Extraction with Catalytic Oxidation", *Ground Water Monitoring Review*, National Ground Water Association, Dublin, OH, USA.

GROUNDWATER CIRCULATION WELL: A SUCCESSFUL CASE STUDY

Benny DeHghi (Honeywell, Torrance, CA)
Mary A. Lucas and Cannon F. Silver, P.E. (Parsons Engineering Science, Pasadena, CA)
Terry H. Feng, Ph.D., P.E., C.Hg. (Parsons Engineering Science, Oakland, CA)

ABSTRACT: An innovative in-well vapor-stripping or groundwater circulation well (GCW) technology was successfully implemented at a former solvent recycling facility as an interim remedial measure (IRM) for the source zone groundwater impacted by halogenated volatile organic compounds (VOCs). Performance data from 11½ months of operation of a pilot and 3 months of operation of a full-scale GCW treatment system have demonstrated the following advantages compared with presumptive pump and treat system at the site; successfully removes contaminants from a hydraulically transmissive, thick aquifer (21 to 25 meters thick); achieves hydraulic containment in the target zone with no above ground groundwater treatment and disposal of treated water; and compliments the previously installed active soil vapor extraction (SVE) soil IRM system. As of February 9, 2001, the system has achieved the following performance results: during 11½ months of operation, the pilot GCW stripped approximately 28 kilograms (kgs) of VOCs from the groundwater; during the full-scale GCW system operation, consisting of four wells, the system has removed approximately 55 kgs of VOCs from the groundwater over a 2.3-month period. No fouling or well clogging has occurred during the 14 months of pilot/full-scale GCW operation. The overall capital, operation and maintenance costs of the system are significantly less than the costs of a comparable pump and treat system. While this technology has received mixed reviews since its introduction in the United States over 10 years ago, the technology appears successful at meeting remedial goals at this site. Variations in this technology's success appear highly dependent on the site-specific hydrogeology and geochemistry. Key parameters for successful implementation of the technology at the site are presented in this paper.

INTRODUCTION

The site (Figure 1) occupies approximately 0.7 acres of land and is located within an industrial park, with commercial and light industrial facilities surrounding the facility. Site-specific data pertaining to the discussion of the groundwater circulation well (GCW) systems is summarized as follows.

Hydrostrategraphy. Five important hydrogeologic units underlie the site as described below in order of occurrence starting from the ground surface, as shown on Figure 2.

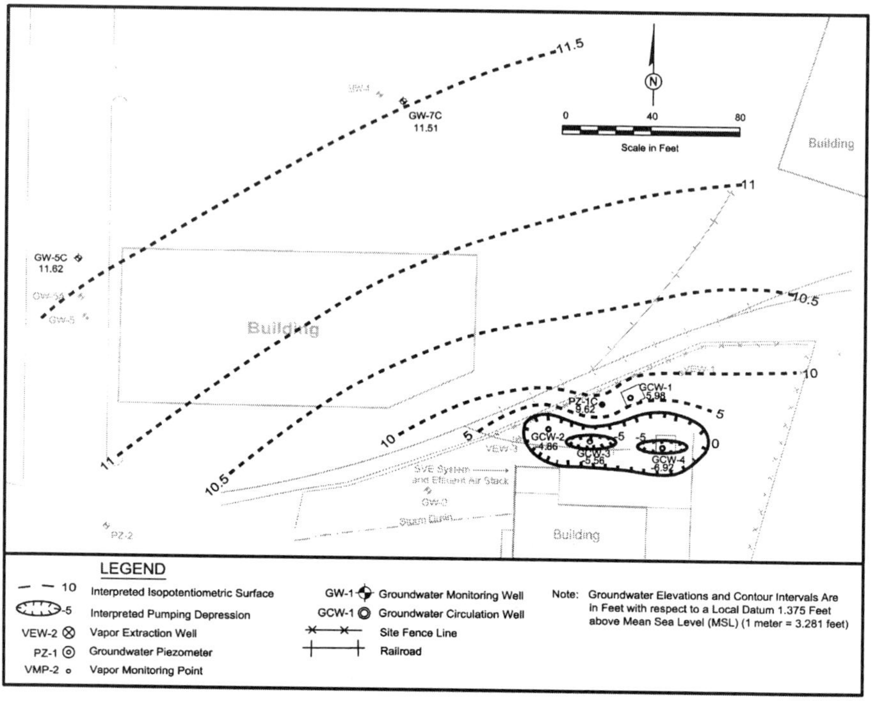

FIGURE 1. Site Layout and GCW extraction induced groundwater flow pattern in lower screen zone (Jan. 26 – Feb.5, 2001).

<u>Near-surface silt and clay</u>: A thin layer of silt constitutes the uppermost stratigraphic unit. This unit lies above the water table and is less than 1.5 meters thick beneath the site, and so does not significantly influence groundwater flow.

<u>Shallow Aquifer</u>: This aquifer unit is a 15 to 25 meter thick layer of coarse sandy gravels and sands, containing large (over 1 meter in diameter) basaltic boulders.

<u>A Local Clay and Gravelly Clay Aquitard:</u> A locally important, discontinuous layer of clay and gravelly clay up to 3 meters thick was encountered in a deep boring at the source area.

<u>Deep Aquifer</u>: A 30 to 36 meters thick section of coarse gravelly sands and clayey gravelly sands that has a base encountered at approximately 55 to 65 meters below ground surface (bgs). This unit cannot always be differentiated from the overlying shallow aquifer where the clay aquitard is absent. VOC plume has not been detected in this aquifer unit.

<u>Confining Unit 1:</u> A regionally extensive confining unit consists primarily of sandy, clayey silt was encountered at the base of the deep aquifer at approximately 65 feet bgs.

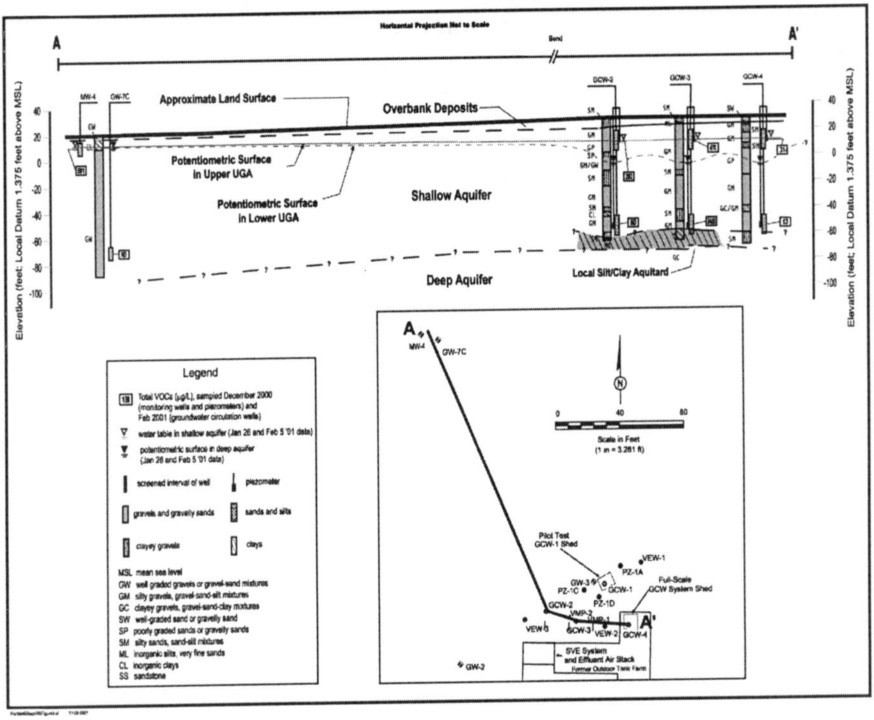

FIGURE 2. Well locations and geologic cross section A-A'.

Groundwater Conditions. Groundwater is encountered at approximately 5 to 7 meters beneath the source area to less than 1 meter bgs in some downgradient areas. Within the shallow aquifer (extending to approximately 25 meters bgs), the groundwater flows in a north-northwesterly direction with a gentle horizontal hydraulic gradient of approximately 0.0003 meter/meter (m/m). This unit has a high permeability (hydraulic conductivity estimated ranging from 0.04 to 0.1 centimeters per second [cm/sec]). The estimated linear horizontal groundwater flow velocity is on the order of 40 meter per year (m/yr) in the vicinity of the site. There is a significant downward flow component as a result of direct infiltration recharge from precipitation and aquifer stratification. Operation of GCWs in the source area results in localized flow condition change as discussed further in later sections.

Geochemical parameters indicate excellent water quality beneath the site. Iron is never detected above detection limit of 0.10 milligrams per liter (mg/L), manganese concentrations range from 0.031 to 0.190 mg/L, alkalinity concentration range from 68.4 to 84.3 mg/L, total dissolve solids concentrations range from 195 to 228 mg/L, and pH ranges from 7.25 to 7.42.

Nature and Extent of Contamination. The chemicals of potential concern in soil and groundwater consist primarily of chlorinated and fluorinated VOCs including 1,1,1-trichloroethane (1,1,1-TCA), tetrachloroethene (PCE), trichloroethene (TCE), 1,1,2-trichloro-1,2,2-trifluorethane (Freon 113), and *cis*-1,2-dichloroethene (*cis*-1,2-DCE). High concentrations of VOCs in soil and soil vapor within the vadose zone immediately beneath the site coincide with the area of highest dissolved VOC concentrations in groundwater. The VOC plume extends vertically approximately 25 meters bgs beneath the soil source area and laterally to at least 700 meters downgradient of the site.

Interim Remedial Measures (IRMs). IRMs for both soil and groundwater have been implemented to contain and remediate the source of VOCs that impacts the groundwater. A SVE system began operation in September 1999 to cleanup contamination in the vadose zone; a pilot GCW system consists of one GCW was put into operation in November 1999 to test the applicability and effectiveness of the recirculation technology; and an expanded GCW system consists of 4 GCWs began operation in November 2000.

MATERIALS AND METHODS

SVE and GCW Systems. Figure 2 shows the layout of both systems. The SVE system consists of three vapor extraction wells (VEWs) with screened interval ranges between 1.67 to 5.18 meters bgs and two vapor monitoring points (VMPs) with screens at 1.67, 3.66-3.81, and 5.18-5.03 meters bgs. The system was installed in July 1999 and has been in operation since then.

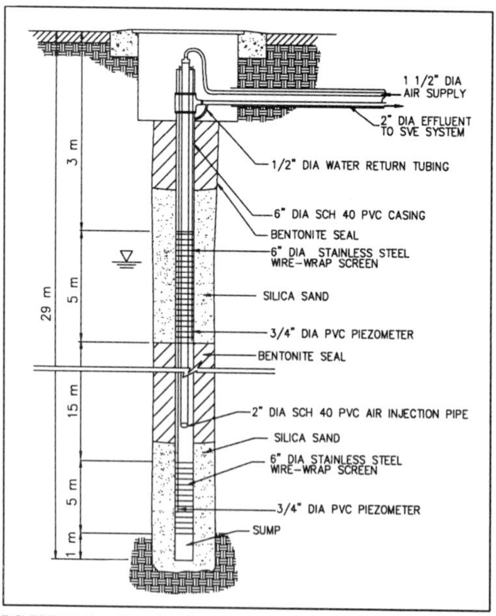

FIGURE 3. Typical groundwater circulation well.

The pilot test GCW system consisted of one well (GCW-1) installed in November 1999. The full-scale GCW system consisted of three additional wells (GCW-2, -3, and -4) installed in October 2000. All GCWs are constructed of 6-inch-diameter, Schedule 40 polyvinylchloride (PVC) well casing with two wire-wrap 15-foot screens (upper and lower) located at depths of 3.05 to 7.62 and 22.86 to 28.95 meter bgs. The upper screen was placed near the top of the shallow aquifer across the water table and the lower screen was placed near the bottom of the shallow aquifer overlying a discontinuous lower permeability layer thought to be a unit separating the shallow and deep aquifers. The average well spacing ranges from 6 to 10 meters. Figure 3 provides a typical GCW design.

Air lines for all four GCWs were connected through pipes placed in trenches and tied to a single air compressor located inside the full-scale GCW system shed at GCW-4 (Figure 2). Air flow and pressure for the total system and for individual GCWs is monitored and controlled at the air supply manifold next to the air compressor. The common 4-inch effluent line serving all four GCWs was connected to the existing SVE effluent line. This approach to system design and construction is intended to enable the SVE system to be used in a vacuum-assist operation for more efficient extraction of vapors stripped into the vadose zone by the four GCWs.

GCW Performance Monitoring. To assess the effectiveness of the GCW system, one round of pre-system startup baseline monitoring and many rounds of post-system startup monitoring were conducted. During the pre-system startup baseline monitoring, water levels and concentrations of pH, temperature, electric conductance, and dissolved oxygen (DO) were measured in the field from all GCWs and monitoring wells located in the vicinity of GCWs. Groundwater samples were also collected from these wells for VOC analysis using the United States Environmental Protection Agency (U.S. EPA) Method 8260B and from selected wells for total dissolved solid (TDS) and general geochemical parameters including major ions and dissolved metals. Soil samples were collected during GCW installation to obtain parameters. The following post-system startup monitoring was performed as part of system optimization and performance evaluation:

- Air flow characteristics monitoring to confirm if optimized bubbling conditions exist at each of the GCWs;
- Flow and pressure measurement at the compressor and individual GCWs;
- Water level and DO were measured in the GCW piezometers and selected monitoring wells;
- Effluent total VOC vapor concentration levels were measured at individual GCWs and at the combined system stack using a PID;
- Effluent air concentrations of VOCs at the discharge stack were measured by collecting air samples from the exhaust point using Summa canisters and analyzed in fixed analytical laboratory using the US EPA Method TO-14 to confirm that all air emission requirements are met;
- Groundwater samples were collected from all GCWs between 10 to 14 days of continuous operation. All groundwater samples were analyzed for VOCs by U.S. EPA Method 8260B; and
- Groundwater samples were also collected and analyzed for TDS, geochemical parameters including dissolved metals (calcium, magnesium, iron, and manganese) and major ions such as alkalinity to evaluate the potential for carbonate or metals fouling in the GCWs.

RESULTS AND DISCUSSION

Since the GCW technology is generally considered an alternative to traditional pump and treat groundwater remediation technology, the technical effectiveness of the GCW is evaluated in terms of hydraulic containment, mass removal, and operation and maintenance issues.

Groundwater Elevations. GCWs extract groundwater from the lower zone through the lower screen, airlift the extracted water up the inside of the well casing to strip VOCs from the water, and recharge the "treated" water back into the formation through the upper well screen. The hydraulic effect to the lower zone is equivalent to traditional pump and treat in which a drawdown cone of depression develops in the lower zone in the vicinity of the lower well screen. The recharge in the upper zone is similar to traditional injection which may result in a water level rise or mounding in the upper zone in the vicinity of the upper well screen. The amount of drawdown in the lower zone or rising in the upper zone is dependent on site hydrogeology and is a direct measure of the GCW system effectiveness. Water level data from this site indicate that the GCW system is very effective in creating not only satisfactory groundwater circulation but also excellent hydraulic control by maintaining a stable hydraulic capture in the lower zone with minimum water level rise in the upper zone, as shown on Figures 1 and 2.

Prior to startup of the expanded system, the depth to groundwater from all GCW piezometers and the existing monitoring wells in the vicinity of GCWs were measured on November 30, 2000. The initial water level elevations ranged between 12.06 and 12.36 feet (3.68 to 3.77 meters) above mean sea level (MSL). Water level in all piezometers responded immediately to the startup of the GCW system. Water levels in all lower zone piezometers dropped significantly (7.6 feet or 2.3 meters in GCW-1 and more than 15 feet or 4.5 meters in GCW-2, -3, and -4) following the GCW system startup. Concurrently, water levels in corresponding upper zone piezometers were either unchanged (GCW-2, -3, and – 4) or increased (GCW-1).

The water level data from GCW piezometers are consistent with the water level responses observed from other groundwater monitoring wells in the vicinity of GCWs (Figures 1 and 2). The water level in the lower zone monitoring well PZ-1C, located approximately 6 meters from GCW-1, decreased immediately after starting the GCW system. Slight water level decreases were also observed in two offsite shallow aquifer monitoring wells, GW-5C and GW-7C, located approximately 60 to 90 meters from the GCWs (Figures 1 and 2), indicating the radius of influence in the lower zone exceeds 90 meters as slight drawdowns are observed at GW-5C and GW-7C. This approximate estimate indicates that the observed hydraulic control effect may be comparable to traditional pump and treat systems.

Although mounding was observed in GCW-1, the effect of mounding is very localized and insignificant as only a 0.04 meter increase in water level was observed in the upper zone monitoring well GCW-3, located less than 3 meters from GCW-1 and no water level rise was observed in all other three GCWs and monitoring wells.

Dissolved Oxygen. DO concentration is an indicator parameter to indirectly assess the effectiveness of the GCW circulation as the GCW process introduces continued supply of DO in the aquifer. DO concentrations were measured in the field in surrounding monitoring wells during GCW operation. Results of the measurements are summarized on Table 1. Prior GCW operation DO measurements showed decreasing concentrations with depth, as expected (Table 1). Once operation of the GCW pilot system began in December 1999, DO concentrations increased in all wells measured. Since startup of the full-scale GCW system in November 30, 2000, an additional increase in DO was observed at the lower zone PZ-1C, which can be attributed to effects from the overlapping radius of influence from other GCWs. Results of the DO measurements agree with the expected pattern of groundwater flow produced by the GCWs, with increases observed in more shallow wells. The observed DO increase in the lower zone monitoring well corresponding to the GCW expansion from pilot to full-scale indicates effective circulation of groundwater between the upper and lower zones.

TABLE 1. Dissolved Oxygen Concentrations During GCW Operation

Date	PZ-1 (mg/L)	PZ-1A (mg/L)	PZ-1C (mg/L)	GW-3 (mg/L)
12/16/99 [1]	8.2	6.3	4.9	7.9
12/16/99 [2]	8.9	6.2	6.5	8.4
12/16/99	10.0	6.3	7.1	9.8
12/16/99	10.8	7.0	7.2	7.6
12/17/99	9.4	7.1	6.0	6.5
12/23/99	12.1	5.5	6.2	9.8
12/30/99	11.8	8.6	6.0	11.3
01/08/00	11.0	8.0	7.5	11.4
01/13/00	11.4	9.1	5.4	9.5
01/20/00	11.5	7.7	5.7	11.6
01/05/01 [3]	NM	6.5	8.7	6.0
12/29/00	NM	8.8	9.3	9.7
12/22/00	NM	5.6	6.2	10.9
12/19/00	NM	5.9	8.7	10.4
12/15/00	NM	9.7	8.7	11.0
12/10/00	NM	9.9	9.1	10.1
12/08/00	NM	10.0	9.0	10.2

NM = not measured.
1. Initial measurement.
2. GCW-1 operation began.
3. Expanded GCW Operation

Groundwater Chemical Analysis. Groundwater samples were collected at various stages of the project to assess the GCW system performance.

Chemical Precipitation and Fouling Potential. The results of field measurement and sampling and analysis of dissolved metals and geochemical parameters indicate that the site groundwater has a neutral pH (7.25 to 7.44), which is within the 6.0 to 7.6 range where carbonate precipitation will not occur (Wasatch, 2000). Within a pH range of 7.7 to 8.5, calcium carbonate would be at equilibrium and may precipitate (Wasatch, 2000). In addition, iron was not detected, and all other metals, major ions and TDS were detected at low levels that do not indicate threat of chemical precipitation and/or fouling. Chemical precipitation and fouling problems have not been encountered during the 14 months of operation of the GCW pilot well and the 3 months operation of the full-scale system.

VOC Removal Efficiencies. Comparison of VOC concentrations in the GCW inflow and outflow over time provides a means of evaluating the groundwater removal efficiency achieved by the GCW. Initial baseline sampling conducted in November 2000, prior to the full-scale system startup, showed that VOC concentrations were generally highest in the lower portion of the treatment zone, except for PCE at GCW-4. At GCW-2 in the upper piezometer, PCE was detected at 870 μg/L and TCE was detected at 96 μg/L and in the lower piezometer, PCE and TCE were both detected at 18,000 μg/L. At GCW-3 in the upper piezometer, PCE was detected at 420 μg/L and TCE was detected at 51 μg/L and in the lower piezometer, PCE and TCE were detected at 530 and 90 μg/L, respectively. At GCW-4 in the upper piezometer, PCE was detected at 3,100 μg/L and TCE was detected at 160 μg/L and in the lower piezometer, PCE and TCE were detected at 810 and 170 μg/L, respectively.

Based on groundwater data collected from the GCW piezometers, reductions in VOC concentrations within the GCWs have been observed as shown in Table 2.

TABLE 2. Calculated VOC Removal Efficiencies

Well	Days of Operation	Upper Zone		Lower Zone	
		PCE	TCE	PCE	TCE
GCW-1	418	60.53%	95.00%	83.72%	60%
GCW-2	70	91.38%	-4.17%	96.89%	98.72%
GCW-3	70	50.00%	-273%	-108%	-811%
GCW-4	70	99.29%	95.31%	96.05%	87.65%

In comparing the groundwater samples collected throughout operation at GCW-3, the concentrations of PCE and TCE initially increased, therefore showing negative values in Table 2, but are slowing decreasing and showing signs of a decreasing trend, likewise at the upper zone at GCW-2 for TCE. VOC concentrations have been reduced and the trend is continuing. It is possible, but further concentration reduction to MCLs will take a long time in the source zone, as leachate from potential DNAPL residuals will continue for a long time. However, this system has been installed in the source zone as an IRM and the objective is mass removal rather than reduction of VOCs to below MCLs.

Concentration Reduction. VOC concentrations in surrounding monitoring wells, specifically PZ-1C, located in close proximity to GCW-1, have been decreasing as a direct result of the GCW operation. In December of 1999, prior to the operation of GCW-1, PCE was detected in PZ-1C at 14,000 μg/L. Following GCW system operation, PCE concentrations in PZ-1C decreased to 2,200 μg/L in December 2000.

Air Emission Monitoring. The air injection rate and pressure were measured during each monitoring round. Air samples were collected for chemical analysis approximately every 10 days of continuous operation and analyzed for VOC by U.S. EPA Method TO-14.

Emissions from the GCWs were calculated using the ideal gas law based on the concentration of individual constituents (1,1-DCE, *cis*-1,2-DCE, Freon 113, 1,1,1-TCA, TCE, and PCE) and the extraction flow rate at the time of sampling. Operating under the full-scale GCW system alone, the VOC removal rate on average was 2.0 pounds per day (0.9 kg/day). Combining the GCW and SVE systems, the VOC removal rate on average was 1.4 pounds per day (0.63 kg/day).

Based on the effluent data, the total VOC mass removed prior to operation of the full-scale GCW system was approximately 1,956 pounds (887 kgs) over 422 operating days, including results from the operation of SVE system (resulting in 1,894 pounds or 859 kgs) and GCW pilot test well (GCW-1) (resulting in 62 pounds or 28 kgs). Since the full-scale GCW system began operation on November 30, 2000, an additional 120 pounds of VOCs have been extracted over 71 days (69 pounds or 31 kgs from the expanded GCW operation and 51 pounds or 23 kgs from the combined SVE/GCW operation). Results of the air emission monitoring demonstrate that the stripping action produced by the GCWs has removed significant quantities of dissolved VOCs from the aquifer.

CONCLUSIONS

Conjunction use of SVE and GCW remedial technologies as interim remedial actions have resulted in remediation of VOC impacted soil and achieve hydraulic containment of groundwater in the source zone. Increases in concentrations and the distribution patterns of DO in surrounding monitoring wells suggest that the groundwater is circulating through the aquifer during GCW operation. Stripping efficiencies based on VOC concentration of PCE and TCE reductions were estimated to range between 88 to 99 percent at GCW-2 and GCW-4. At GCW-3, PCE and TCE concentrations initially increased but have showed a decreasing trend after 70 days of operation. Significant reductions in VOC concentrations have been observed in PZ-1C since the start of the GCW operation. This is consistent with the hydraulic observation that the GCW operation has created local capture zone in the low screen interval and PZ-1C is located at the downgradient side of the capture zone. Removal of VOCs from the aquifer was demonstrated by analysis of air emission samples and calculated removal rates of approximately 1.95 pounds (0.9 kgs) per day. Based on the geochemistry data, there is limited potential for precipitation or fouling at this site, and none has been observed in over 14 months of GCW operation.

Although different views have been reported within the technical community on applicability of GCW technology, the technology has proven to be fairly effective at this site due to a good match between the technology and the site remedial objectives. Appropriate aquifer characteristics with proper inorganic water chemistry and limited stratification at the site are considered the key conditions that enabled successful application of this technology. Incorporating the GCW systems to an existing SVE system simplifies the design and reduces the capital cost.

REFERENCES

Wasatch Environmental, Inc. 2000. *Draft DDC Pilot Test Report Baron-Blakeslee Incorporated Site Portland, Oregon.* February.

USE OF IN SITU FLOW METERS AROUND A GROUNDWATER CIRCULATION WELL

Michelle A. Simon, United States Environmental Protection Agency, USA
Tong Li, Tetra Tech EM, Inc., USA
Benjamin J. Hough, Tetra Tech EM, Inc., USA
Richard L. Johnson, Oregon Graduate Institute, USA

ABSTRACT

The primary benefit of groundwater circulation well (GCW) technology is the development of strong vertical flows surrounding the treatment well. The extent of significant vertical flow surrounding a circulation well is difficult to establish from traditional groundwater elevation measurements. In June 2000, in situ flow meters were buried in the aquifer around a GCW at Cape Canaveral Air Station, Florida (CCAS), to measure the magnitude and direction of the surrounding groundwater flow. This aquifer is contaminated with trichloroethylene and other halogenated volatile organic compounds. The HydroTechnics® in situ flow sensors use a thermal perturbation technique to directly measure the direction and magnitude of groundwater flow in velocity ranges from 5×10^{-6} to 1×10^{-3} cm/s. The in situ flow meter measurements were collected under non-pumping, groundwater extraction, and groundwater circulation conditions. The magnitude and extent of vertical flow around a groundwater circulation well were greater than originally anticipated.

INTRODUCTION

Typically the concentrations of the contaminants in the influent of a GCW and the immediate surrounding aquifer initially increase soon after the startup of the GCW. GCWs are thought to create strong vertical flows which facilitate the desorption of contaminants from low permeability zones and dense non-aqueous phase liquid (DNAPL) pools in the aquifer. However, the presence of strong vertical flows is often difficult to observe in the field. As with many sites, water level measures, and even pressure transducers placed in monitoring wells surrounding the GCW at CCAS, could not definitively confirm the presence of strong vertical flows. The manufacturer has developed an in situ permeable flow sensor that uses a thermal perturbation technique to measure the direction and magnitude of groundwater flow in saturated porous media (Ballard 1996). The \$2500 sensors are permanently buried into the formation in order to have direct contact with the formation. The sensors can be placed into the formation via drilling rig or by direct push technology. The sensors are connected to a data acquisition system that can be programmed and accessed remotely, via modem and telephone connection, for up to two years time.

Objective

The objective of this study was to use the in situ permeable flow sensors to determine the presence, as well as magnitude and direction, of the presumed vertical recirculation cell around a GCW installed near Facility 1381 at CCAS.

The GCW was installed in November 1999 and operated intermittently from February 29 to August 1, 2000. The results of the GCW study can be found in Parsons (2001). Step and long term constant rate extraction tests, as well as step and long term constant rate dipole flow aquifer tests (Simon et al. 2000), were conducted during September 2000. This paper discusses the flow sensor measurements made during these September 2000 aquifer tests.

Site Description

Facility 1381 is located in the central portion of CCAS. Chlorinated solvent contamination at the site is suspected to have resulted from past metal cleaning operations and dumping of solvents. Lithology at the site consists of poorly sorted, coarse to fine sands and shell material from the ground surface to approximately 35 feet (ft) (10.7 m) below ground surface (bgs). Groundwater is first encountered at depth of 5 to 8 ft (1.5 to 3.5 m) bgs. Vertical hydraulic conductivity data, measured in intact cores from the site, decreased with depth from 30 m/day at the water table to 0.03 m/day at a depth of 40 ft (11 m) bgs (Parsons 2001).

The transmissivity of the aquifer, as calculated from the aquifer tests, ranged from 1790 to 2190 ft^2/day(165 to 205 m^2/day). Assuming an aquifer thickness of 42 ft (12.8 m), this results in an aquifer hydraulic conductivity of 43 to 53 ft/day (13-16 m/day). Calculated from the dipole test, the transmissivity of the deep aquifer, i.e., deeper than 5 m bgs, is substantially lower than that of the upper, 196 to 337 ft^2/day (18.2 to 31.3 m^2/day). The anisotropy ratio (Kh:Kv) was 5:1. The specific yield values were found to be 0.06 to 0.09 and the storativity values ranged from 0.006 to 0.007. The water table gradient across the site is 0.007 (Tetra Tech 2001).

MATERIALS AND METHODS

The CCAS GCW is shown schematically in Figure 1. Seven in situ flow sensors were installed at this site. The shallow flow sensors were completed at a depth of 11.5 ft (3.5 m) bgs and the deep flow sensors were completed at a depth of 22 ft (6.7 m) bgs. Figure 2 is the hydrograph of the pump and dipole tests that were performed in September 2000.

The manufacturer recommends that the sensors be installed with greater than 5 ft (1.5 m) of groundwater submergence. Because the water levels were lower than anticipated, the shallow sensors had less than the recommended submergence of groundwater and their results were more erratic than the deeper sensors.

RESULTS AND DISCUSSION

Figures 3, 4 and 5 illustrate the flow sensor results from cross-section A-O - B under non-pumping, pumping and GCW recirculating conditions, respectively. Figure 3 illustrates groundwater flow at the site under natural conditions. Figure 3 shows a vertical flow in the shallow sensors that is probably not real. As the water table surface heats during the day, the upper portion of the aquifer is hotter than the deeper. The sensor's calibration software computes this temperature gradient as a vertical flow but the water is not actually flowing upward.

Figure 4 shows the resultant groundwater flow under pumping conditions by subtracting the natural flow from the actual sensor data. Similarly, Figure 5 shows the resultant groundwater flow under recirculating conditions by subtracting the natural flow from the actual sensor data. The flow sensors measure the groundwater flow in approximately one cubic meter of the immediate aquifer and can show local irregularities. Extraction of groundwater at this particular GCW at a rate of 10 gallons per minute (gpm) (37.9 L/min) produced vertical flow patterns comparable to those under recirculating conditions at 12.5 gpm (47.3 L/min). The shallow sensors in Figure 5 show groundwater moving away from the upper screen under recirculating conditions, but not under non-pumping or pumping conditions (Figures 3 and 4). The deep sensors show similar flow of groundwater toward the extraction screen of the GCW, under pumping and recirculating conditions.

ACKNOWLEDGMENTS

The authors appreciate the support of the United States Air Force Center for Environmental Excellence, United States Environmental Protection Agency, Tetra Tech EM Inc., and Oregon Graduate Institute in the performance of this study. Although the research described in this paper has been funded in part by the U.S. EPA, it has not been subjected to Agency review. Therefore, it does not reflect the official views of the Agency. The use of product or trade names does not constitute Federal Government endorsement of product.

REFERENCES

Ballard, Sanford. 1996. "The In Situ Permeable Flow Sensor: A Ground-Water Flow Velocity Meter." *Ground Water 34*(2): 231-240.

Parsons. 2001. "Groundwater Circulation Well Technology Evaluation at Facility 1381, Cape Canaveral Air Station, Florida." Prepared for United States Air Force Center of Environmental Excellence.

Simon, Michelle A., Benjamin Hough, Tong Li, Zbigniew J. Kabala. 2000. "Field Evaluation of Dipole Method to Measure Formation Anisotropy" In *Phys., Chem., Therm. Technol., Int. Conf. Rem. Chlorinated Recalcitrant Compd., 2st (2000)*, Editor(s): Wickramanayake, Godage B.; Gavaskar, Arun R.. Publisher: Battelle Press, Columbus, Ohio. 43-47.

Tetra Tech. 2001. "Data Summary Report for Groundwater Circulation Well Technology Evaluation at the Ordnance Support Facility, Facility 1381, Cape Canaveral Air Station, Cape Canaveral, Florida." Prepared for the United States Environmental Protection Agency.

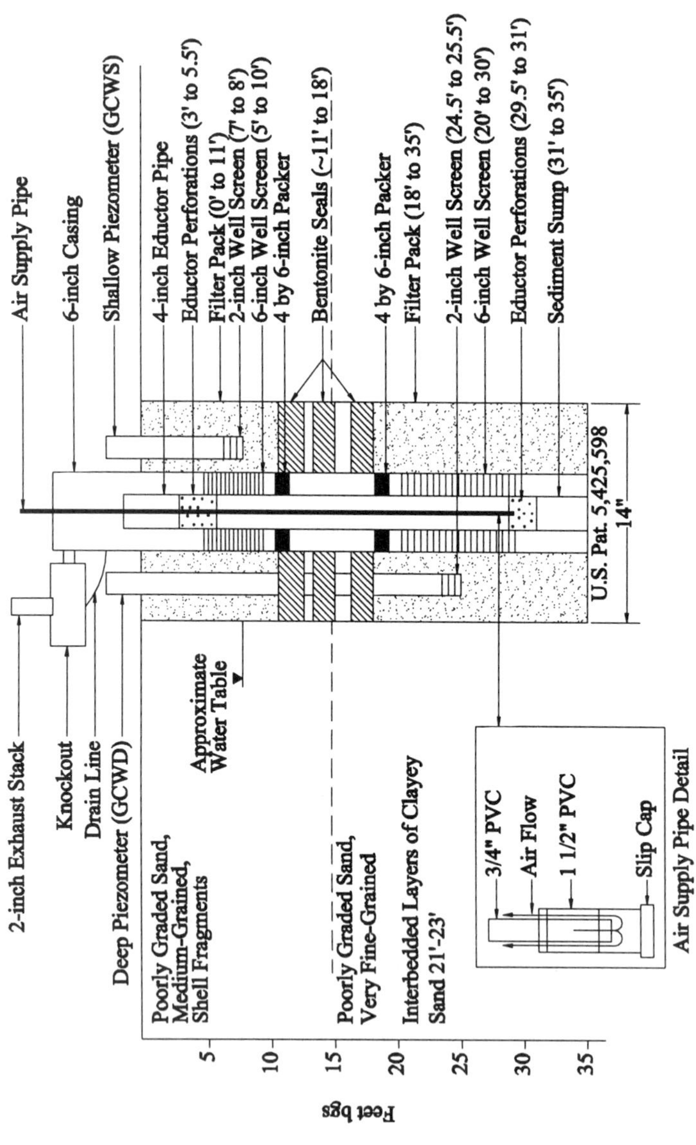

FIGURE 1 Cape Canaveral Air Station Groundwater Circulation Well Schematic (from Parsons 2001)

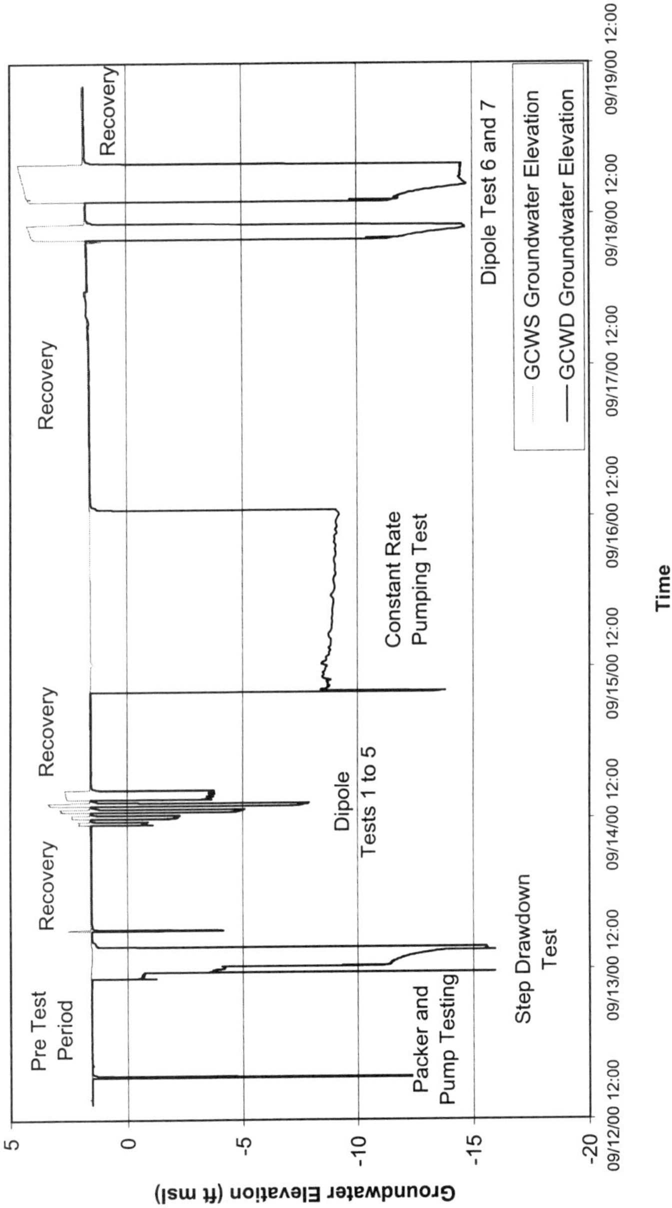

FIGURE 2 Groundwater Circulation Well Shallow and Deep Piezometer Hydrographs from September 2000 Aquifer Testing

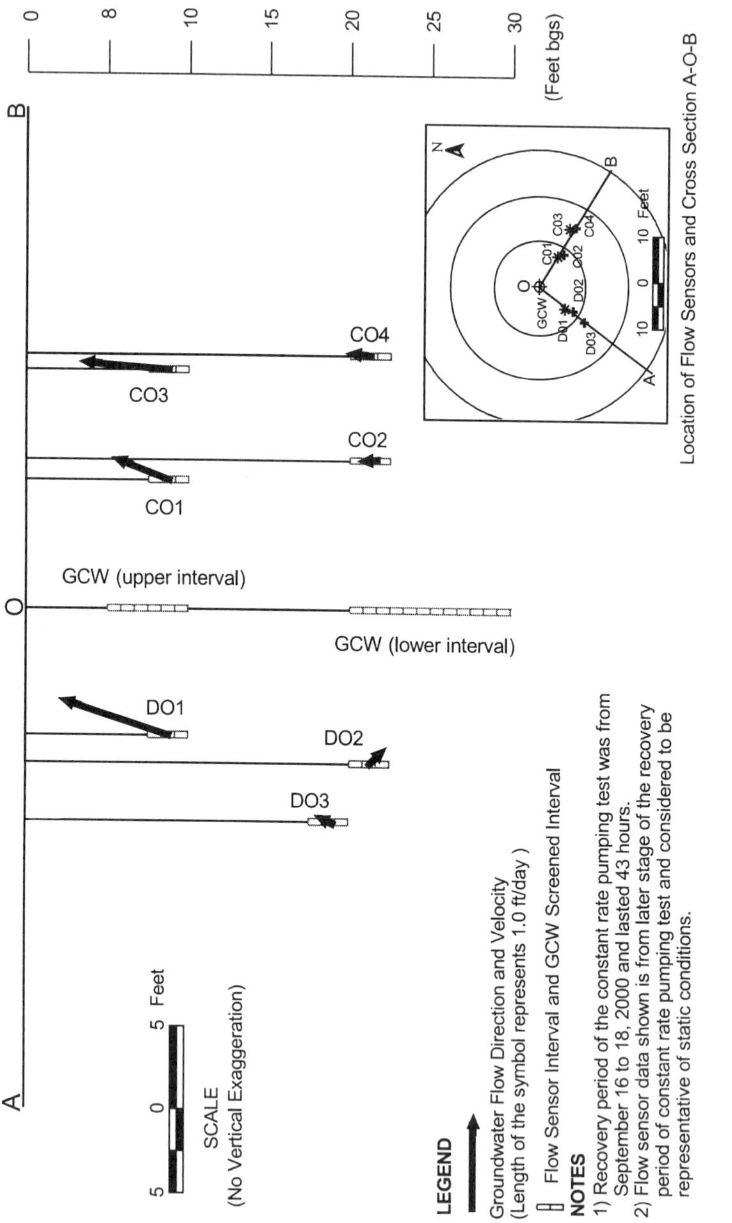

FIGURE 3 Cross Section A-O-B Groundwater Flow Under Non-pumping (Natural) Conditions (Actual Data)

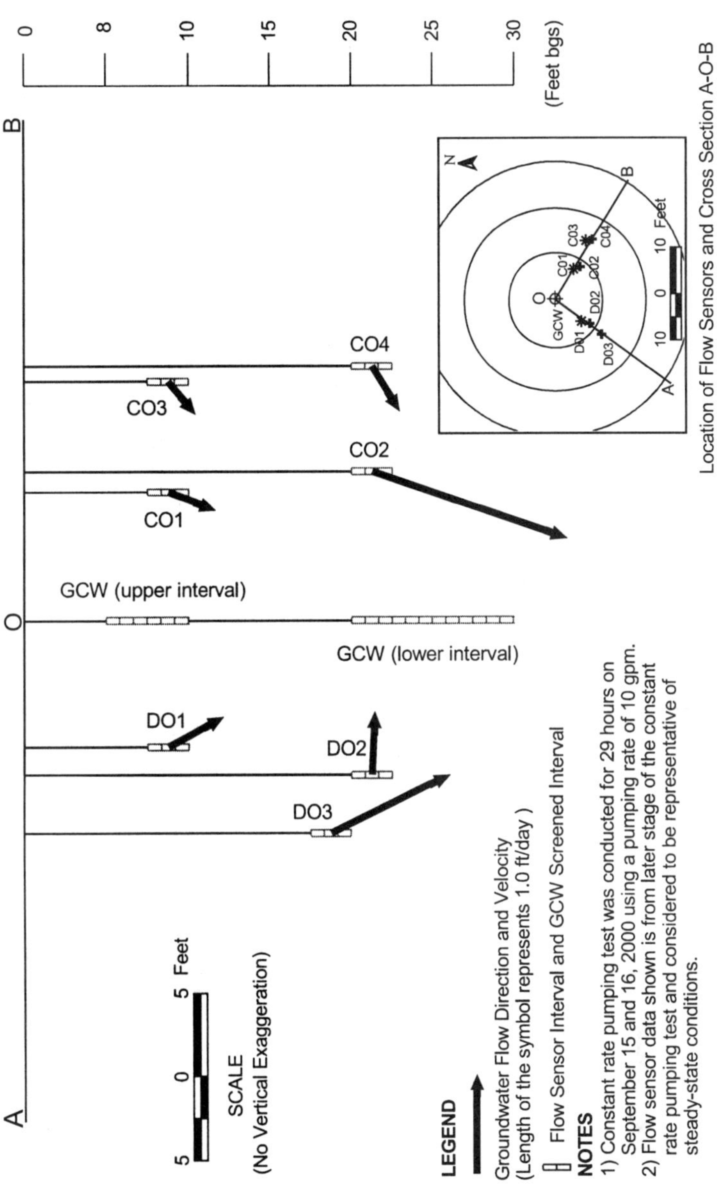

FIGURE 4 Cross Section A-O-B Resultant Groundwater Flow Under Pumping Conditions (Actual Data minus Background Natural Flow)

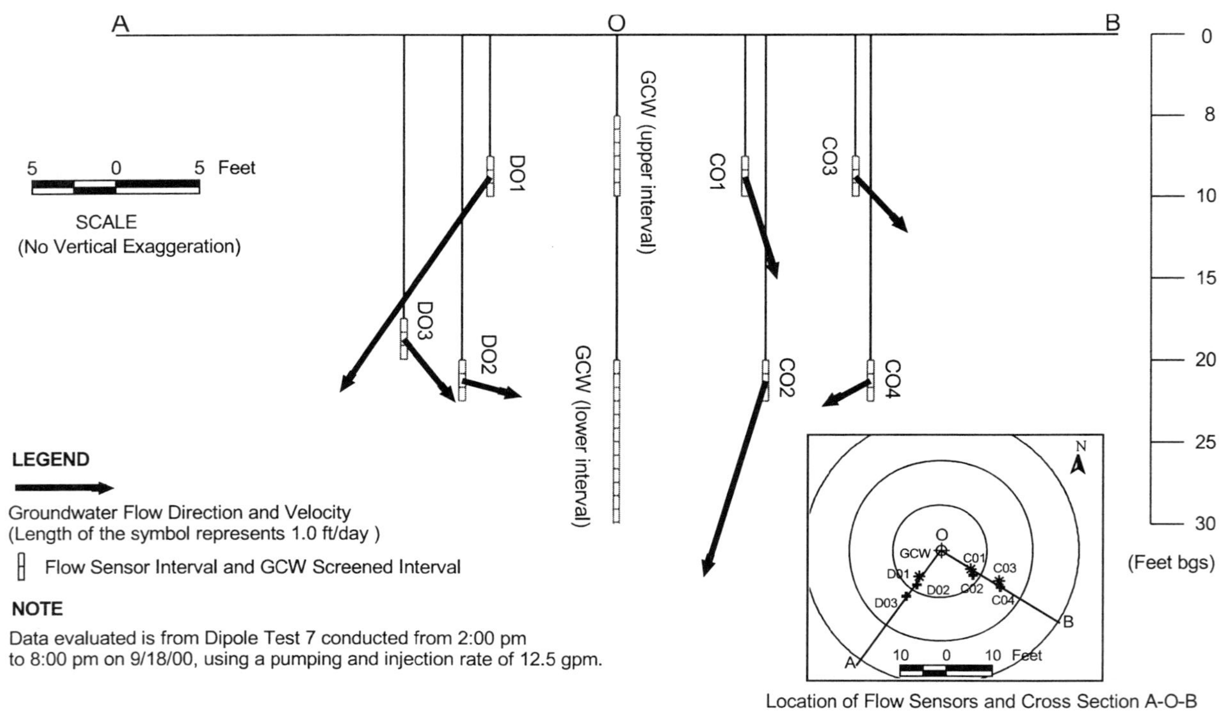

FIGURE 5 **Cross Section A-O-B Resultant Groundwater Flow Under Recirculation Conditions (Actual Data minus Background Natural Flow)**

ENHANCEMENT OF BIODEGRADATION OF DICHLORODIETHYL ETHER BY PRE-OXIDATION

Milica Kaludjerski, Fatih Büyüksönmez, Tuba Turan Ertas, Mirat D. Gurol
San Diego State University, San Diego, CA, USA

ABSTRACT: Wastewater from chemical manufacturing plants often contains hazardous xenobiotic (man made) compounds whose discharge into municipal treatment systems is prohibited because these compounds can persist in biological treatment systems and pass through to the effluent or absorb to the biosolids without any transformation. Some of these compounds can upset the biological process through toxicity. Dichlorodiethyl ether (DCDE) is such a compound. DCDE is widely used in the United States in the manufacture of pesticides and pharmaceuticals, as a solvent and cleaning fluid, as a constituent of paints and varnishes, and in the purification of oils and gasoline. DCDE is non-biodegradable in conventional biological treatment systems (Lewis, 1991). It is reactive, flammable, a mutagen and a probable carcinogen. Its inhalation can cause central nervous system depression (Agency for Toxic Substances and Disease Registry [ATSDR], 1989). Because of these properties, DCDE is present on the "Toxic Substances" list and the "Community Right to Know" list.

The objective of this study was to examine the feasibility of enhancing biodegradability through the pre-oxidation. The oxidation processes applied in this study were ozonation. Biodegradability was examined at different oxidation levels of DCDE: 25, 50, 75 and 100 %. The amount of non-oxidized DCDE was confirmed by gas chromatography, while total organic carbon (TOC) was used as surrogate parameters for oxidation products and the remaining DCDE. Activated sludge cultures were grown in steady-state conditions in our laboratories as non-acclimated or acclimated to DCDE, and were used in our experiments to test the biodegradability of oxidized DCDE solutions.

The results indicated that the biodegradability of pre-oxidized DCDE solutions improved very significantly compared to non-oxidized solutions. The best degradation rate was at 75% oxidation of DCDE. There was no statistical difference in the rate of biodegradation of oxidized products by either acclimated or non-acclimated bacteria. These results are the first to show that DCDE can be degraded by an integrated treatment scheme of sequential chemical and biological oxidation processes.

INTRODUCTION

Wastewater from chemical manufacturing plants often contains hazardous xenobiotic (man made) compounds whose discharge into municipal treatment systems is prohibited because these compounds can persist in biological treatment systems and pass through to the effluent or absorb to the biosolids without any transformation. Some of these compounds can upset the biological process through toxicity. As a result, industries are forced to implement costly alternatives such as chemical treatment and off-site land disposal. Dichlorodiethyl ether (DCDE) is such a compound.

DCDE is widely used in the United States in the manufacture of pesticides and pharmaceuticals, as a solvent and cleaning fluid, as a constituent of paints and varnishes, and in the purification of oils and gasoline. Its production in 1986 was reported to be 1.2 tons (ATSDR, 1989). DCDE is non-biodegradable in conventional biological treatment systems (Lewis, 1991). It is reactive, flammable, a mutagen and a probable carcinogen. Its inhalation can cause central nervous system depression (ATSDR, 1989). Because of these properties, DCDE is present on the "Extremely Hazardous Substances" list and the "Community Right to Know" list. In addition, DCDE is classified as a hazardous chemical under the Resource Conservation and Recovery Act (RCRA), which mandates that wastes containing DCDE be disposed of in an authorized RCRA facility.

OBJECTIVE

The objective of this study was to investigate the feasibility of enhancing the biodegradability of DCDE by chemical oxidation using ozone. The goal of oxidative treatment of wastewater that contain recalcitrant compounds is often not to mineralize the compounds completely to CO_2 and H_2O, but rather to convert the recalcitrant compounds to by-products that are readily biodegradable in conventional biological treatment processes. The purpose of coupling chemical and biological processes is to allow removal of the target pollutants using minimum yet optimum amounts of costly chemical oxidants, followed by relatively inexpensive, often already existing, biological processes such as activated sludge.

First, several experiments were conducted to investigate the potential biodegradability of DCDE as a single carbon source. Second, after being exposed to various degrees of chemical oxidation by an oxidation technique, the extent of biodegradability of the oxidized DCDE samples was determined, and the rates of biodegradation were measured. Third, the responses of acclimated and non-acclimated cultures to non-oxidized DCDE were compared. Microorganisms were acclimated to the presence of DCDE to allow production of specific enzymes that can degrade DCDE and its intermediates.

MATERIALS AND METHOD

Ozonation of DCDE. The experimental set-up used in these experiments is shown in Figure 1. It consists of an ozone generator, a pressurized oxygen gas cylinder, monitoring devices, and a 4L glass reaction vessel. The reaction vessel, which has openings for ozone inlet, outlet and for liquid sampling, was operated in semi-batch mode. Changing the power input into the generator controlled the amount of ozone produced. Ozone gas flow was controlled by a gas flow meter that was calibrated by the soap film technique.

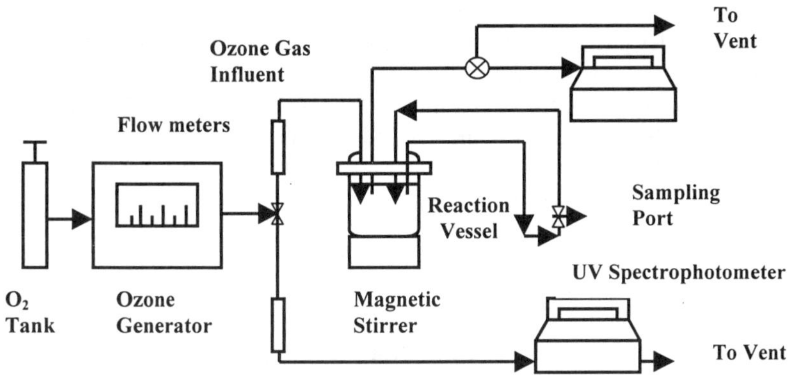

FIGURE 1. Experimental set-up for ozonation.

A glass diffuser was used to sparge ozone gas into the solution in the reaction vessel. The three liters of solution used for each test was mixed by a Teflon™ coated stirrer at high speed to maximize the mass-transfer and provide complete mixing. Teflon™ tubing was used for all the connections from the ozone generator to the reaction vessel. A piston pump with ceramic liner and piston was used to take samples from the reactor at various time intervals. Since natural waters contain bicarbonate ion, 5 mM sodium bicarbonate was added into solution. All ozonation experiments were conducted at pH 7.0 after adjustment with 0.1N sulfuric acid. The ozone gas concentration and the flow rate of the influent gas for all the experiments were kept at 29.5 mg/L and 1.5 L/min, respectively. The experiments were started with 200 mg/L of DCDE, and ozonated to provide 25, 50, 75 and 100 percent oxidation of DCDE (with an acceptable error range of ±5%). These solutions were made available for biodegradation studies.

Acclimated and Non-acclimated Sludge Cultures. Heterogeneous microbial populations used in this study were grown in our laboratory in 5L reactors, operated in semi-batch mode. The original biomass was obtained from the return line of the activated sludge reactor (RAS) in the North City Reclamation Plant, San Diego, CA. The mixed liquor suspended solid concentration in the reactor (MLSS) was adjusted to and maintained at 3,200 mg/L. The biomass was fed with a mixture of filtered primary effluent and chemical oxygen demand (COD) supplement solution to adjust the strength of the feed to 1,500 mg/L as COD. The COD supplement solution was prepared using the modified recipe of Henze (1992). The recipe contained the following chemicals per liter of distilled water: acetic acid, 2.5 mL; butyric acid, 2.5 mL; propionic acid, 2.0 mL; ethanol, 1.0 mL; glutamic acid monosodium salt, 5 g; glucose, 3.75 g; dibasic potassium phosphate, 80 g; monobasic potassium phosphate, 40 g. The primary effluent was collected once a week from the same plant, and the COD supplement solution was prepared in our laboratory. The mixture of filtered primary effluent and COD supplement solution was added to the reactor daily at a volume of 4 liters. Upon arrival at the SDSU laboratories, the primary effluent was filtered through a high porosity filter paper and then through a glass-fiber filter paper with a pore size of 0.45 μm. The filtrate was refrigerated to slow down the biodegradation of the organic compounds. Prior to its use, the COD concentration of the primary effluent was measured by the Closed Reflux, Colorimetric method (Standard Methods, 1995). Twice a day, 2L of the supernatant were withdrawn from the reactor to maintain the desired liquid and solid residence times. The operating parameters of the reactor consisted of sludge residence time of 5 days, food to microorganism ratio (F/M ratio) of 0.375 day-1 and a hydraulic residence time (HRT) of 1.25 days. The reactors were aerated with compressed air (1 L/min), which maintained about 4 mg/L of dissolved oxygen at ambient temperature (22°C±2). Once a week, one liter from each reactor was replaced by fresh RAS from the same plant in order to maintain diversity and to prevent major shifts of microbial community. Prior to experimentation, non-acclimated cultures were grown in our laboratories for at least 2 weeks. Also, in order to acclimate the microorganisms, one of the two reactors was treated with 0.4 mg/L of DCDE for at least 4 weeks in addition to the mixture of primary effluent and the COD supplement solution.

Biomass Harvesting. A standardized biomass harvesting protocol was used for each experiment. One liter of the biomass was withdrawn from the reactor and the biomass harvested by centrifugation at 35,280 × g for 15 minutes. The pellet was recovered and re-suspended in one liter of mineral salt solution (MSS). The MSS contained per liter of distilled water: dibasic potassium phosphate, 86 mg; monobasic potassium phosphate, 217.5 mg; monobasic sodium phosphate, 26.24 mg; ammonium chloride, 20 mg; magnesium sulfate, 22.5 mg; calcium chloride, 27.5 mg; ferric chloride, 0.25 mg; manganese sulfate, 0.04 mg; boric acid, 0.06 mg; zinc sulfate, 0.04 mg; and ammonium molybdate, 0.03 mg. After harvesting, the biomass was aerated at airflow of 1.65 L/min for two hours to achieve a stable rate of endogenous respiration during which suspended solids analysis was performed. The biomass was adjusted with MSS to yield 2000 mg/L as suspended solids.

Experimental Protocol for Biodegradation Studies. A modification of the Organization of Economic Cooperation (OECD) Method 209 (OECD, 1987) was followed as the experimental protocol for short-term studies. The method was based on the measurement of endogenous and exogenous respiration rates with dissolved oxygen (DO) probes (Orion, Model #97-08). The experiments were conducted in BOD bottles. A personal computer, with Pentium II processor, equipped with the data acquisition board (IO Tech, Model DynaRes-8U) was set up to acquire the dissolved oxygen data during the biodegradation experiments and then to calculate the kinetic constants of the biodegradation rate. Yeast extract (0.15 mg/L) was added to each BOD bottle before starting the experiments to provide the necessary micronutrients for the microorganisms. The oxidized solutions of DCDE to 25, 50, 75 and 100% oxidation, and a non-oxidized DCDE solution containing before oxidation 200 mg/L of DCDE were tested in triplicate. Oxygen consumption in the bottles was monitored as an indicator of bacterial degradation of substrate in this model activated sludge system. The duration of the experiments was kept at 30 minutes in order to prevent further acclimation of microorganisms as well as prevent depletion of oxygen. Using DO probes attached to a highly sensitive data acquisition system allowed the collection of data at a very high speed: one data point per second per bottle. The respiration rate of the microorganisms in each bottle was determined by measuring the oxygen concentration as a function of time, and by calculating the corresponding slopes.

The endogenous rate of respiration of microorganisms was obtained by running a control experiment with microorganisms alone. The net exogenous respiration rates for DCDE solutions were obtained by subtracting the endogenous rate from the observed exogenous respiration rates. After acknowledging the values of the respiration rates, it was possible to calculate the kinetic parameters of the Monod equation that is presented below:

$$\mu = \frac{\mu_{max} \cdot S}{S + K_S}$$

where μ is the specific growth rate (time-1), μ_{max} is the maximum specific growth rate constant (time-1), K_S is the substrate concentration that gives one-half μ_{max} (mg/L), and S is the substrate concentration (mg/L).The constants, μ_{max} and K_S were determined through non-linear curve fitting. The calculations of kinetic constants and statistical significance tests were performed by using Sigma Plot software for 95% confidence level. Comparison of the kinetic constants obtained for the ozonated samples and the non-ozonated samples allowed the proper assessment of the extent of improvement of biodegradation as well as the optimum ozonation levels.

Analytical Methods

Ozone Measurement

The influent and effluent gas concentration of ozone was measured by a Shimadzu UV-1601 UV-Visible Spectrophotometer at a wavelength of 258 nm. Potassium Iodide method was used to calibrate the spectrophotometer (Standard Methods, 1995). For the calibration, ozone gas was dissolved in a 2% KI solution. The iodine formed was titrated with 0.1 N standardized sodium thiosulfate under acidic conditions using a starch indicator. The liquid phase ozone concentration was determined by the Indigo method (Bader and Hoigne, 1981).

DCDE Measurement

The DCDE concentration in solution was determined by Shimadzu Gas Chromatograph equipped with a flame ionization detector. The aqueous samples were injected directly on a Nukol column (15 m x 0.53 mm with 0.5 mm film) that was obtained from Supelco (Bellefonte, PA). Helium was used as a carrier gas at 15 ml/min, while hydrogen and air are used to fuel the detector. The injector and detector were operated at temperatures of 150°C and 250°C respectively. The initial column temperature was 80°C, and it was increased to 200°C at rate of 5°C/min.

TOC Measurement

TOC concentration can be obtained by subtracting the IC from the TC measured. Both total carbon (TC) and inorganic carbon (IC) were measured using Schimadzu TOC Analyzer (5000A). TC and IC are measured on a same principle by measuring the amount of CO_2, which is detected by non-dispersive infrared gas analyzer. TC sample is combusted and IC is acidified to decompose to become CO_2.

RESEARCH RESULTS AND SIGNIFICANCE

The DCDE removal by ozonation under the stated conditions is presented in Figure 2. Ozone was found effective in oxidizing DCDE, removing 200 mg/L within 120 minutes, consuming about 3.3 moles of ozone per mole of DCDE. During this time period, removal of DCDE by stripping was found to be negligible.

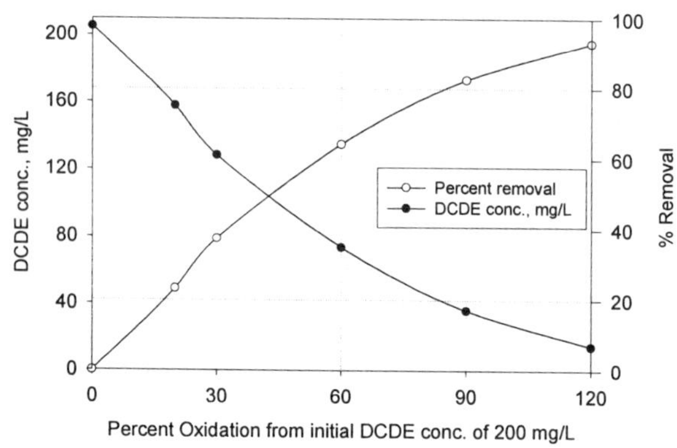

FIGURE 2. Removal of DCDE by ozonation

Samples of ozonated solutions at 25, 50, 75 and 100% (±5%) were collected and subjected to biodegradation tests, as described above. Since the ozonated solutions contained a mixture of the oxidation products and the remaining DCDE, the total organic carbon (TOC) was used as a surrogate measurement for total organic material present in the solutions. The amount of remaining DCDE was measured by gas chromatography. The TOC of the oxidation end-products was estimated by taking the difference between total TOC and the TOC attributable to DCDE; it was then referred to as "non-parent TOC".

The kinetic constants were determined based on "non-parent TOC". Figure 3 shows the net exogenous respiration rates of acclimated culture of ozonated DCDE solutions as a function of substrate concentration presented as non-parent TOC. It should be noted that non-ozonated DCDE solution was not biodegradable, i.e., the net exogenous respiration rate was observed to be zero. This solution, which also corresponds to zero amount of non-parent TOC, is represented in these plots as the point of origin. Furthermore, different levels of non-parent TOC were obtained by dilution of the solutions ozonated at 25, 50, 75 and 100%. The results in Figure 4 clearly show that for ozonated solutions, an increase in the substrate concentration is reflected by a rapid increase in respiration rates, reaching a plateau at higher concentrations, as would be expected. The increase in respiration rates for all levels of ozonation is significantly higher than the rate of non-ozonated DCDE, which is zero. This is clearly due to formation of biodegradable reaction products upon ozonation of DCDE. Some of these products were identified in our laboratories by other researchers as mainly carboxylic acids and aldehydes (Ertas and Gurol, 2001).

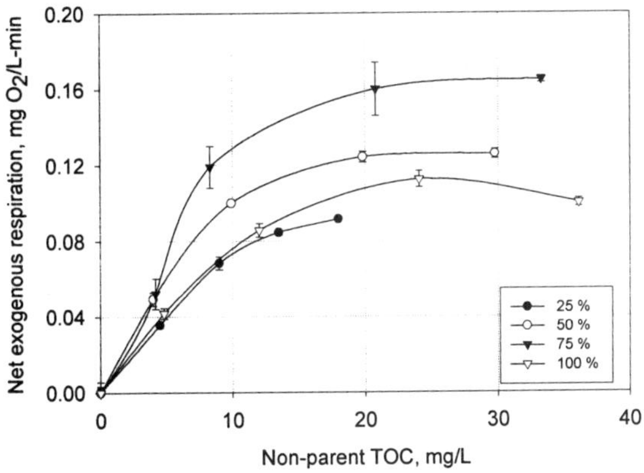

FIGURE 3. Biodegradability of ozonated DCDE by acclimated cultures.

The results also show that different levels of ozonation exerts different plateau, equaling maximum respiration rates, suggesting that the nature of non-parent TOC is different for each level of ozonation. Increasing the level of ozonation, for a given concentration of non-parent TOC, causes an increase in the respiration rates only up to 75% of ozonation level. A 100% ozonation produces a significant drop in the respiration rate. This behavior could be due to formation of more inhibitory oxidation products at 100% ozonation level.

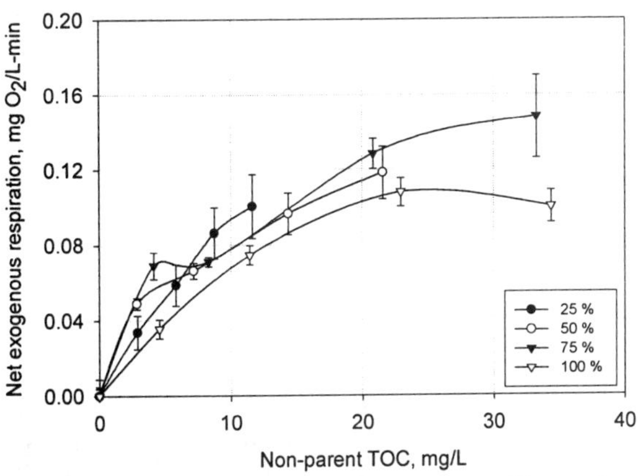

FIGURE 4. Biodegradability of ozonated DCDE of non-acclimated cultures.

The results obtained with non-acclimated culture, as presented in Figure 4, agree very well with those obtained with acclimated culture. Again the highest respiration rate was obtained for 75% ozonation level.

The μ_{max} values estimated for acclimated and non-acclimated cultures are plotted in Figure 5 as a function of different ozonation levels. Again, zero percent ozonation corresponds to the non-ozonated solution. It is clear that 75 % oxidation of DCDE by ozone produces a significantly higher μ_{max} value compared to other oxidation levels. The kinetic constants obtained for two cultures are not statistically different from each other at 95% confidence interval. Figure 6 presents the K_S values estimated based on non-parent TOC as the substrate versus different ozonation levels. The K_S values were found to be relatively insensitive to the level of ozonation, considering the large error predicted by the statistical program at 95% confidence interval.

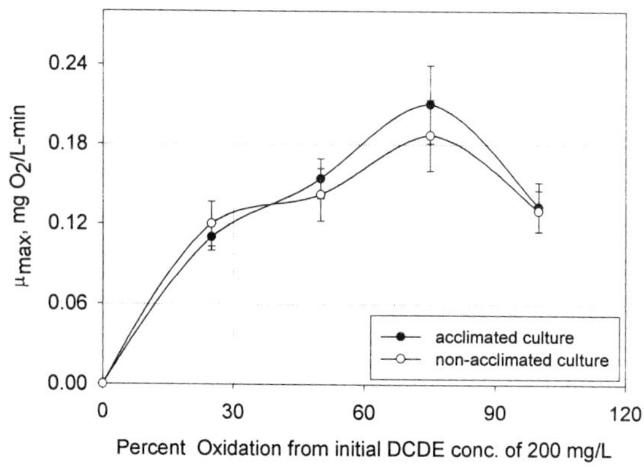

FIGURE 5. Comparison of μ_{max} values for acclimated and non-acclimated cultures.

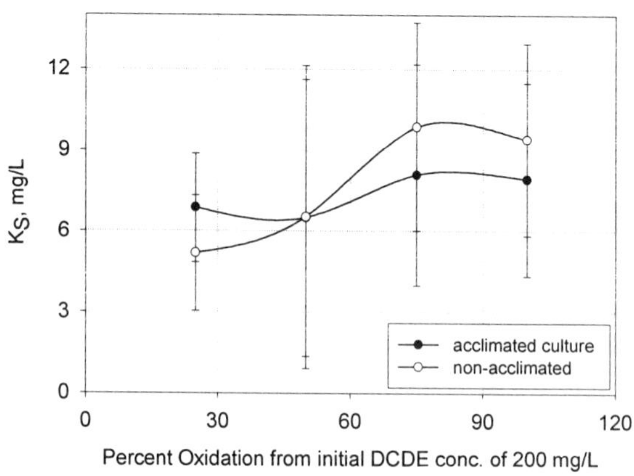

FIGURE 6. Comparison of K_S for acclimated and non-acclimated cultures.

CONCLUSIONS

The results indicated that DCDE was basically non-biodegradable by the activated sludge cultures maintained in our laboratories. Acclimating the cultures to DCDE did not cause any biodegradation of DCDE. However, ozonation was able to remove considerable amount of DCDE in relatively short time, consuming about 3 moles of ozone per mole of oxidized DCDE. Hence, it can be concluded that ozone alone can be used to oxidize all DCDE. Ozonation can also be used to oxidize DCDE to biodegradable compounds, such as carboxylic acids and aldehydes that can further be treated in biological treatment facilities. Our results indicated that such ozonation by-products were significantly more biodegradable with cultures both non-acclimated and acclimated to DCDE. Furthermore, reaction products of 75% oxidation of DCDE by ozone induce higher bioactivity for both acclimated and non-acclimated cultures, compared to other oxidation levels investigated in this study, including 100% oxidation. This result implies that some of the reaction byproducts at 100% oxidation might be inhibitory to the microorganisms. Our results imply that ozonation might also improve the biodegradability of DCDE in contaminated groundwater during in-situ or pump-and-treat remediation efforts.

REFERENCES

Agency for Toxic Substances and Disease Registry (ATSDR), 1989. *Toxicological Profile for Bis (2-chloroethyl) Ether*, ATSDR/TP-89/02.

Bader, H. and Hoigne, J., 1981. "Determination of Ozone in Water by the Indigo Method", *Water Resources*, Vol. 15, No. 4, pp. 449-456.

Ertas, T. T. and Gurol, M. D., 2001. "Oxidation of Diethylene Glycol (DEG) with Ozone and Fenton Processes", Submitted to *Chemosphere*.

Henze, M., 1992. "Characterization of Wastewater for Modeling of Activated Sludge Processes", *Water Science and Technology*, Vol.25, No 6, pp. 1-15.

Lewis, R. J. Sr., 1991. *Hazardous Chemicals Desk Reference*. Van Nostrand Reinhold, second edition, New York, NY.

OECD, Organization of Economic Cooperation, 1987. *Guidelines for Testing of Chemicals*. OECD, Paris, France.

Standard Methods. 1995. APHA, AWWA, WEF, nineteen edition, Washington, DC.

Verschueren, K., 1983. *Handbook of Environmental Data on Organic Chemicals*. Van Nostrand Reinhold, second edition, New York, NY.

ENHANCED BIODEGRADATION OF HYDROCARBON CONTAMINANTS USING PHOTOCATALYTIC PRETREATMENT

Chad E. Green (University of California, Davis)
Anita Stansberry, Jeff Sczechowski and Yarrow Nelson
(California Polytechnic State Univ., San Luis Obispo, CA)

ABSTRACT: Photocatalysis was investigated as a pretreatment for hydrocarbon groundwater contamination at an abandoned oil field in Guadalupe, CA. A laboratory photoreactor, with a UV fluorescent lamp submerged in a suspension of TiO_2 catalyst particles, was used to determine oxidation rates and the increase in biodegradability of the hydrocarbon mixture. Chemical oxidation demand (COD) and 5-day biological oxygen demand (BOD_5) were measured as a function of exposure time using standard methods. During eight-hour photoreactor runs in a nitrogen atmosphere, the COD of diluent solutions in de-ionized water and diluent solutions in simulated groundwater both decreased by approximately 50%, while the BOD_5 increased by a factor of 2 to 3. The biodegradability quotient, BOD_5/COD, increased from initial values of between 0.03 and 0.06 up to final values of between 0.27 and 0.29 during the eight-hour photoreactor runs. Little or no decrease in the chemical oxygen demand was observed in a control with no UV light and a control with no TiO_2, confirming the role of photocatalysis in these experiments. Based on the experimental oxidation and biodegradation kinetics, design parameters for *ex-situ* and *in-situ* remediation applications are presented.

INTRODUCTION

Improper handling and disposal of hazardous wastes has led to widespread groundwater contamination at thousands of sites throughout the world (Ollis et al., 1989). Ideally, groundwater contaminants could be removed via natural attenuation processes mediated by indigenous microorganisms. Unfortunately, some contaminants are not readily biodegradable or the kinetics of biodegradation are so slow that concentration limits designated by federal and local agencies cannot be met in a reasonable time period. Thus there has been a tremendous amount of research conducted to develop technologies that accelerate natural attenuation by enhancing or augmenting biodegradation via the introduction of additional nutrients, electron acceptors, or microorganisms to the subsurface environment. A more novel approach, explored in the present research, is to increase the biodegradability of the contaminants by physical-chemical pretreatment using photocatalysis. This approach could conceivably be applied to both above ground treatment schemes and in-situ remediation.

Photocatalysis is a promising advanced oxidation processes (AOP) that has proven useful for the oxidation of many organic hazardous waste compounds, including alkanes, alkenes, haloalkanes, haloalkenes, haloaromatics, aromatics (e.g. phenol, nitrobenzene, nitrotoluenes), surfactants, herbicides, and even some toxic metals (Ollis et al., 1989). Photocatalysis or photocatalytic oxidation (PCO) is a type of heterogeneous catalysis in which a semiconductor material is irradiated with photons of ultraviolet wavelength ($\lambda < 385$ nm), creating energetic electrons and holes on the surface of the semiconductor. In an aqueous environment, these charged carriers then react with dissolved oxygen and/or water to produce highly reactive hydroxyl radicals that readily oxidize the contaminants. PCO effectively oxidizes a wide variety of environmental contaminants, yet the most economically efficient means of remediation may be a combination of PCO and natural attenuation. Through oxidation, chemical contaminant structure is changed, often resulting in compounds that are much more biodegradable. In the present study, the feasibility of such a hybrid physical-biological treatment scheme was investigated for remediation of hydrocarbon-contaminated groundwater at a site on the central coast of California. The field site is an abandoned oil field that was contaminated by a mixture of C_{10} to C_{30} hydrocarbons (similar to diesel fuel), which was used as a diluent for pumping viscous crude oil. Groundwater contaminated with this diluent was collected and treated with a laboratory photoreactor using suspended TiO_2 catalyst. The kinetics of photo-oxidation and biodegradation determined in this study were then used to design potential *ex-situ* and *in-situ* hybrid remediation systems.

MATERIALS AND METHODS

Preparation of Contaminated Water. Samples of hydrocarbon diluent were collected during groundwater sampling from the Guadalupe Dunes Oil Field. A stock solution of contaminant was prepared by continuously mixing 200 mL of raw diluent with 3 L of filtered (5 µm) de-ionized water or simulated groundwater. Simulated groundwater contained 50 mg/L $CaCl_2$ (Fisher Chemical, Fisher Scientific, Fair Lawn, NJ) and 200 mg/L $NaHCO_3$ (Spectrum Chemical Co., Redondo Beach, CA). Contaminated water samples were allowed to come to equilibrium for 5 days before being used in the photoreactor.

Photoreactor Apparatus. The photoreactor used for these experiments was an annular re-circulating batch reactor as previously described (Turchi and Ollis, 1989). Briefly, the system is modular in design, with a pump, a slurry vessel and an annular flow tube. The slurry vessel is a 2 L, jacketed, round bottom flask with three threaded sampling ports. The annular reactor column is composed of two 18-inch long concentric Pyrex® tubes 30 mm and 54 mm in diameter. Within the inner concentric tube was an 18-inch, 15-watt fluorescent ultraviolet lamp ($\lambda \approx$ 365 nm, Visual Effects, Inc., New York, NY). Slurry flow was maintained at 6.4 L/min using a 2.3 amp sealed bearing centrifugal pump (March Mfg, Inc., Glenview, CA). The temperature of the photoreactor was kept at 25°C by pumping ice water into the jacket of the slurry vessel using a 1-100 rpm adjustable peristaltic pump (Cole Parmer, Chicago, IL). Experiments were conducted with either air or nitrogen sparging (A & R Welding, San Luis Obispo, CA) through a porous stone into the reaction vessel.

Photoreactor Operation and Sampling. Prior to the photoreactor runs, titanium dioxide (P25, Degussa Corp.) was added to the contaminated water at a concentration of 1 g/L and manually stirred. Once the TiO_2 was thoroughly mixed with the contaminated water, it was poured into the reaction vessel. 40 mL samples were collected at 1 hr intervals and filtered through a 0.45 μm Nylon membrane (Whatman, Maidstone, England, Cat. No. 7404002) into collection vials.

Chemical Oxygen Demand (COD) Analysis. COD was analyzed using the Accutest mercury-free Micro-COD System® (Bioscience, Inc., Bethlehem, PA). Potassium biphthalate (KHP) (Spectrum Chemical Co., Redondo Beach, CA), was used as a COD standard. Absorbance of KHP standards of known concentrations were measured and the COD in mg O_2/L was calculated using the stoichiometric relation between KHP and oxygen (Eqn. 1). A calibration curve was prepared to determine COD levels in the samples taken from the reactor.

$$4\ K\,C_8H_5O_4 + 29\ O_2 \rightarrow 32\ CO_2 + 10\ H_2O + 4\ K \qquad (1)$$

Biological Oxygen Demand (BOD) Analysis. The initial microbial seed for BOD analyses was prepared using Polyseed® BOD Seed Inoculum (Interbio, Woodland, TX). An enrichment culture acclimated to growth on diluent was prepared from Polyseed® and used as the seed for all BOD measurements. Each 60 mL BOD bottle (Wheaton, Millville, NJ) contained 10 mL of reactor effluent sample, 2 mL acclimated seed inoculum, 12 μL of concentrated buffered mineral solution (Hach, Loveland, CO) and dilution water. The dissolved oxygen (DO) concentration in each bottle was measured using a portable DO meter with a self-stirring, polarometric DO probe (YSI, Yellow Springs, OH). The BOD bottles were incubated at 20°C for five days. DO measurements were repeated every twenty-four hours for five days, and the BOD_5 was calculated from the observed O_2 depletion.

RESULTS

To investigate the effectiveness of photocatalysis as a pretreatment to biodegradation, photoreactor experiments were run for eight hours, analyzing COD and BOD_5 of reactor effluent at 1 hr intervals. The feed used in the photoreactor experiments was either de-ionized water saturated with diluent or simulated groundwater saturated with diluent. Diluent-saturated de-ionized water was tested to evaluate the effect of photocatalysis on biodegradability without the influence of common inorganic ions. Simulated groundwater was tested to determine if common groundwater ions, particularly bicarbonate ions, had any adverse effects on the rate of photocatalytic oxidation of diluent. The ratio of BOD_5/COD was used as a measure of the biodegradability of the photooxidized oxidized diluent. Preliminary experiments indicated that photocatalysis rates were 50% lower with nitrogen sparging compared to air sparging (data not shown). This agrees with a previous study that investigated the effects of dissolved oxygen on photocatalytic reaction rates (Hsiao et al., 1983). Although photocatalysis was slower under a nitrogen atmosphere, these results indicate that oxygen is not absolutely necessary for photocatalysis under these conditions. Since groundwater conditions are likely to be anoxic, all subsequent experiments were conducted with nitrogen sparging.

For diluent dissolved in de-ionized water, the COD of the photoreactor effluent decreased significantly (Figure 1) while the BOD_5 increased significantly (Figure 2) over the course of the eight-hour experiments. The BOD_5/COD ratio increased substantially over the eight-hour photoreactor run (Figure 3). Results for diluent in simulated groundwater were similar to the de-ionized water experiments, with COD of the photoreactor effluent decreasing significantly (Figure 1) and the BOD_5 increasing significantly (Figure 2). For the simulated groundwater experiment, the BOD_5/COD ratio increased linearly to over 9 times its initial value (Figure 3).

First order kinetics accurately modeled the COD data for both the de-ionized water and simulated groundwater experiments ($R^2 = 0.93$, $p<0.0001$), while zero order kinetics more accurately modeled the BOD_5 data measured during the first 5 days ($R^2 = 0.85$, $p<0.0004$) for each of the experiments. Table 1 summarizes the first order and zero order rate constants from the COD and BOD_5 analyses of the photoreactor runs. COD rate constants from a control without UV light and a control without TiO_2 are also shown for comparison. The rate constants from the controls are very low compared to the rate constants from the photoreactor experiments with both UV and TiO_2. BOD_5/COD data followed zero order kinetics ($R^2 = 0.95$, $p<0.00001$). The zero order BOD_5/COD rate constant for diluent in simulated groundwater was slightly higher than that for diluent in de-ionized water, indicating that the common inorganic ions present in the simulated groundwater had no detrimental effect on the rate of photocatalytic oxidation. These rate constants are also given in Table 1.

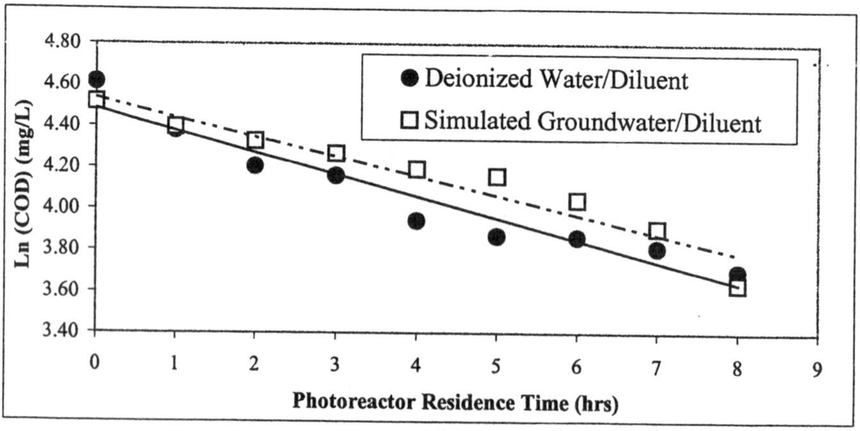

FIGURE 1. COD of photoreactor effluent for experiments with diluent in deionized water and simulated groundwater. Note the log scale for COD. These semi-log plots were used for the determination of first order rate constants.

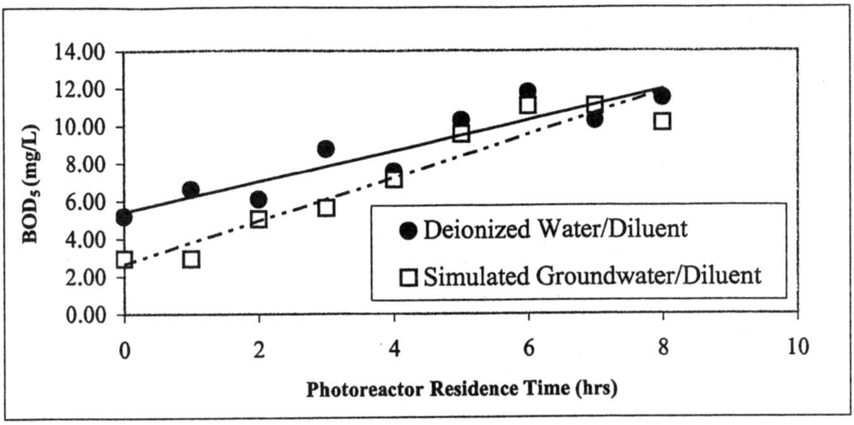

FIGURE 2. BOD_5 Analysis of Photoreactor Effluent from Diluent Saturated De-ionized Water and Simulated Groundwater Experiments.

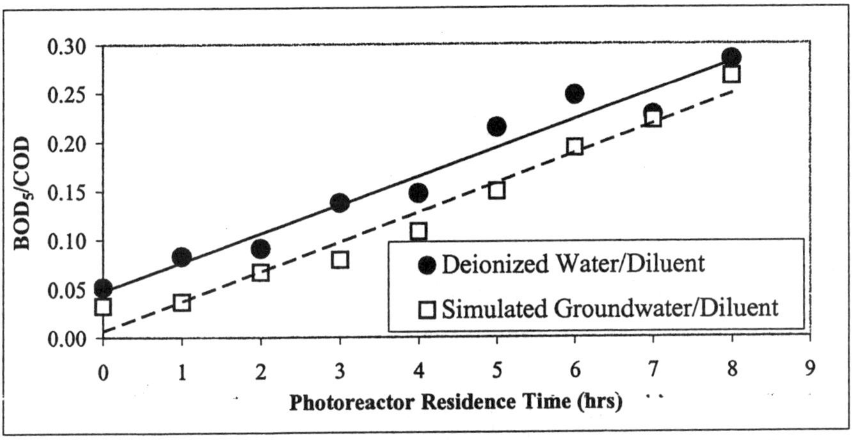

FIGURE 3. BOD_5/COD Analysis of Photoreactor Effluent from Diluent Saturated De-ionized Water and Simulated Groundwater Experiments.

TABLE 1. Summary of COD, BOD_5 and BOD_5/COD Rate Constants for Each Photoreactor Experiment.

Photoreactor Experiment with Diluent Saturated Water	COD First Order Rate Constants (mg/L*hr)	BOD_5 Zero Order Rate Constants (mg/L*hr)	BOD_5/COD Zero Order Rate Constants (1/hr)
De-ionized Water	0.1053	0.818	0.0292
Simulated Groundwater	0.094	1.15	0.0303
No UV Control	0.0261	-------	-------
No TiO_2 Control	0.0125	-------	-------

DISCUSSION

Gilbert (1987) studied the effect of prior ozonation on the biodegradability of twenty-eight organic compounds and concluded that oxidation was sufficient for rapid biodegradation when the BOD_5/COD quotient reached at least 0.4. Thus, an optimally designed photocatalytic pretreatment would utilize a reaction time capable of increasing the BOD_5/COD ratio to 0.4. In the current study, the BOD_5/COD ratio for the diluent increased five fold over the eight-hour experiment, but only reached a maximum of 0.3 during this time frame. Using the kinetic parameters obtained in this study, the data were extrapolated and it was estimated that it would take approximately 12 hours to reach a BOD_5/COD ratio of 0.4 under these conditions.

Results of the photocatalysis experiments are encouraging because they indicate that the process is feasible under typical groundwater conditions. The observation that photocatalysis is only inhibited by 50% under nitrogen compared to air is an indication that the process is viable even under the anoxic conditions that exist in the subsurface. Also, it is apparent that groundwater ions, such as bicarbonate, do not inhibit the photocatalytic degradation of hydrocarbon contaminants present in diluent contaminated groundwater, as confirmed with the simulated groundwater experiment. In fact, the photocatalysis rates and final biodegradability quotient (BOD_5/COD) were higher for the simulated groundwater than for diluent dissolved in de-ionized water.

It is important to note that these experiments were carried out in a slurry reactor with TiO_2 particles in suspension. In such a reactor, the overall reaction rate is limited by surface transformations and not mass transfer. In other reactor configurations or in an *in-situ* treatment application, reaction rates may be limited by mass transfer and, thus the time required to reach a BOD_5/COD ratio of 0.4 could be higher.

The kinetic data obtained in these experiments can be used to estimate design parameters for *ex-situ* (pump and treat) and *in-situ* remediation applications. An *ex-situ* system can be envisioned that consists of a packed-bed solar array photoreactor followed by an attached-film bioreactor for final biodegradation of the hydrocarbon contaminant. The proposed photoreactor is a fixed-bed reactor packed with TiO_2 coated silica gel, as described previously (Crittenden et al., 1996). The UV flux in such a system is about 600 times greater than that employed in these laboratory experiments. Based on this UV flux and first-order COD kinetics from the simulated groundwater experiments, the photoreactor retention time required to increase the BOD_5/COD ratio to 0.4 would be 1.4 min (based on an initial BOD_5COD ratio of 0.03). Thus, for example, a pump and treat system for groundwater saturated with diluent and operating at a flow rate of 20 L/min would require an effective photoreactor volume of 28 L. The bioreactor stage can be designed based on the experimentally observed first order BOD degradation kinetics (k = 0.096 mg/L-day). Assuming a maximum biomass concentration of 14 g/L (Shieh and Keenan, 1986), the bioreactor retention time required to reduce the contaminant concentration below 5 ppb would be 6 hr. Thus the example design flow rate of 20 L/min would require a bioreactor volume of 7200 L.

In-situ application of this hybrid technology would require bringing contaminated water into contact with UV light in the saturated zone of the contaminated aquifer. One way to accomplish this would be to bury fiber optic cables coated with TiO_2 in a subsurface reactive barrier as described by Peill and Hoffmann (Peill and Hoffmann, 1995). The required thickness of this reactive barrier will depend on photocatalysis kinetics and groundwater velocity. For a groundwater velocity of 50 m/yr (as observed at the field site), the required barrier thickness would be extremely thin (approx. 0.1 cm) based on the calculated retention time. However, a practical design would have to account for the lack of solar radiation for up to 16 hrs/day. Thus, a prudent design would employ a retention time of 16 hrs, increasing the required barrier thickness to 8 cm for this site. A practical remediation scheme could use a funnel and gate system with impermeable barriers funneling contaminated groundwater into the photoreactive barrier. Subsequent biodegradation of oxidized contaminant would be mediated by natural soil microbes down-gradient from the reactive barrier. Based on BOD degradation rates observed in the laboratory, the 50 m/yr groundwater velocity, and a soil microorganism concentration of 1 g/L soil, the distance required to reach 5 ppb would be only 40 cm. Of course, actual biodegradation rates will be considerably slower in the anaerobic groundwater environment. These illustrative calculations are very approximate, but indicate that the proposed hybrid treatment system is potentially feasible for both *ex-situ* and *in-situ* remediation.

CONCLUSIONS

The objective of this project was to determine the feasibility of a hybrid photocatalysis/biodegradation treatment approach to remediating groundwater contaminated with hydrocarbon diluent. Results indicated that photocatalytic pretreatment was successful in increasing the biodegradability of the hydrocarbon diluent in contaminated de-ionized water and in contaminated simulated groundwater. Chemical oxygen demand of the photoreactor effluent decreased by an average of 50% over the course of the eight-hour experiments, while the biological oxygen demand of the effluent rose by 2-3 times its original value for both deionized water and simulated groundwater experiments. The BOD_5/COD ratio, which was used as a measure of biodegradability, increased linearly from initial values of 0.03 and 0.06 to final values after eight hours of 0.27 and 0.29 for the simulated groundwater and de-ionized water experiments, respectively. Extrapolation of the linear regression equation was used to estimate the reaction time required to reach a BOD_5/COD ratio of 0.4. It was estimated that a ratio of 0.4 would be reached after a photoreactor retention time of 12-13 hours. It is at this point that biodegradability is considered acceptable for bioremediation. Additional control experiments were done without titanium dioxide and without ultraviolet light. As expected, contaminant oxidation was very slow in these experiments, confirming that the observed reductions in COD were the result of TiO_2-catalyzed photo-oxidation.

Photo-oxidation rates with nitrogen sparging were 50% lower than rates observed with air sparging. Although this reduction in rate is significant, these results indicate that photocatalytic oxidation under anoxic groundwater conditions is possible. Furthermore, dissolved ions at concentrations typical in groundwater did not inhibit photocatalytic oxidation. The success of these experiments indicates that *in situ* groundwater treatment using photocatalysis could be accomplished without the need for air sparging of the groundwater.

Finally, potential *ex-situ* and *in-situ* remediation designs were presented based on the observed photocatalysis and biodegradation kinetics. While these designs are preliminary, they indicate that photocatalysis pretreatment has potential in both scenarios, thus warranting further research in the application of this hybrid technology.

REFERENCES

Crittenden, J. C., Y. Zhang, D. W. Hand, D. L. Perram, and E. G. Marchand. 1996. "Solar Detoxification of Fuel-Contaminated Groundwater Using Fixed-Bed Photocatalysis." *Wat. Environ. Res. 68*(3): 270-278.

Gilbert, E. 1987. "Biodegradability of Ozonation Products as a Function of COD and DOC Elimination by Example of Substituted Aromatic Substances." *Wat. Res. 21*(10): 1273-1278.

Hsiao, C. Y., C. L. Lee, and D. F. Ollis. 1983. "Heterogeneous Photocatalysis: Degradation of Dilute Solutions of Dichloromethane (CH_2Cl_2), Chloroform ($CHCl_3$), and Carbon Tetrachloride (CCl_4) with illuminated TiO_2 Photocatalyst." *J. Catalysis. 82*: 418-423.

Ollis, D. F., E. Pelizzeti, and N. Serpone. 1989. "Heterogeneous Photocatalysis in the Environment: Application to Water Purification." In N. Serpone and E. Pelizzeti (Eds.), *Photocatalysis: Fundamentals and Applications,* pp. 604-637. Wiley, New York, NY.

Peill, N. J., and M. R. Hoffman. 1995. "Development and Optimization of a TiO_2-Coated Fiber-Optic Cable Reactor: Photocatalytic Degradation of 4-Chlorophenol." *Environ. Sci. Technol. 29*: 2974-2981.

Shieh, W. K., and J. D. Keenan. 1986. "Fluidized Bed Biofilm Reactor for Wastewater Treatment." In A. Fiechter (Eds.), *Advances in Biochemical Engineering/Biotechnology*, pp. 131-169. Springer-Verlag, New York, NY.

Turchi, C. S., and D. F. Ollis. 1989. "Mixed Reactant Photocatalysis: Intermediates and Mutual Rate Inhibition." *J. Catalysis. 119*: 483-496.

EFFECT OF CHEMICAL OXIDATION ON IN-SITU BIODEGRADATION OF CHLORINATED SOLVENTS

David S. Lipson, CPG – BBL (Golden, CO, USA)
John L. Persico, PG – BBL (Cranbury, NJ, USA)

ABSTRACT: The effect of chemical oxidation pilot studies on in-situ biodegradation of chlorinated solvents in groundwater was evaluated at a dense, non-aqueous phase liquid (DNAPL) in fractured bedrock site in New Jersey, USA. In-situ biodegradation is a component of the site-wide groundwater remedy, to be used as a follow-up to chemical oxidation (otherwise known as a polishing strategy). To evaluate the effect of chemical oxidation on in-situ biodegradation, site data were collected before, during, and after chemical oxidation pilot studies. Results show that short-duration chemical oxidation pilot studies did not adversely affect in-situ biodegradation in the short term, and in-situ solvent biodegradation was enhanced in the fractured bedrock aquifer. We hypothesize that one of the chemical oxidation injectants (i.e., acetic acid) was used as a carbon source by subsurface microorganisms during solvent biodegradation reactions.

INTRODUCTION

The site is an industrial property where historic waste disposal practices led to subsurface releases of the chlorinated solvents perchloroethylene (PCE) and trichloroethylene (TCE). DNAPL has been observed at only one location, a deep overburden groundwater monitoring well, but the DNAPL zone spans an area of approximately 5,000 square feet as inferred by maximum PCE and TCE groundwater concentrations up to 300 parts per million. The resulting plumes have been the focus of ongoing site remedial activities during recent decades. Remediation of bedrock beneath the site has been particularly difficult due to the effect of matrix diffusion, in which PCE and TCE are absorbed into the bedrock matrix and essentially immobilized there due to the impermeable nature of the rock. "Matrix diffusion" refers to a transport mechanism where chemical mass migrates from areas of high concentration to areas of low concentration, in this case from the bedrock fractures to the bedrock matrix. Matrix diffusion has rendered a significant proportion of the PCE and TCE at the site inaccessible to all practicable in-situ remedial technologies. The remedial strategy discussed herein involves reducing contaminant mass within bedrock fractures using in-situ chemical oxidation, reversing the direction of matrix diffusion from the bedrock matrix back into the fractures, and monitoring in-situ biodegradation of the residual plume.

Bedrock at the site consists of fractured basalt units of the Preakness Formation. The Preakness Formation was formed as a series of horizontal basalt flows separated by sedimentary rocks deposited during erosional periods between flow events. Of particular interest at the site is the first inter-basaltic layer of sedimentary rock, which is a thin siltstone layer approximately 100 feet below ground surface (Figure 1). The siltstone layer and a portion of the basalt immediately above and below it were confirmed to be hydraulically significant based on tracer test results and borehole image processing methods. Groundwater and soil sampling results also show that the inter-basaltic siltstone layer is a preferred flow pathway for site contaminants in groundwater.

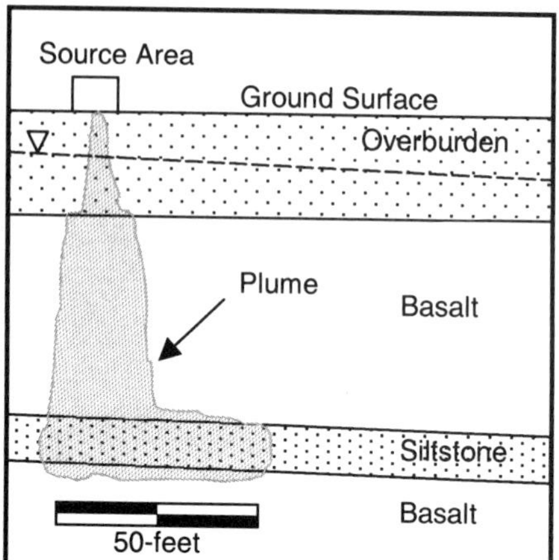

FIGURE 1. Site schematic.

Chemical oxidation technology was selected for pilot-scale evaluation because it could potentially destroy a significant mass of solvents in-situ. Two short-duration (two to four week) pilot studies were performed, one for overburden soils and the other for bedrock. The pilot studies generally consisted of injecting acetic acid, ferrous sulfate, and hydrogen peroxide into the groundwater through appropriately screened wells, and collecting data at the site monitoring network. Acetic acid was used to reduce groundwater pH levels to within an optimal range for chemical oxidation because it is relatively innocuous compared with strong acids such as hydrochloric and sulfuric. Concerns were raised about the possibility of chemical oxidation inhibiting the rate of in-situ biodegradation reactions or, worse, killing off a fraction of the subsurface microflora. These concerns were addressed by analyzing pre-pilot test groundwater samples for geomicrobiological indicator parameters and monitoring in-situ geochemical conditions during the chemical oxidation pilot studies.

METHODS

Monitoring Networks. The groundwater monitoring networks for overburden and bedrock are illustrated conceptually in Figures 2 and 3, respectively. Generally, the network consisted of two-inch diameter wells located at varying distances from the injection wells. The wells were screened at appropriate depth intervals in order to monitor hydrogeologic units of interest. As shown in Figure 2, the monitoring network for the overburden pilot study

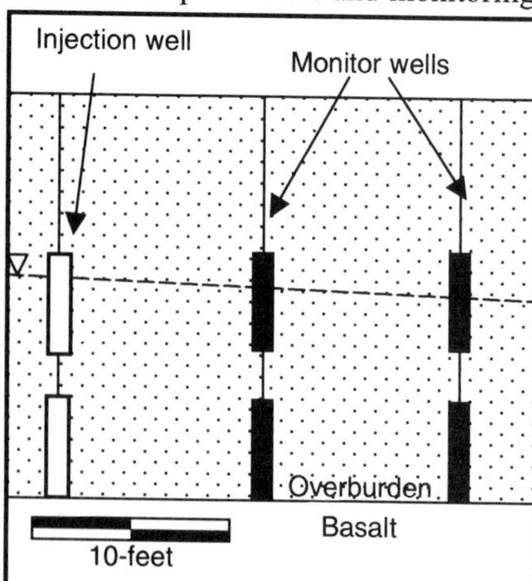

FIGURE 2. Conceptual setup of overburden pilot study.

consisted of two injection and four monitoring wells screening the top and bottom of the saturated soil zone. The well clusters used in the overburden pilot study were located at distances of 10 and 20 feet from the injection well. As shown in Figure 3, the monitoring network for the bedrock pilot study consisted of one injection well and four monitoring wells screened across the siltstone layer. The monitoring wells used in the bedrock pilot study were located at distances of approximately 15, 30, 65, and 80 feet from the injection well.

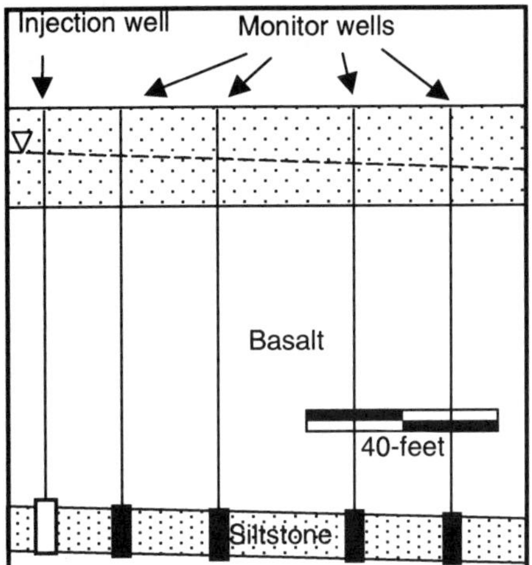

FIGURE 3. Conceptual setup of bedrock pilot test.

Sample Collection. The site monitoring network was sampled before, during, and after the chemical oxidation pilot tests. Samples were collected at monitoring wells using low-flow groundwater sampling methods based on the principle that low-stress well purging rates would provide samples reasonably representative of the groundwater in the geologic units of interest. All samples were collected, preserved, documented, and shipped according to standard remedial investigation/feasibility study (RI/FS) methods typically acceptable to the United States Environmental Protection Agency (USEPA) and the New Jersey Department of Environmental Protection (NJDEP).

Sample Analyses. A one-time sampling event was conducted before implementing the chemical oxidation pilot studies in order to characterize baseline geochemical and microbiological conditions in site groundwater. The samples were analyzed for the parameters listed in Table 1.

Table 1 – Sample analytical parameters for groundwater

Parameter	Analytical Method
Volatile organic compounds (VOCs)	SW-846
pH, temperature, conductivity, oxidation-reduction potential (ORP)	Field measurements
Nutrients	SW-846
Dissolved gases	GC/MS
Inorganic byproducts	SW-846
Dissolved organic carbon (DOC)	SW-846
Phospholipid fatty acids (PLFAs)	GC/MS

Additional sampling events were conducted during and after the pilot studies in order to evaluate the effectiveness of chemical oxidation, and to monitor potential geochemical changes that might adversely influence in-situ biodegradation reactions. Effectiveness monitoring samples were analyzed for VOCs using SW-846 methods and geochemical indicator parameters (pH, temperature, conductivity, and ORP) using field instruments.

Pilot Studies. Two short-duration (two to four weeks) chemical oxidation pilot studies were performed: The first was performed in overburden soils and the second was performed in bedrock. Approximately four months elapsed between the pilot studies in order to allow groundwater conditions to stabilize. The pilot studies were performed by first injecting acetic acid to lower groundwater pH, then injecting ferrous sulfate to serve as a catalyst, and finally injecting hydrogen peroxide as an oxidizer. It is important to note this sequence of events because, as discussed in greater detail below, this allowed the acetic acid to migrate beyond the chemical oxidation reaction zone. Chemical dosages were predetermined based on VOC concentrations and geochemical characteristics of the groundwater. Immediately after injection into groundwater, the hydrogen peroxide and ferrous sulfate caused a nearly instantaneous chemical oxidation reaction to occur in which hydroxyl radicals were generated and VOCs and other organic compounds were oxidized. This reaction occurred in a matter of minutes and resulted in the generation of byproduct gases (e.g., carbon dioxide) below the water table that occupied of portion of the porosity. The chemical oxidation reaction and generation of gases was readily observed in the injection and monitoring wells as bubbling, water column rises, and pressure buildups.

During the bedrock pilot study, approximately 950 pounds of acetic acid, 100 pounds of ferrous sulfate, and 5,561 pounds of hydrogen peroxide were injected into groundwater over a four-week period. Similar quantities were used during the overburden pilot study.

RESULTS AND DISCUSSION

Baseline Sampling Event. Results for both soil and bedrock groundwater zones indicated the presence of:

- Normal pH and sufficient buffering capacity;
- Anaerobic microorganisms (e.g., iron-reducing bacteria, sulfate-reducing bacteria, and methanogens);
- Reducing conditions (e.g., negative ORP readings);
- Macronutrients (e.g., nitrogen, phosphorus);
- Alternative electron acceptors (e.g., sulfate, carbon dioxide);
- Alternative carbon sources (e.g., DOC); and
- Elevated concentrations of biodegradation byproducts (e.g., ethenes).

These results show that prior to the chemical oxidation pilot studies, solvents were being biodegraded in groundwater by means of reductive dechlorination reactions associated with sulfate reduction and methanogenesis. The biodegradation capacity of the overburden appeared to be greater than the capacity of the bedrock zone. This was based on the PLFA data, which indicate that overburden samples had higher biomass and greater microbial diversity than the bedrock samples, and geochemical data indicating that bedrock zone samples were nearly depleted of alternative carbon sources. Thus, the overburden had a more robust microbial community adapted to the presence of VOCs in groundwater before the chemical oxidation pilot studies, while the bedrock had the potential to support in-situ biodegradation of VOCs by nutrient enhancements.

Additional Sampling Events. Results of the geochemical monitoring data are summarized in Figures 4 through 7. These figures show time-series plots of the geochemical data with time before (baseline), during, and after the pilot studies. Measurements plotted with time include groundwater pH, temperature, conductivity, and ORP. DO and chloride results were not plotted due to analytical interference associated with chemical oxidation.

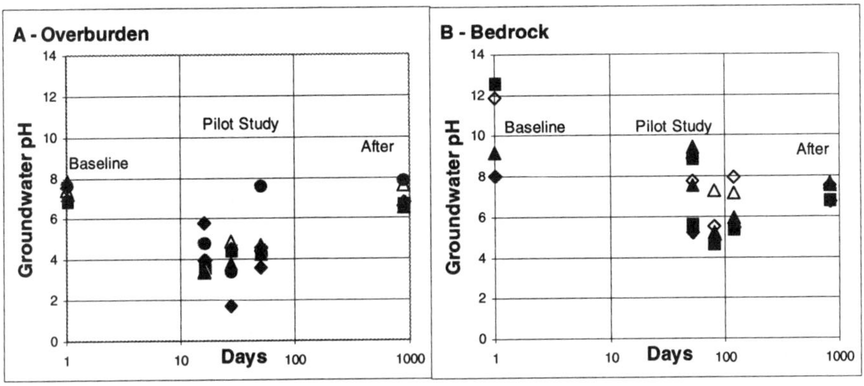

Figure 4. Groundwater pH with time before, during, and after chemical oxidation pilot studies.

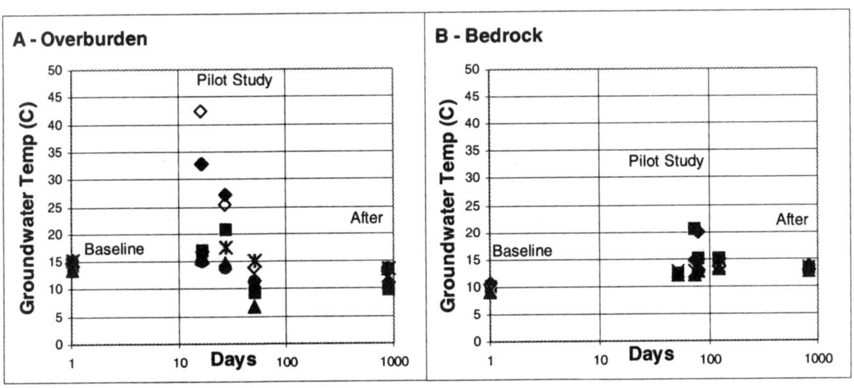

Figure 5. Groundwater temperature with time before, during, and after chemical oxidation pilot studies.

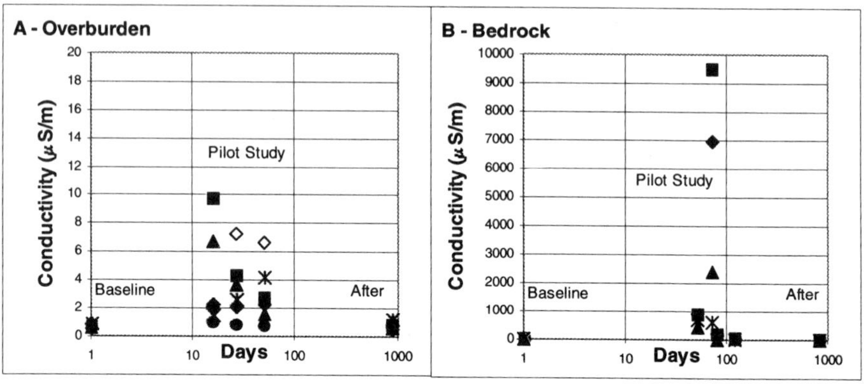

Figure 6. Groundwater electrical conductivity with time before, during, and after chemical oxidation pilot studies.

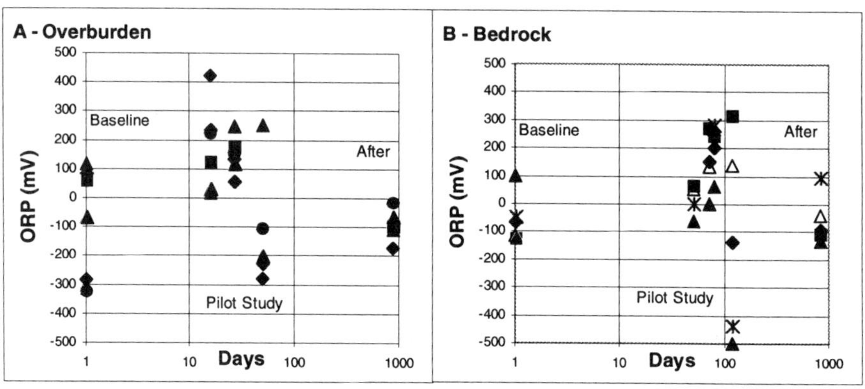

Figure 7. Groundwater oxidation-reduction potential (ORP) with time before, during, and after chemical oxidation pilot studies.

Results show that groundwater geochemistry changed from reducing conditions to oxidizing conditions during the pilot studies and rebounded back to reducing conditions after the pilot studies were complete. Figure 4 shows that groundwater pH decreased during the pilot studies due to the addition of acetic acid and returned to baseline conditions afterward. Elevated pH readings in bedrock groundwater prior to the pilot studies were likely influenced by monitor well grouting materials. Post-pilot study pH values for bedrock groundwater were consistent with the ambient pH of Preakness Formation groundwater. It was also observed that the pH of bedrock groundwater did not decrease below 4 standard units likely due to the buffering capacity of the bedrock. A pH range from 2 to 4 standard units is considered optimal for Fenton's reaction to occur because this range is needed to maintain iron in the ferrous state (Oberle and Schroder, 2000). Figure 5 shows that groundwater temperature increased during the pilot studies due to chemical oxidation (an exothermic reaction) and returned to baseline conditions afterward. Figure 6 shows that groundwater conductivity increased during the pilot studies due to the injection of ferrous sulfate, and returned to baseline conditions afterward. Groundwater conductivity at some bedrock sampling locations increased by several orders of magnitude immediately after injection of ferrous sulfate, while overburden groundwater conductivity values exhibited a much smaller increase (compare scales used on Figures 6A and 6B). It is probable that the increased conductivity values were also influenced by the lower groundwater pH, which would result in dissolution of aquifer minerals. Figure 7 shows that groundwater ORP increased during the chemical oxidation pilot studies and returned to baseline conditions afterward. These results are consistent with the presence of oxidizing conditions.

Figures 8 and 9 show the ratio of TCE to PCE concentrations (TCE/PCE ratio) and the ratio of DCE to TCE concentrations (DCE/TCE ratio) in groundwater samples collected before, during, and after the pilot studies. Concentration ratios are a useful way of evaluating analytical results for a series of compounds that undergoes sequential degradation, such as the PCE, TCE, and cis-1,2-dichloroethylene (DCE) series. For TCE and PCE concentrations, a ratio less than one indicates that the PCE concentration is greater than the TCE concentration; a ratio greater than one indicates that the TCE concentration is greater than the PCE concentration; and a ratio of one indicates that TCE and PCE have approximately the same concentration. The same is true of DCE and TCE concentrations. By plotting these ratios with time it is possible to discern changes in degradation rates.

As shown in Figure 8A the TCE/PCE concentration ratio increased with time from less than one before the overburden pilot study to greater than one after the pilot study at most of the monitoring wells. These results show that PCE concentrations initially exceeded TCE concentrations prior to the pilot study, and that TCE concentrations increased relative to PCE during and after the pilot study. These changes cannot be attributed solely to chemical oxidation because it is non-selective with regard to VOCs (Suthersan, 1997). The information shown in Figure 8A is evidence that the PCE dechlorination rate increased significantly in overburden groundwater as a result of the chemical oxidation pilot study.

In contrast, the TCE/PCE concentration ratio in bedrock groundwater remained less than one throughout and after the chemical oxidation pilot study (Figure 8B). In fact, the data show that TCE/PCE concentration ratios decreased slightly with time during the bedrock pilot study. This is due to the increased TCE dechlorination rate, as discussed below, and the presence of DNAPL which acts as a constant concentration source of PCE.

Figure 9 shows the ratio of DCE to TCE concentrations in groundwater samples collected before, during, and after the pilot studies. As shown in Figure 9A, the DCE/TCE concentration ratio increased with time from less than one before the pilot study to greater than one after the pilot study in only two of six overburden monitoring wells sampled. The information shown in Figure 9A is evidence that the TCE dechlorination rate increased slightly in overburden groundwater as a result of the chemical oxidation pilot study.

As shown in Figure 9B, the DCE/TCE concentration ratio in bedrock groundwater increased with time from less than one before the pilot study to much greater than one after the pilot study at a majority of monitoring wells. In fact, the DCE/TCE concentration ratio in bedrock groundwater at one monitoring location increased with time from less than one to more than 200. The information shown in Figure 9B is evidence that the TCE dechlorination rate increased significantly in bedrock groundwater as a result of the chemical oxidation pilot study.

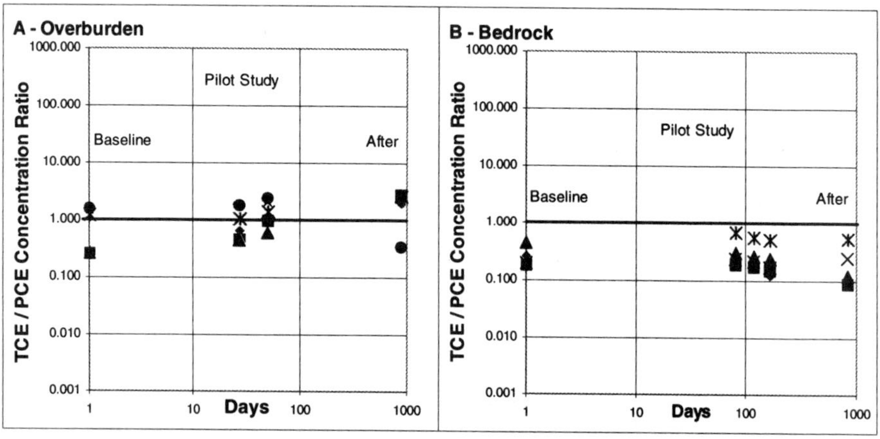

Figure 8. Ratio of TCE to PCE concentrations in groundwater samples before, during, and after chemical oxidation pilot studies.

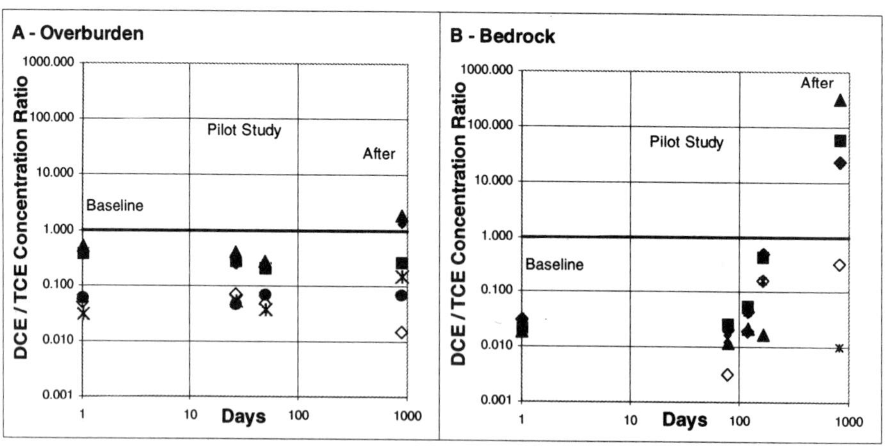

Figure 9. Ratio of DCE to TCE concentrations in groundwater samples before, during, and after chemical oxidation pilot studies.

CONCLUSIONS

In-situ biodegradation of solvents was occurring in overburden and bedrock groundwater prior to the chemical oxidation pilot tests as indicated by the baseline geochemical and microbiological data. Results demonstrated the presence of anaerobic microorganisms such as sulfate reducing bacteria and methanogens, macronutrients, alternative electron acceptors, and low levels of intermediate byproducts. Organic carbon appeared to be a limiting factor that inhibited biodegradation rates.

During the chemical oxidation pilot studies, geochemistry of both overburden and bedrock groundwater was temporarily changed from reducing to oxidizing conditions, as shown by the information in Figures 4 through 7. Although VOC concentrations decreased slightly, the chemical oxidation pilot studies did not appear to be effective at reducing a significant mass of VOCs in groundwater. Groundwater geochemistry rebounded after the pilot studies and reducing conditions were enhanced in the fractured bedrock aquifer (Figure 7).

In-situ biodegradation rates of solvents were increased in overburden and bedrock groundwater as a result of the chemical oxidation pilot tests (Figures 8 and 9). Significant increases in the TCE/PCE and DCE/TCE concentration ratios in overburden and bedrock groundwater (respectively) provide evidence of the increased biodegradation rates. These changes cannot be attributed solely to chemical oxidation and natural aquifer processes. Further evidence of enhanced biodegradation is shown by the three-to-six month lag time required for the TCE/PCE and DCE/TCE concentration ratios to increase at this site. Enhancements to biodegradation reactions exhibit a characteristic lag phase, or acclimation phase, that can span approximately three to six months in anaerobic environments (Chappelle, 1993).

We hypothesize that a portion of the acetic acid injected during the chemical oxidation pilot studies was used by indigenous groundwater microorganisms as a carbon source. Acetic acid dissociates in groundwater to acetate and hydrogen ions, thereby lowering groundwater pH. Acetate can be metabolized by a variety of microorganisms and has been observed by others to act as an electron donor during biologically mediated reductive dechlorination reactions involving PCE, TCE, and DCE (Vogel and McCarty, 1985; Freedman and Gossett, 1989).

Since acetic acid was the first chemical injected during the pilot studies, it was probably transported in groundwater a longer distance away from the injection wells compared with ferrous sulfate and hydrogen peroxide. This is because chemical oxidation was limited to an area relatively close to the injection wells due to a combination of the nearly instantaneous reaction rate and a reduction in permeability associated with generated gases. Therefore, the more distant areas in the bedrock must have received a dose of acetic acid (and acetate) but were not influenced by chemical oxidation reactions. After the chemical oxidation pilot studies, reductive dechlorination of solvents was likely enhanced in portions of the un-oxidized bedrock as indigenous microorganisms adapted to the use of acetic acid and acetate as a source of organic carbon. Furthermore, acetic acid and acetate likely diffused into the bedrock matrix and could serve as a continuing source of organic carbon for some time due to the process of reverse diffusion, whereby a chemical can diffuse from the rock matrix into fractures. On this basis, enhanced in-situ biodegradation by means of injecting innocuous dissolved organic compounds into bedrock may be a viable component of the groundwater remedy at this and other DNAPL in fractured rock sites. Results of this work support the growing body of evidence that in-situ bioremediation methods can be an appropriate remedial strategy for removing chlorinated solvents from DNAPL source zones (e.g., Cope and Hughes, 2001).

More detailed analyses of the pilot test data are warranted. Stoichiometric evaluations of the mass balance between PCE, TCE, and DCE concentrations before, during, and after the pilot studies could be used to verify our observations and conclusions. These evaluations are currently being performed and may be provided in future publications.

REFERENCES

Chappelle, F. H. 1993. *Ground-Water Microbiology and Geochemistry*. John Wiley & Sons, Inc. New York.

Cope, N. and J. B. Hughes. 2001. "Biologically-Enhanced Removal of PCE from NAPL Source Zones." *Environ. Sci. Technol. 35*: 2014-2021.

Freedman, D. L. and J. M. Gossett. 1989. "Biological Reductive Dechlorination of Tetrachloroethylene and Trichloroethylene to Ethylene Under Methanogenic Conditions." *Appl. Environ. Microbiol. 55*: 2144-2151.

Oberle, D. W. and D. L. Schroder. 2000. "Design Considerations for In-Situ Chemical Oxidation." In G. B. Wickramanayake, A. R. Gavaskar, and A. S. C. Chen (Eds.), *Chemical Oxidation and Reactive Barriers: Remediation of Chlorinated and Recalcitrant Compounds*, pp. 91-99. Battelle Press. Columbus, Ohio.

Suthersan, S. S. 1997. *Remediation Engineering: Design Concepts*. CRC Press, Inc. Boca Raton, Florida.

Vogel, T. M. and P. L. McCarty. 1985. "Biotransformation of Tetrachloroethylene to Trichloroethylene, Dichloroethylene, Vinyl Chloride, and Carbon Dioxide under Methanogenic Conditions." *Appl. Environ. Microbiol. 49*: 1080-1083.

BIOREMEDIATION OF OIL-COATED STONES

Prayad Pokethitiyook (Mahidol University, Bangkok, Thailand)
Michael A. Connor (University of Melbourne, Parkville, Vic., Australia)

ABSTRACT: The suitability of bioremediation techniques for on-site cleaning of oil-coated gravels was investigated. The oil was present as a 1 mm thick, paste-like layer containing small sand particles. Two types of fixed-film reactors were studied. In one, stones were placed in an aerated tank of water; in the other, water was continuously recirculated over a packed bed of stones. The reactors were inoculated with a culture acclimatized to the oil on the stones. Comparable oil removal and degradation rates were achieved in both reactors. The maximum extent of oil removal achieved was around 60%, by which point the residual oil had changed color, become flaky and detached readily from the stones. The effects of various factors on reactor performance were studied. Physical processes were found to play a significant role in oil removal. Nutrient supplementation led to increased initial oil removal and degradation rates but had less effect on the amount of oil finally removed. As expected, decreases in operating temperature lowered degradation rates; they also caused a marked and unexpected decrease in levels of oil removal. Except at the start, rates of bioremediation appeared to be controlled by diffusion processes within the oil itself.

INTRODUCTION

ESSO Australia Pty Ltd operates a Crude Oil and Gas Processing Facility at Longford, in Victoria, Australia. This facility uses gravel as a ground cover beneath its process equipment. Oil from minor leaks and spills collects on the gravel stones where it weathers to a highly viscous, paste-like oily residue. These oil-coated stones have to be replaced periodically; this is costly as they are classified as a prescribed waste and have to be disposed of in an off-site licensed waste treatment facility. In the early 1990's, work began to see if bioremediation would be a cost effective method of cleaning the oily stones on-site. Early results were encouraging and provided the impetus for the work described in this paper.

In this study, two types of fixed-film batch reactor bioremediation processes were studied. In the first type (referred to here as an AS (aerated submerged) reactor), the stones, supported on a stainless steel mesh tray (300 mm long and 190 mm wide), were submerged in a water tank aerated using diffusers. In the second, packed bed (PB) type, the stones were packed into a cylindrical wire mesh holder to form a bed approximately 100 mm deep. The holder was suspended in a 100 mm diameter vertical glass tube. Water was recirculated continuously over the bed. Both reactors were seeded with a microbial culture derived from moist soil collected at the Longford plant and adapted to using oil from the stones as its sole carbon source. Within a few days after inoculation, biofilms were visible on the surfaces of the stones. More details of the experimental equipment can be found in Pokethitiyook (1999).

The aim of the study was to obtain information of practical relevance to on-site operation of these reactors. Factors investigated included:

- The effect of temperature on rates of hydrocarbon removal
- The effect on removal rates of adding supplementary nutrients to the water
- The contribution of purely physical processes to oil removal rates
- Which of the reactor types was the better.

MATERIALS AND METHODS

Our experimental program was strongly influenced by the lack of uniformity among the stones. Although reasonably isometric, they still varied somewhat in shape, in size (sizes ranged from 7.5 to 25mm) and in oil layer thickness (around 1mm). To minimize the effect of these variations, stones were sorted into batches with fairly well matched size and shape characteristics.

Chromatographic analysis showed the weathered oil residues on the stones to consist of a highly complex mixture of hydrocarbons containing reduced levels of more volatile components compared with the original oils. Embedded in the residues were numerous small soil and sand particles, referred to here as 'grits'. Their presence complicated the procedures needed to establish carbon and oil mass balances over the reactors during experiments. Traces of moisture were also present. Therefore stones had to be dried to constant weight, at room temperature, before being used in experiments; this usually took several days.

Initially, oil removal rates during bioremediation experiments were going to be found by periodically sampling stones from each reactor and measuring how much oil they still retained. It soon became apparent that differences between individual stones were too great for this to work; to obtain statistically meaningful results, an impractically large number of stones would be needed in each sample. Therefore, a different approach was tried. For each set of experimental conditions being investigated, a series of five or six separate experiments was undertaken. These experiments ran for different times but were conducted under conditions that were as near as possible identical. Each experiment contributed one point on a plot of extent or rate of oil removal versus time. Though very time-consuming, this approach of using a separate batch of stones for each experimental point largely overcame the problems caused by variations in stone characteristics.

The experimental procedure was as follows. A pre-sorted batch of oil-coated stones was dried, weighed and placed in the reactor. Reactor operation then commenced, with the stones being kept under a predetermined set of conditions. After a preset time (usually 5, 10, 20, 30 or 40 days) the experiment was stopped, and the stones carefully removed. Any residual moisture was removed by drying the stones at room temperature. The stones were then reweighed. The following information was also obtained:

- The mass of any grits exposed during the biodegradation process that had detached from the stones and fallen into the water tank
- The mass of oil on the stones removed from the reactors; this was determined by first extracting the oil from the weighed stones with freon (1,1,2-trichloro-1,2,2-tri-fluoroethane) (or, later, hexane), then weighing

the dry, oil-free stones; allowance also had to be made for the mass of grits released during the extraction process

- The mass of small amounts of undegraded oil that on occasions detached from the stones and accumulated on reactor surfaces.

At the end of each experiment the volatile suspended solids (VSS) and (filtered) chemical oxygen demand (COD) levels in the water were also measured. The VSS results provided a measure of microbial cell concentrations in the bulk water phase while the COD showed how much dissolved organic matter was present.

From the above results one could determine both the amount of oil lost by the stones and the fraction of this oil that had been biodegraded. Average rates of oil removal could also be found for each experimental run. Because batches of stones were not identical, these rates had to be normalised before comparing results from different experiments. Stone surface area was felt to be the most appropriate normalising parameter.

Because of the range of stone sizes and shapes, determining stone surface area was not a simple matter. A procedure was devised that enabled the surface areas of each of a representative sample of 45 cleaned, oil-free stones to be found. This involved tightly wrapping each (pre-weighed) stone in thin plastic film, painting the exposed areas of the film, removing and unfolding the film, and using a computerised image analysis program to determine the area of painted surface. The procedure was calibrated using cubes and spheres; it consistently over-predicted surface areas by 8-9% (Pokethitiyook, 1999). It was found that surface area was related to mass by the following relationship:

$$Y = 1.057X + 5.795$$

where Y is a stone's surface area (cm^2) and X is its mass (g). (The correlation coefficient was 0.9 at a 95% confidence level (Pokethitiyook, 1999).)

After each experiment, the mass of each stone (oil-free) was measured. Its surface area was estimated using the above relationship and the areas of the individual stones summed to give the overall batch surface area. This latter area was then used to compute rates of oil removal on an area basis.

Because of the labour- and time-intensive nature of the experimental procedures, we decided to focus our efforts on AS-type reactors (which were easier to make and set up) and limit the experiments carried out on the PB reactor.

The results from one series of experiments on the AS reactor were used as a reference data set. The six runs in this experimental series were stopped after 5, 10, 20, 30, 40 and 100 days respectively; all were carried out at 20°C in an aqueous medium to which both an inoculum and extra nutrients had been added. The nutrient solution comprised a mixture of salts that would supply any N, P, S and trace metal needs of the microbial community. In general, runs were not extended beyond 40 days. This was because rates of biodegradation slowed markedly after 30 to 40 days and bare patches started to appear on some stones. In addition, by this stage much of the remaining oil had become much paler in colour, lost its pasty consistency and become flaky (and hence, from a practical standpoint, easier to remove by physical rather than microbial processes).

RESULTS AND DISCUSSION

Comparisons between AS and PB Reactors. The nature of the biofilm community differed between the two reactors. In the PB reactor, fungal mycelia could be seen on the stones, but these were absent in the AS reactors. In addition, oil removal in the PB reactor occurred preferentially in the topmost part of the bed and on the upper surfaces of individual stones; in the AS reactors on the other hand, oil removal occurred more uniformly over the stone surfaces.

Despite these differences, the pattern of oil removal in a PB reactor run at 20°C with nutrient supplementation (the only complete experimental run carried out on the PB reactor) corresponded closely with that for an AS reactor under similar operating conditions – see Figure 1. The extent of oil removal (expressed as a percentage of the initial level) appeared slightly higher for the AS reactor but, as can be seen from Figure 2, the PB reactor looked better when rates of oil removal per unit area were compared. Hence neither reactor type can be said to have a clear advantage over the other.

Figure 2 shows that rates of oil removal in the PB reactor were initially above those in the AS reactor. At this stage VSS and COD levels for the two reactor types also differed markedly, with those in the PB reactor being much higher. These differences appear to be related to flow patterns at the water/oil boundary. In the PB reactor, water droplets strike the upper layer of stones in the packed bed before streaming over the stones beneath. In both cases, shear forces at the water/oil interface would be high, and rapid transfer of the more accessible (and soluble) components of the oil into the water phase could be expected. This would result in a high initial COD in the bulk water phase. Development of a sizeable microbial community could be expected to follow, leading to a high VSS.

In the AS reactor, the only water movements are those caused by air rising from the diffusers. Shear forces at the oil/water interface would be much lower than in the PB reactor and transfer of oil components into the water would be commensurately slower. This slow release rate would also enable the rapidly developing biofilm community to intercept a larger fraction of the solubilized hydrocarbons. Much smaller amounts of hydrocarbon would reach the bulk water phase than in the PB reactor, hence the much lower COD and VSS levels. Support for the idea that water flow patterns have a sizeable effect on oil removal rates came from experiments run at 10°C in an AS reactor; when the air diffusers were switched off, oil removal rates were 20% lower than when they were on.

Figure 2 also shows that the rate of oil removal decreased markedly over time, in both reactors. This occurred despite the fact that microbial populations had had a chance to build up and that there was a more than adequate supply of nutrients. This implies that the limiting factor in the removal process is the movement of hydrocarbon from the oil layer into the water.

The Contribution of Physical Processes to Oil Removal. In several experiments using the AS reactors enough sodium azide (NaN_3) was added to the water to raise its concentration to 0.05 molar. This is well above the concentration regularly used in microbiological experiments to suppress microbial activity.

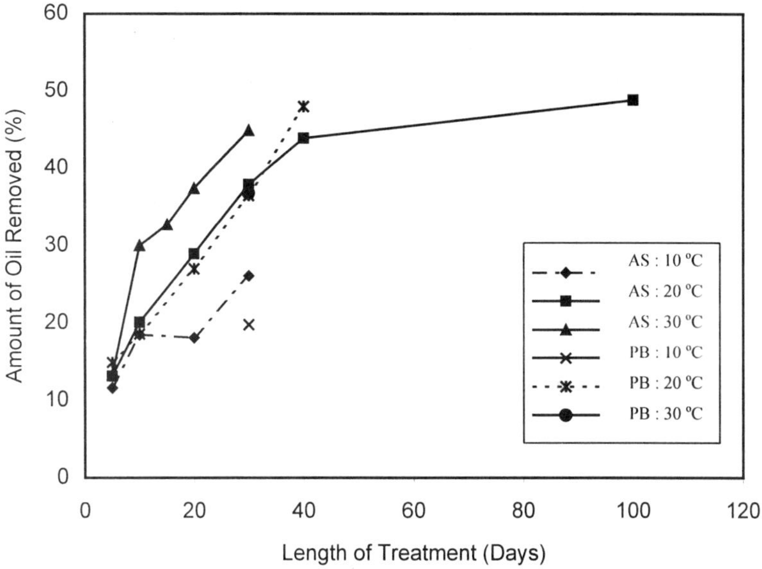

FIGURE 1 : Changes over time of the percentage of oil removed and biodegraded, at different temperatures, in thc AS and PB reactors.

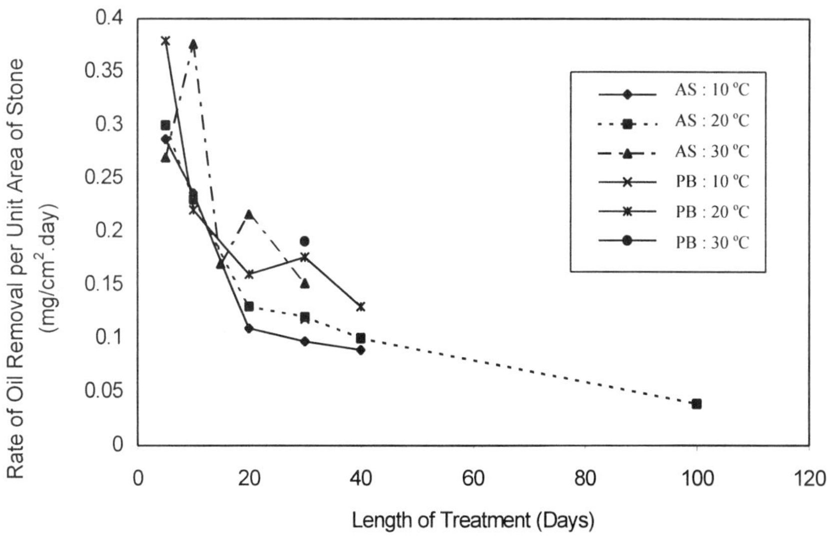

FIGURE 2 : Changes over time of rates of oil biodegradation per unit area of stone surface, at different temperatures, for the AS and PB reactors.

Despite the absence of microbial activity in these experiments, loss of oil from the stones still took place; after 40 days, the amount removed was around half that removed in experiments where microbiological activity was not suppressed. However, this oil was not degraded. It appeared that oil was being solubilized or emulsified at the oil/water boundary and then transported through the water to other parts of the reactor. Once in contact with equipment surfaces or the air, this oil tended to come out of solution, forming droplets that either floated in clusters on the water or coalesced and accumulated on reactor surfaces.

The above evidence shows that physical and interfacial processes have a strong influence on oil removal rates. However, whether or not these processes also affect the extent of oil removal that is ultimately attainable remains uncertain.

The Role of Nutrients. The role played by nutrients in bioremediation is still poorly understood. Other researchers have found that in some cases addition of supplementary nutrients appears to accelerate bioremediation processes whereas in others it seems to have little or no effect. In this study experiments were undertaken in AS reactors at 20°C with and without addition of supplementary nutrients. The results obtained are shown in Table 1; for each pair of values, the upper one gives the percentage of the original oil content removed by the end of the experiment while the lower one (in brackets) gives the percentage of the original oil content that has been totally biodegraded.

TABLE 1. The percentages of oil removed, and biodegraded (in brackets).

Categories	Experimental Period (Days)						
	5	10	15	20	30	40	100
No nutrients added (20°C)	10.1 (9.5)	18.2 (16.9)		30.5 (26.1)	35.2* (30.8)	36.0 (34.4)	
Nutrients added (10°C)	21.9 (11.6)	21.1* (18.6)		19.6 (18.1)	29.5 (26.1)		
Nutrients added (20°C)	22.5 (13.1)	30.4 (20.1)		34.2 (28.9)	40.0 (37.9)	45.4 (43.9)	53.5 (48.8)
Nutrients added (30°C)	15.0 (13.3)	35.0* (30.0)	37.7 (32.7)	40.3* (37.4)	49.2 (44.9)		

* Mean of two sets of results

In the early stages, oil was removed from the stones much more rapidly when nutrients were added. However, the amounts of oil fully biodegraded differed much less in the two cases. After 30 to 40 days the extent of oil removal was around 20% greater in the case where nutrients were added and much of the oil removed had been fully biodegraded.

The addition of nutrients also had a marked effect on COD and VSS levels in the bulk water phase. When extra nutrients were added, COD levels showed a rapid initial increase, followed 20 to 30 days later by a substantial decrease. This contrasted with the pattern in the non-supplemented reactor, where the COD built up slowly over the course of the experiment. Patterns of VSS development also differed. When extra nutrients were added, the VSS peaked quickly, then fell away before starting to increase again. When no extra nutrients were added, the VSS increased relatively slowly, peaking only after 30 days and then falling away.

The above observations suggest that the main effect of adding extra nutrients is to stimulate biofilm development. Since biofilms secrete biochemical agents that promote oil solubilization, flows of solubilized oil components into the water could be expected to increase correspondingly. This would bring about an upsurge in COD levels, and an increase in bulk water phase microbial activity (manifested as an increase in VSS levels) could be expected to follow soon afterwards. As the biofilm thickened and the accessibility of readily solubilized components in the oil layer decreased, it would be expected that less dissolved organic matter would find its way through to the bulk water phase. This would result in both COD and VSS levels falling, as was observed in practice.

Whilst the main effect of nutrient supplementation is to promote biofilm growth and activity, it also influences the overall level of oil removal achieved. When extra nutrients were added there was a small but significant increase in both the total amount of oil removed and the amount of oil that was biodegraded.

The Effects of Temperature. Temperature is generally assumed to affect bioremediation by influencing chemical or biochemical reaction rates. For example, it has been reported that reductions in temperature reduce enzyme activity (Atlas and Bartha, 1972) and that increases in temperature increase rates of hydrocarbon metabolism (Bossert and Bartha, 1984).

In this study, biodegradation experiments were conducted in AS reactors at temperatures of 10°C, 20°C and 30°C. Extra nutrients were added in all cases. The effect on biodegradation rates was as expected, with rates increasing as the temperature increased. What was unexpected was that, as is evident from Table 1, the overall extent of oil removal was also strongly influenced by temperature; in particular, after 30 days the level achieved at 10°C was well below that achieved at the higher temperatures. The failure of the removal curves for different temperatures to converge as the length of experimental runs increased implied that factors other than reaction rates were also being influenced by temperature.

Oil Composition and Properties. The above and other observations led us to consider whether it was the changes in oil composition and properties that took place as experiments progressed that were the main factors limiting the extent of oil removal achieved in our runs. During experiments the physical appearance of the oil changed markedly, going from a highly viscous, pasty liquid to a flaky solid. (A similar change was observed by Pritchard *et al.* (1991,1992) in their studies of the biodegradation of residues from the Exxon Valdez oil spill.) Such a change would be consistent with preferential removal of more mobile components from the oil. This would not be unexpected since it is the lower molecular weight, more mobile hydrocarbons that are usually the more readily solubilized and biodegraded. This idea was tested by modelling the internal diffusion processes within the oil (Pokethitiyook, 1999) and the results obtained supported the idea that it is these internal diffusion processes that control the extent of oil removal ultimately achievable.

CONCLUSIONS

On-site clean-up of the oil-coated stones used as ground cover beneath process equipment at the Longford Crude Oil and Gas Processing Facility is in principle feasible. The most appropriate approach would appear to be to use a two-step process. The first step would be one of bioremediation; the stones would be treated in an AS type reactor for around 40 days at ambient temperatures to transform the paste-like residual oil layer to a flaky surface layer. The second step would involve the use of some physical process to abrade the flaky surface layer and produce stones clean enough for re-use as ground cover. There would still be a small amount of waxy hydrocarbon residue to dispose of since only half of the original oil present is removed in the bioremediation process. Whether such a process would be economic is highly questionable, however.

Insights were also gained into the effects of various factors on the bioremediation of weathered, highly viscous oil residues. There is strong evidence to suggest that, except in the initial stages, the rate of bioremediation is controlled by diffusion processes within the oil layer itself. This would account for the fact that at low temperatures not only the rate but also the extent of oil removal and degradation is markedly decreased. It also helps explain why adding extra nutrients enhances oil removal and degradation at the start of experiments but has a rapidly diminishing influence in the later stages.

ACKNOWLEDGEMENTS

The help of staff at the Longford Facility and of Ms C. Buchanan (whose work led to the present study) is gratefully acknowledged.

REFERENCES

Atlas, R. M. and R. Bartha. 1972. "Biodegradation of Petroleum in Seawater at Low Temperatures." *Canadian Journal of Microbiology. 18*: 1851-1855.

Bossert, I. and R. Bartha. 1984. "The Fate of Petroleum in Soil Ecosystems." In R. M. Atlas (Ed.), *Petroleum Microbiology*, pp. 435-473. Macmillan, New York.

Pokethitiyook, P. 1999. *Bioremediation of Oil-coated Gravels*. PhD thesis, University of Melbourne, Australia.

Pritchard, P. H. and C. F. Costa. 1991. "EPA's Alaska Oil Spill Bioremediation Project." *Environmental Science and Technology. 25*: 372-379.

Pritchard, P. H., Mueller, J. G., Rogers, J. C., Kremer, F. V. and J. A. Glaser. 1992. "Oil Spill Bioremediation: Experiences, Lessons and Results from the *Exxon Valdez* Oil Spill in Alaska." *Biodegradation. 3*: 315-335.

EFFECT OF OIL SPILL BIOREMEDIATION ON THE BENTHIC FAUNA

Sarah Macnaughton, Fabien Daniel, David Mitchell, Richard Swannell, (AEA Technology, Culham, UK), John Patrick (University of Reading, Reading, UK), Michaela Schratzberger, Caroline Wall, Rebecca Kilbride, Sian Boyd, Hubert Rees (CEFAS Laboratory, Burnham on Crouch, UK), Kenneth Lee (Fisheries and Oceans, Dartmouth, Canada).

ABSTRACT: This project was designed to determine the impact of different bioremediation strategies on oil biodegradation and benthic meio- and macrofaunal communities. A field trial simulating an oil spill was established at a hydrodymanically active shoreline site in England. Microbiological and chemical investigations indicated that fertilizer stimulated oil biodegradation with treated oiled sediment showing a significant (P<0.05) reduction in oil compared to the unfertilized but oiled sediment. After 11 weeks, bioremediation had not stimulated faster recolonization rates of fertilized vs non-fertilized oiled sediments by benthic fauna. Multivariate analyses suggested that the fertilizer caused changes in species distribution patterns of benthic fauna. Fertilizer did not effect trophic relationships in nematode assemblages, but the impact of oiling and nutrient addition on macrofauna biomass was detected within 3 weeks of the fertilizer addition. These preliminary results suggest that although fertilizer treatment had no adverse effect on the benthic fauna, it also did not promote more rapid recovery of the benthic invertebrate community following the oil spill.

INTRODUCTION

Large oil spills can be both environmentally damaging and costly to clean-up. Responders have developed a range of techniques for treating oil spilled at sea that has subsequently been deposited on the shoreline. The environmental impacts of these treatments are being investigated, and while it may be argued that the use of some can bring a net environmental benefit, others have been shown to have at best little and sometimes significant negative effects on the shoreline environment (Sell et al. 1995). The majority of spilled oil that does not evaporate or is removed by human intervention is consumed by microbes (Prince, 1993). Oil biodegradation often is limited by sub-optimal levels of nutrients, such as N and P (Swannell et al., 1996). Bioremediation of oil spills has focused on alleviating this limitation by adding fertilizers to contaminated shorelines (Head and Swannell, 1999). Its use for cleaning shorelines has been investigated through experimental oil spills (Swannell et al. 1999a, Venosa et al. 1996) and after real spill incidents (Bragg et al. 1992; Swannell et al. 1999b). However, the effect of the fertilizer treatment on shoreline ecosystems is not clear. The intertidal and shallow water invertebrate communities in soft sediments are commonly the most seriously affected parts of the ecosystem (Jewett et al., 1999). The recovery of these communities after an oil spill is dependent on the recruitment pattern of affected species, their life history traits, and their tolerance to toxic compounds in the sediment. The community recovery is thus a function of mortality rates of the settled organisms caused by the pollutant, the species immigration or recolonisation rates and their ability to select different polluted habitats (Christie and Berge 1995).

Meio- (63-500 μm) and macrofauna (>500 μm) are vital links in the food webs of the marine and estuarine environment, yet are relatively immobile. These organisms can serve as indicators of the amount and extent of damage caused by an oil spill and the effects of bioremediation treatments on the recolonisation process of oiled sediments. Meiofauna are mainly dispersed by suspension (Palmer 1988) whereas macrofauna disperse by larval settlement and adult migration (Christie and Berge, 1995). The effects of oil on meio- and macrofauna have been studied following real oil spills (Jewett et al., 1999); in field experiments (Christie and Berge, 1995); experimental mesocosms (Frithsen et al., 1985); and laboratory microcosms (Grassle et al., 1980). However, there is no published information on field studies assessing both the effects of oil contamination and bioremediation treatments on benthic infauna.

Herein is presented an investigation of the impact of the use of fertilizers for bioremediation on the shoreline environment, specifically the meio- and macrobenthic assemblages recolonising and living in the oiled sediments.

MATERIALS AND METHODS

Field trial design. A randomised block design utilising Nitex mesh enclosures modified from that described in Swannell et al. (1999a) was used. Each block contained five treatments coded as follows : Untreated Control (no oil, no bioremediation treatment) (**UC**); Oiled Control (oil only, no bioremediation treatment) (**OC**); Test Plot 1 (oil and slow release fertilizer) (**SF**); Test Plot 2 (oil and regular addition of liquid inorganic fertilizer) (**LF**); Defaunated control (no oil, no bioremediation treatment with the natural benthic fauna removed at the start of the experiment) (**DF**). The additional defaunated control enabled separation of the effects of experimental set-up from possible treatment effects. To ensure that the oiled sediment was retained while not restricting the recolonisation of the macrofauna, a mesh size of 1000 μm was used.

Sediment from three of the plots was defaunated by drying the sediment above the high tide mark overnight. Nine oiled plots were filled with oiled beach sediment (weathered and emulsified (with 25% sea water) Fortiers crude oil) at an initial concentration of 3.5 l/m^2, and buried at each of the allocated plot locations at a depth of 0.15 m, 4 m apart (to avoid cross contamination). The final 3 plots were used as unoiled controls. Sampling was conducted 0, 2, 3, 7, 11, 16 and 45 weeks after the first nutrient addition.

Chemical and microbiological analyses. Sediment samples (3x50 g) were collected for oil chemistry analysis (total petroleum hydrocarbons, TPH). The initial oil concentrations ranged from 30,000 to 38,000 mg/kg. Recalcitrant compounds from the hopane family were quantified and compared with the remainder of the oil components thus providing a determination of oil biodegradation. During the first 2 weeks of the study *in situ* CO_2 evolution was assessed on a daily basis by *in situ* respirometry. The methods used for these chemical and microbiological analyses are given in Swannell et al. (1999a).

Meio- and macrofauna. Samples for meiofauna analysis were taken to a depth of 20 cm. Cores were separated into 5 cm layers (0 – 5, 5 – 10, 10 – 15, and 15 – 20). Samples for macrofauna analysis were collected from within a 0.1 m^2 quadrat. Macrofauna samples were separated into 5 cm layers ((0 – 5, 5 – 10, 10 – 15, and 15 – 20). Meiofauna and macrofauna samples were fixed in 7 % formaldehyde in filtered seawater (63 μm filter pore size) prior to analysis. Meiofauna samples were washed onto a 63 μm pore size sieve to remove the fine silt fraction and the formalin. After decanting the samples five times onto the sieve in order to separate the meiofauna and lighter sediment fractions from the coarser material, meiofauna was extracted with Ludox™ 40 with a specific gravity of 1.15 (Somerfield and Warwick, 1996). The extraction process was repeated three times. The extracts were evaporated slowly in anhydrous glycerol and spread on microscope slides. Nematodes were identified to genus or species level by reference to descriptions in Platt and Warwick (1983; 1988) and Warwick et al. (1998). Macrofauna samples were washed through a 500 μm sieve and retained individuals counted and identified to family or species level. Macrofauna species were transferred into vials for biomass determination by weight. Biomass from headless polychaete fragments was arbitrarily allocated across the weights for all identified species, in proportion to the amount of biomass represented by each.

Statistical Analysis (Univariate). Total individuals, total species, Shannon-Wiener diversity (H'; Shannon and Weaver, 1949) and evenness (Pielou's J'; Pielou, 1975) were calculated to describe benthic assemblage structure. Significance of differences between treatments was tested using a one-way ANOVA. Tukey's HSD multiple comparisons test was used in pairwise comparisons of controls and treatments (STATGRAPHICS Plus, Version 3.3).

Free-living nematodes were divided into groups according to the structure of their buccal cavity (Wieser, 1953): **Group 1A**: selective deposit feeders; **Group 1B**: non-selective deposit feeders; **Group 2A:** epigrowth or diatom feeders; and **Group 2B**: predators.

Statistical Analysis (Multivariate). Multi-dimensional scaling (MDS) ordination using the Bray-Curtis similarity measure was applied to species abundance data (Clarke and Warwick 1994). In an MDS ordination, the samples are mapped in such a way that distances between pairs of samples reflect their relative dissimilarity in species composition. The MDS ordination can therefore be used to identify groups of samples having similar faunal assemblages. A stress value gives an indication of how well the two-dimensional plot represents the multi-dimensional sample relationship. Values between 0.05 and 0.2 generally correspond to a good representation of sample similarities (Clarke and Warwick, 1994). Analysis of similarities (ANOSIM) was performed to test the significance of differences in benthic assemblage composition between treatments. Based on the number of possible permutations when three replicates per treatment are collected as in this study, the next lowest *P*-value to zero is $P = 0.1$. Community groupings identified in the MDS ordinations were explored further by applying the similarity percentages program (SIMPER) to determine the contribution of individual species to the average dissimilarity between samples. Multivariate analyses were performed using PRIMER (Clarke and Warwick, 1994).

RESULTS AND DISCUSSION

Bioremediation. Significantly less TPH remained in the bioremediated plots at the end of the study then in the oiled control (Figure 1A). The decrease in the ratio of TPH/hopane showed that over the course of the experiment, bioremediation increased oil biodegradation, which accounts, at least in part, for the reductions in TPH (Figure 2B). In all plots there was an unexplained increase in the ratio at week 11. Carbon dioxide emissions indicated a significant difference in biodegradation rates dependent on treatment, with both the slow release (~1800 mmol C) and liquid fertilizers (~1000 mmol C) stimulating increased levels of degradation in comparison to the oiled control (~300 mmol C; $P<0.05$).

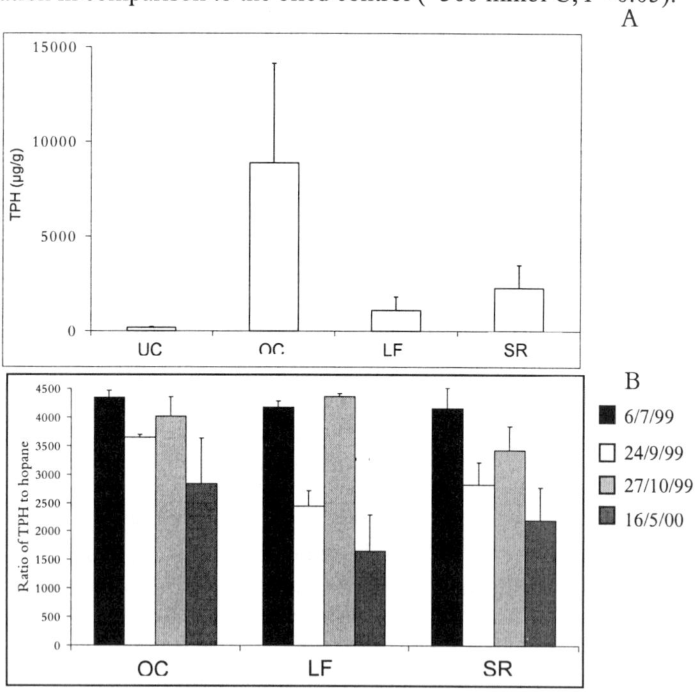

FIGURE 1. A) TPH concentration at the end of the study (the decrease in TPH was significantly greater in the fertilizer treated plots, *P*<0.05); B) Mean ratios of TPH to hopane following fertilizer treatment. Error bars represent Standard Deviation.

Meio- and Macrofauna. Results are presented from samples collected pre-experiment, 3, 7, and 11 weeks after the first nutrient addition to a depth of 0-5 cm. A total of 84 nematode species and 18 macrofauna species were identified.

Total nematode abundance was significantly ($P<0.05$) lower in all oiled sediments compared to the unoiled and defaunated controls 11 weeks after fertilizer addition (Figure 2). Total macrofauna abundance was significantly lower in the oiled control and liquid fertilizer treatments compared to the unoiled controls ($P < 0.05$). Nematode assemblages were dominated by predators (60 % of total abundance) > non-selective deposit feeders (19 %) and epigrowth feeders (18 %) > bacteria feeders (3 %). However, there was no significant difference in the proportions of nematode feeding types following the different fertilizer treatments after 11 weeks ($P>0.05$).

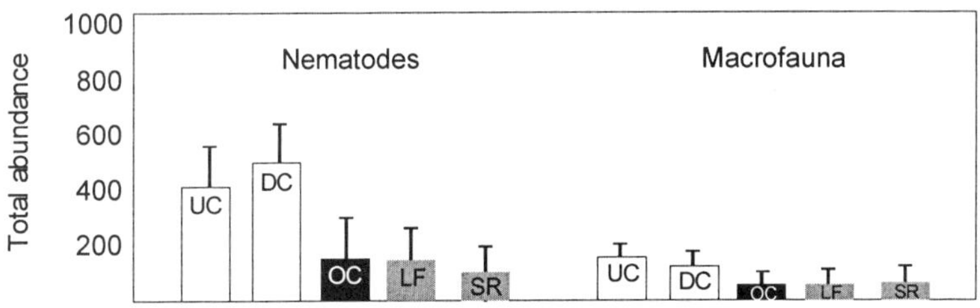

FIGURE 2. Means (+/- 95 % pooled confidence intervals) of total nematode and macrofauna abundance of experimental treatments and controls (Univariate analysis).

Results from the multivariate MDS ordinations, ANOSIM and SIMPER analyses revealed a number of trends in the response of nematode and macrofauna assemblages to the effects of the experimental set-up, oil and fertilizer after 11 weeks. Samples collected in the defaunated and unoiled control plots cluster closely together, indicating that there were no significant differences between the two types of controls. Assemblages from the unoiled plots were more similar to each other than to those from the oiled plots. Nematode assemblages collected in the non-fertilized oiled plots were more similar to those from the unoiled plots than to those sampled in the fertilized oiled plots (Figure 3). Macrofauna assemblages collected from the oiled sediments treated with the liquid fertilizer were more similar to the unoiled sediments than to those sampled from the other experimental treatments (Figure 3).

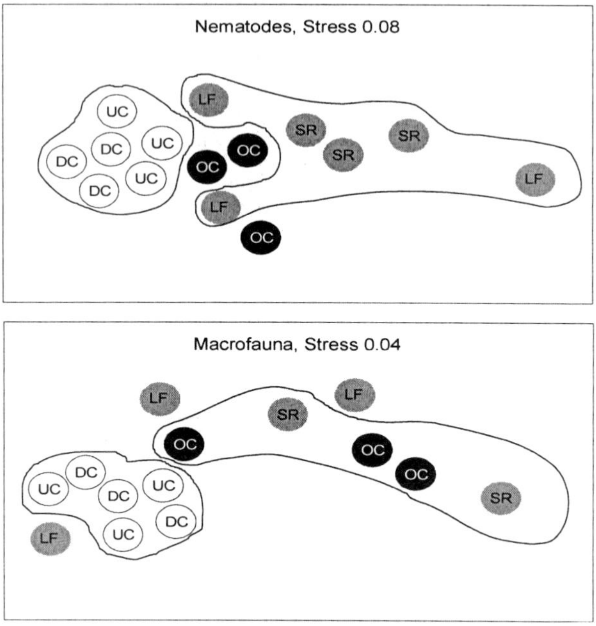

FIGURE 3. MDS ordination based on square-root transformed abundance data of nematode and macrofauna species 11 weeks after fertilizer application (circled treatments are significantly different at *P* < 0.10).

CONCLUSIONS

Oil concentration in all oiled samples decreased over time. By the end of the study, both fertilizer treatments stimulated a significantly greater decrease in TPH content of the sediment than that observed in the oiled controls ($P<0.05$). However, there was no significant difference in the efficacy of the slow release and liquid fertilizer treatments ($P>0.05$) supporting earlier observations (Swannell et al., 1999b). Use of slow release fertilizers offers a simple and cost effective approach to shoreline bioremediation. After 11 weeks, the stimulation of biodegradation had not resulted in faster recolonisation rates of fertilized versus non-fertilized oiled sediments by benthic meio- and macrofauna. Results from multivariate analyses indicated that the species composition in the unoiled control plots was more similar to the non-fertilized oiled sediments than to the fertilized treatments. These data, together with the lack of any statistically significant differences between untreated and fertilized oiled sediments in terms of assemblage structure, may reflect the generally high ecological tolerance of estuarine benthic infauna to perturbations. Analysis of samples obtained 45 weeks after nutrient treatment is continuing and will provide further information regarding the benthic response to fertilizer treatment. Results obtained to date provide important information for further second field trials at a sheltered sand flat which is currently underway. Both habitats are sufficiently different in terms of the physiographic environment (salinity, exposure to wave action etc.) to support quantitatively and qualitatively different meio- and macrofauna assemblages and to offer the potential for contrasting insights into the consequences of oiling and bioremediation treatments on benthic recovery.

ACKNOWLEDGEMENTS

This work was supported by grants from the Maritime and Coastguard Agency (UK), the Ministry for Agriculture, Fisheries and Food (UK) and the UK Environment Agency.

REFERENCES

Bragg, J. R., R. C. Prince, J. B. Wilkinson, and R. M. Atlas. 1992. "Bioremediation for shoreline cleanup following the 1989 Alaskan oil spill." Exxon Co., USA, Houston.

Christie, H. and J. A. Berge. 1995. "*In situ* Experiments on Recolonisation of Intertidal Mudflat Fauna to Sediment Contaminated with Different Concentrations of Oil." *Sarsia. 80*: 175-185.

Clarke, K. R. and R. M. Warwick. 1994. "Change in Marine Communities: an Approach to Statistical Analysis and Interpretation." pp. 144. Plymouth Marine Laboratory, Plymouth, UK.

Frithsen, J. B., R. Elmgren, and D. T. Rudnick. 1985. "Responses of Benthic Meiofauna to Long-Term, Low-Level Additions of No. 2 Fuel Oil." *Mar. Ecol. Prog. Series. 23*: 1-14.

Grassle, F. J., R. Elmgren, and J. P. Grassle. 1980. "Response of Benthic Communities in Merl Experimental Ecosystems to Low Level, Chronic Additions of no. 2 Fuel Oil." *Mar. Environ. Res.. 4*: 279-297.

Head, I. M. and R. P. J. Swannell. 1999. "Bioremediation of Petroleum Hydrocarbon Contaminants in Marine Habitats." *Cur. Opin. Biotech. 10*: 234-239.
Jewett, S. C., T. A. Dean, R. O. Smith, and A. Blanchard. 1999. "'Exxon Valdez' Oil Spill: Impacts and Recovery in the Soft Bottom Benthic Community in and Adjacent to Eelgrass Beds." *Mar. Ecol. Prog. Series. 185*: 59-83.

Jewett, S. C., T. A. Dean, R. O. Smith, and A. Blanchard. 1999. "Exxon Valdez oil spill: impacts and recovery in the soft bottom benthic community in and adjacent to eelgrass beds." *Mar. Ecol. Prog. Series 185*: 59-83.

Palmer, M. A. 1988. "Dispersal of Marine Meiofauna: a Review and Conceptual Model Explaining Passive Transport and Active Emergence with Implications for Recruitment." *Mar. Ecol. Prog. Series. 48*: 81-91.

Pielou, E. C. 1975. Ecological Diversity, Wiley-Interscience, New York.

Platt, H. M. and R. M. Warwick. 1983. "Free-Living Marine Nematodes. Part I. British Enoploids." pp. 307. Cambridge University Press, Cambridge, UK.

Platt, H. M. and R. M. Warwick. 1988. "Free-Living Marine Nematodes. Part II." British Chromadorids. pp. 502. Cambridge University Press, Cambridge, UK.

Prince, R. C. 1993. "Petroleum Spill bioremediation in marine environments." *Crit. Rev. Microbiol. 19*: 217-242.

Sell D., L. Conway, T. Clark, G. B. Picken, J. M. Baker, G. M. Dunnet, A. D. McIntyre and R. B. Clark. 1995. "Scientific Criteria to Optimize Oil Spill Cleanup" In *Proc. of the 1995 Oil Spill Conference*, pp. 595-610.

Shannon, C.E., and W. Weaver. 1949. The Mathematical Theory of Communication. Univ. of Illinois Press, Urbana, IL.

Somerfield, P. J. and R. M. Warwick.1996. "Meiofauna in Marine Pollution Monitoring Programmes. A Laboratory Manual." Ministry of Agriculture, Fisheries and Food, Directorate of Fisheries Research, pp. 71. Lowestoft, UK.

Swannell, R. P. J, K. Lee, and M. McDonagh. 1996. "Field evaluations of oil spill bioremediation." *Microbiol. Rev. 60*: 342-365.

Swannell, R. P. J., D. Mitchell, S. Petch, I. M. Head, A. Willis, K. Lee, and J. Lepo. 1999a. "Bioremediation of Oil Contaminated Fine Sediments". In *Proc. Of the 1999 International Oil Spill Conference,* pp. 751-756.

Swannell, R. P. J., D. J. Mitchell, G. Lethbridge, D. Jones, D. Heath, M. Hagley, D. M. Jones, S. Petch, R. Milne, R. Croxford and K. Lee. 1999b. "The Use of Bioremediation to Treat an Oiled Shoreline Following the *Sea Empress* Incident." *Environ. Technol. 20*: 863-873.

Venosa, M. T. Suidan, B. A. Wrenn, K. L. Strohmeier, J. R. Haines, B. L. Eberhart, D. King and E. Holder. 1996. "Bioremediation of Experimental Spill on the Shoreline of Delaware bay." *Environ. Sci. Tech. 30*: 1764-1777.

Warwick, R. M., H. M. Platt, and P. J. Somerfield. 1998. "Free-Living Marine Nematodes. Part III. British Monhysterids." pp. 296.Field Studies Council, Shrewsbury, UK

Wieser, W. 1953. "Die Beziehung zwischen Mundhoehlengestalt, Ernaehrungsweise und Vorkommen bei freilebenden marinen Nematoden." *Arkiv foer Zoologi. 4*: 439-484.

FIELD TESTS OF BIOREMEDIATION AGENT WITH MICROBIOLOGICAL CULTURES

Hiroaki Tsutsumi (Prefectural University of Kumamoto, Japan), Masakazu Kono, Toyoharu Hozumi (Oppenheimer Biotechnology Japan, Kobe, Japan), Shouji Hirashima (Yakult Pharmaceutical Industry Co., Tokyo, Japan), Koji Takai (Fuji Packing, Co., Iwami-machi, Tottori-ken, Japan), Takehiko Manabe (Hyogo Prefectural Fisheries Research Institute, Akashi, Japan), Makoo Haraguchi, Isao Yamamoto (Sumitomo Marine Research Institute, Inc., Tokyo, Japan) and Carl Oppenheimer (Oppenheimer Biotechnology, Austin)

ABSTRACT: A field test was conducted to examine the effectiveness of a seed culture of petroleum degrading bacteria on enhancing biodegradation of the heavy oil spilt on the shore by an accident of a tanker in the Sea of Japan in January 1997. In this study, we developed a new method to quantitatively evaluate the impact of a bioremediation agent treatment on the oil resulting from the oil spill that had coated the hard substrates on the shore, such as rocks and concrete. This method uses a technique of digital photographic image analysis with a personal computer with a procedure. The results of the analysis detail, both visually and graphically, the quantitative changes of the oil on the rock surfaces relative to controls. The results of the field test clearly show that the bioremediation agent with mircrobiological cultures is able to significantly enhance the biodegradation of oil in the field.

INTRODUCTION

On January 2, 1997, a Russian tanker, the *Nakhodka*, broke apart and sank in the Sea of Japan. Approximately 5,000 tons of heavy oil was spilled from the ship, and carried by winds to the coast of Japan. The oil had impacted the coast of approximately 1,200 linear km. Such large-scale oil pollution on the shoreline has not occurred in Japan since the oil spill from the oil tank in Mizushima, Okayama prefecture, in 1975 (Okaichi and Tatsumi, 1975).

The oil spilled from the tanker was removed from the shore by the hands of innumerable volunteers, treated with chemical agents (oil dispersants). In a small town in Kasumi-cho, Kinosaki-gun, Hyogo prefecture, surprisingly, a bioremediation agent with microbiological cultures, TerraZyme™ (Oppenheimer Biotechnology, Inc.), was applied to enhance the bio-degradation of oil on the shore by the members of a local fishery cooperative association. They scattered approximately 250 kg of the bioremediation agent on the oil attached to shoreline rocks and concrete in the small fishing harbor of the town (Fig. 1).

Bioremediation agents including inorganic or organic fertilizers and oil degradation bacteria have recently been applied to clean-up efforts (Atlas, 1995; Hoff, 1993; Swannell et al., 1996). However, in the earlier studies on the bioremediation of oil, through the 1970s and 1980s, only the fertilizers were tried to use for the field experiments to demonstrate the feasibility of bioremediation in enhancing the degradation of oil (Oliveiri et al., 1978; Sendstad, 1980; Sendstad et al., 1984; Sveum and Ladousse, 1989). Therefore, the application of TerraZyme™ on the shore in Kasumi-cho may be the first trial to treat the heavy oil on the shore after oil spill with microbiological cultures

In this study, we developed a technique to quantitatively measure the impact of treatment on the substrate covered by oil with a digital photographic image analysis running on a personal computer. We conducted field tests with TerraZyme™ to enhance the biodegradation of the heavy oil from the *Nakhodka* that adhered to the rocks and concrete blocks on the shore around the Kasumi-cho area, and evaluated the effectiveness of the bioremediation agent with microbiological cultures on the enhancement of biodegradation of the heavy oil.

(a)

(b)

FIGURE 1. Treatment of heavy oil on the shore with a bioremediation agent with microbiological cultures, TerraZyme™. (a) Treatment of oil adhered on the rock in a port of Shibayama, Kasumi-cho, Hyogo, Japan, in Feb. 7, 1997. (b) Treated oil on the rock with TerraZyme™. One day after the treatment, the oil was easily removed from the rock by hand.

Objective. The objective of this study is to describe a method to quantitatively evaluate the clean-up impact of the bioremediation agent in the field, to repoort the enhanced biodegradation of oil after the treatment with TerraZyme™ on the shore, and to and discuss the use of seed cultures of oil degrading bacteria for the treatment of oil pollution following oil spills.

MATERIALS AND METHODS

The study was conducted on the shore in Kasumi-cho, Kinosaki-gun, Hyogo prefecture, Japan (134° 43' E, 35° 42'), where a large amount of oil was spilled from the *Nakhodka* in January 1997. We prepared three sets of polluted rocks, having a flat face over 200 cm^2. These rocks were placed within the wire cages (62 x 42 x 32 cm) and anchored at the intertidal zone on the beach. The experiments were carried out for eight weeks, from July 9, 1997 to September 4, 1997. Once a week, the wire cages with the rocks and concrete blocks were carried to a higher point on the beach. Each of the rocks was photographed, treated with a bioremediation agent with microbiological cultures, TerraZyme™ (Oppenheimer Biotechnology, Inc.), by sprinkling the powder of TerraZyme™ over them at a density of 100 g/m^2, and returned to the wire cages. The cages set at the water's edge on the beach again. Throuth the experiments, no nutrients were added to the wire cages with the rocks and concrete blocks.

TerraZyme™ is made up of a bentonite substrate and various petroleum-degrading microorganisms, which have been collected from many different geographical localities under different environmental conditions, shown extensive clean-up impacts on oil pollution under various environmental conditions. 1 g of TerraZyme™ contains approximately 10^8 to 10^9 cells of bacteria. TerraZyme™ is noted in the latest NCP of the U.S. EPA as one of the bioremediation agents with microbiological cultures for treating the oil pollution (U.S. EPA, 1999).

The photographs of the rocks were scanned for later computer analysis. The scanned pictures of the rocks were processed, and then analyzed with a digital photographic image analysis system, using a personal computer.

The personal computer system for image analysis is made up of a personal computer (Power Macintosh 7300, Apple), a 13 inch color monitor (Apple), a flatbed image scanner (GT6500, EPSON), software for image analysis (NIH Image Ver. 1.60), which is a freeware digital image analysis program, provided by National Institute of Health (NIH), U.S., and Plugin software (EPSON Scanner Plugin, EPSON) originally provided for Adobe Photoshop, digital image processing software, which can be used to control the scanner directly from "NIH Image". It enables to import a digital image of a photograph through the flatbed image scanner directly to the monitor of the personal computer that "NIH Image" is runing on.

RESULTS

The general procedures of digital photographic image analysis:

1) Photographs of the polluted faces of the rocks were taken, developed, and printed.
2) The digital image was transferred to a personal computer system through a flatbed scanner.

3) The digital image was converted to a gray color scale on the computer monitor.
4) All of the pixels that formed the image were classified into three groups according to the darkness levels (0 to 255) (Fig. 2).
5) The number of pixels in each group was counted, and calculated for the 'percent coverage' of these pixels on the substrate.

In the sample-image of the rock shown in Fig. 2, the pixels of Group 1 consisted of the darkest ones (over Level 225) in gray-scale. All of these pixels looked black in the image, which indicated the oil adhered to the rock. The pixels of Group 3 had the brightest colors, which were ranged between Level 120 and 200 in gray-scale. These colors were derived from the ground colors of the rock itself. The pixels of Group 2 had intermediate gray colors between Level 200 and 225 in gray-scale. These pixels indicated that a relatively thin layer of oil remained on the rock.

The digital image of the rock was processed to one with two-tone color of pixels at the boundary level in gray-scale between two groups of the pixels. Fig. 3(a) indicates an image of the rock with the pixels of Group 1 only, whose colors are darker than Level 225 in gray-scale. Fig. 3(b) indicates an image of the rock with the pixels of Group 1 and 2, whose colors are darker than Level 200 in gray-scale. Fig. 3(c) indicates the pixels of Group 1, 2 and 3, that is, the entire image of the rock. The number of the pixels of each group were counted on the image, and its 'percent coverage' on the substrate was calculated with the scale information.

Thus, by this method, the areas covered by the oil on the rock surface can be expressed as numerical data. If oil was lost from the rock surface from the treatment experiments, the loss of oil should be expressed as decrease of the area of Group 1 and increase of the areas of Group 2 and 3.

Fig. 4 shows the photographic digital images of (A) the treated rocks with TerraZyme™ and (B) untreaed rocks, which were processed and expressed in three

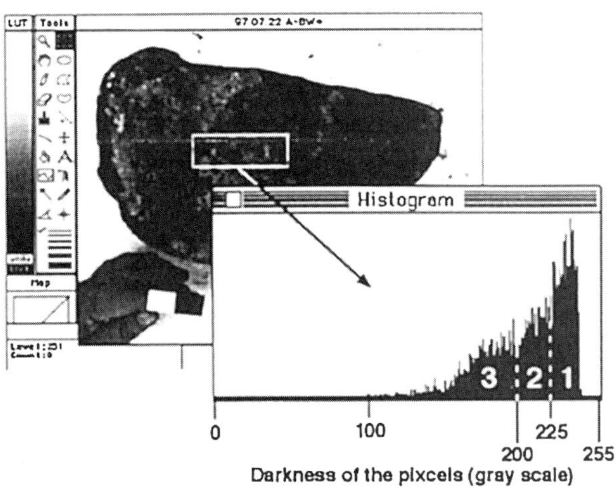

FIGURE 2. A frequency distribution of the pixels arranged in darkness (of gray color) on the digital image. The pixels of the digital image were classified into three different groups in darkness.

(a)

(b)

(c)

Group	Gray-scale (level)	Area(cm^2)
1	100 to 199	40
2	200 to 224	72
3	225 to 237	160

FIGURE 3. Processed digital images with two-tone pixel-colors at the boundary- levels in gray-scale between two groups of the pixels. (a) an image of the rock with the pixels of only Group 1, of which colors are darker than Level 225 in gray scale, (b) an image of the rock with the pixels of Group 1 and 2, in which the colors are darker than Level 200 in gray scale, (c) image of the whole rock with the pixels of Group 1, 2 and 3.

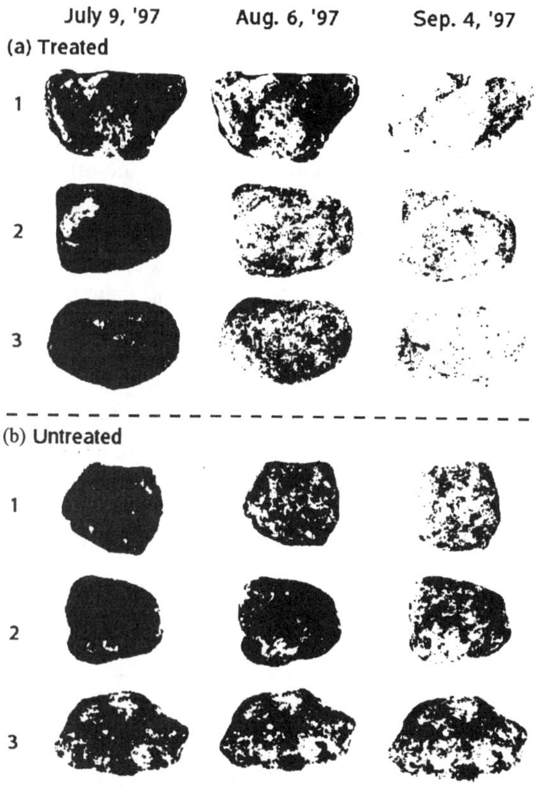

FIGURE 4. The photographic digital images of the rocks s expressed with three different groups of pixels according to the darkness of their gray colors. (a) treated rocks with TerraZyme™ (b) untreated rocks

gray-scale colors (white, gray and black) to clarly show the distribution of oil on them. At the start of the oil treatment test, the images of both of the treated and untreated rocks were mostly black (Group 1). However, eight weeks later, there were distinct differences in the distribution of three groups of pixels in the images between the treated and untreated rocks. The majority of the images of the treaed rocks were replaced by gray pixels of Group 2 or white ones of Group 3, while those of the untreated rocks remained black pixels.

Table 1 notes the percentage coverage of three different gray-color pixel groups (Group 1 to 3) on the images of both of the treated and untreated rocks. At the start of the oil treatment test on July 9, the percent coverage of Group 1 with oil pollution was ranged between 77.7 and 94.2% on all of the rocks. Eight weeks later, the percent coverage of Group 1 decreased to 14.0% (mean) on the treated rocks, while that of the untreated rocks remained 51.2% (mean). Contrastivelly, the percent coverage of Group 2 increased from 0% to 51.0% (mean) on the treated rocks, while that of the untreated rocks only slightly increased to 18.2% (mean).

Thus, the marked decrease of the percent coverage of Group 1 and marked increase of Group 2 on the treated rocks demonstrates the effectiveness of TerraZyme™ for the treatment of oil on the substrate in the field.

DISCUSSION

In the monitoring of oil pollution in the field, it is very difficult to completely collect the oil adhered to the substrates for quantitative measurement of the remaining oil. Even if we succeeded in taking the samples of the oil, we would still be unable to follow sequential changes of oil pollution over time at the sampling sites after the treatment. In field ecological studies, "percent coverage" that is, the percentage of coverage of plants or animals on the substrate, has been widely used to evaluate the abundance of organisms without disturbing them or their habitats, i.e. the amounts of algae that covered the sediments on the shore (Tsutsumi, 1990). Recently, this 'percent coverage' has been measured from the photographic images with PC-based graphic digitizers or image analysis systems (Meese and Tomich, 1992). In this study, we also applied the digital image analysis system with a personal computer to convert the visual impression of the coverage to digitized quantitative data that were obtained from objective and standardized procedures. The 'per-

TABLE 1. The areas covered by three different groups of pixels on the treated and untreated rocks.

	Group 1 (Black) (%)						Group 2 (Gray) (%)						Group 3 (White) (%)					
	Treated			Untreated			Treated			Untreated			Treated			Untreated		
Time (week)	1	2	3	1	2	3	1	2	3	1	2	3	1	2	3	1	2	3
0	84.2	92.1	97.1	94.2	89.4	77.7	0.0	0.0	0.0	0.0	0.0	0.0	15.8	7.9	2.9	5.8	10.6	22.3
1	68.4	87.5	88.7	74.5	92.0	80.7	17.3	0.0	0.0	0.0	2.8	6.1	14.3	12.5	11.3	25.5	5.2	13.2
2	58.8	79.2	58.3	72.1	84.3	74.4	17.6	10.6	35.1	15.8	5.8	0.0	23.5	10.2	6.7	12.1	9.9	25.6
3	62.5	67.9	44.3	81.2	79.3	66.1	13.6	23.4	38.3	5.2	4.7	0.0	23.9	8.7	17.4	13.6	15.9	33.9
4	66.9	37.7	50.1	65.8	81.7	71.3	10.3	51.3	38.3	14.2	6.5	0.0	22.8	10.9	11.6	20.0	11.9	28.7
5	***	***	***	***	***	***	***	***	***	***	***	***	***	***	***	***	***	***
6	23.5	26.0	15.7	67.9	67.5	61.2	35.7	47.9	73.6	9.1	20.0	0.0	40.8	26.0	10.7	23.0	12.5	38.8
7	18.4	38.9	13.6	63.6	51.7	50.7	39.3	24.9	61.7	11.8	30.0	0.0	42.3	36.2	24.6	24.5	18.3	49.3
8	16.2	17.4	8.4	33.0	64.2	56.2	41.9	30.9	80.0	32.1	22.4	0.0	41.9	23.0	11.6	34.8	17.5	43.8
Ave.	14.0			51.2			51.0			18.2			25.5			32.0		

cent coverage' of oil on the rock should reflect the amount of oil residues on the rock. Thus, this technique enables us to quantitatively follow the changes of percent coverage of oil on the substrate after the treatment.

In previous studies on the application of the bioremediation agents to oil pollution, adding seed-cultures of petroleum-degrading bacteria have proven less promising for the promotion of biodegradation of oil than adding fertilizers such as Inipol EAP22™ and ensuring adequate aeration (Atlas, 1995), partly due to the technical difficulties in evaluating their impacts. However, TerraZyme™ has an excellent potential to biodegrade all of the ingredients of heavy oil including the fraction of resin and asphaltum (Hozumi et al., 2000), which was not degraded by the treatement with Inipol EAP22™.

In this study, we did not follow the changes of the number of bacteria both on the treated and untreated rocks. Therefore, we did not confirm the growth of petroleum-degrading bacteria on the oil of the rocks. However, TerraZyme™ is made up of only a bentonite substrate and petroleum-degrading bacteria. The results of the field application tests of TerraZyme™ on the polluted shore in this study clearly demonstrate the promotion of removal of oil from the rocks by the treatment, and predict the usefulness of seed-cultures of petroleum-degrading bacteria for bioremediation of oil in the field.

REFERENCES

Atlas, R. M. 1995. "Petroleum biodegradation and oil spill bioremediation." *Mar. Pollut. Bull. 31*: 178-182.

Hoff, R. Z. 1993. "Bioremediation: an review of its development and use for oil spill clean-up." *Mar. Pollut. Bull. 26*: 476-481.

Hozumi, T., H. Tsutsumi, and M. Kono 2000. "Bioremediation on the shore after an oil spill from the Nakhodka in the Sea of Japan. I. Chemistry and characteristics of the heavy oil loaded on the Nakhodka and biodegradation tests of the oil by a bioremediation agent with microbial cultures in the laboratory." *Mar. Pollut. Bull. 40*: 308-14.

Meese, R. J., and P. A. Tomich 1992. "Dots on the rocks, a comparison of percent cover estimation methods." *J. exp. mar. Biol. Ecol. 165*: 59-73.

Okaichi, T., and S. Tatsumi (ed.) 1975. *Oil spill in the Seto Inland Sea (Setonaikai no juyu osen)*. Asia Planning, Takamatsu, 182pp.

Oliveiri, R., A. Robertielllo, and L. Degen 1978. "Enhancement of microbial degradation of oil pollutants using lipophilic fertilizers." *Mar. Pollut. Bull. 9*: 217-220.

Sendstad, E. 1980. "Accelerated biodegradation of crude oil on Arctic shorelines." *In Proc. 3rd Arctic Mar. Oil Spill Program Tech. Semi.*, pp. 402-416, Environment Canada, Ottawa, Canada.

Sendstad, E., P. Sveum, L. J. Endal, Y. Brattbakk, and O. Ronning 1984. "Studies on a seven year old seashore crude oil spill on Spitsbergen." *In Proc. 7th Arctic Mar. Oil Spill Program Tech. Semi.*, pp. 60-74, Environment Canada, Ottawa, Canada.

Sveum, P., and A. Ladousse 1989. "Biodegradation of oil in the Arctic: enhancement by oil-soluble fertilizer application." *In Proc. 1989 Oil Spill Conf.*, pp. 439-446. American Petroleum Institute, Washington, DC.

Swannell, R. P. J., Lee, K., and M. McDonagh 1996. "Field evaluations of marine oil spill bioremediation." *Microbiol. Rev. 60*: 342-365.

Tsutsumi, H. 1990. "Population persistence of *Capitella* sp. (Polychaeta; Capitellidae) on a mud flat subject to environmental disturbance by organic enrichment." *Mar. Ecol. Prog. Ser. 63*: 147-156.

Tsutsumi, H., M. Kono, K. Takai, T. Manabe, M. Haraguchi, I. Yamamoto and C. Oppenheimer 2000. "Bioremediation on the shore after an oil spill from the Nakhodka in the Sea of Japan. III. Field tests of a bioremediation agent with microbiological cultures for the treatment of an oil spill." *Mar. Pollut. Bull. 40*: 320-324.

U.S. EPA 1999. *Environmental protection agency national contingency plan product schedule*. U.S. Environmental Protection Agency, Feb. 1999, 8pp.

2001 AUTHOR INDEX

This index contains names, affiliations, and volume/page citations for all authors who contributed to the ten-volume proceedings of the Sixth International In Situ and On-Site Bioremediation Symposium (San Diego, California, June 4-7, 2001). Ordering information is provided on the back cover of this book.

The citations reference the ten volumes as follows:

6(1): Magar, V.S., J.T. Gibbs, K.T. O'Reilly, M.R. Hyman, and A. Leeson (Eds.), *Bioremediation of MTBE, Alcohols, and Ethers*. Battelle Press, Columbus, OH, 2001. 249 pp.

6(2): Leeson, A., M.E. Kelley, H.S. Rifai, and V.S. Magar (Eds.), *Natural Attenuation of Environmental Contaminants*. Battelle Press, Columbus, OH, 2001. 307 pp.

6(3): Magar, V.S., G. Johnson, S.K. Ong, and A. Leeson (Eds.), *Bioremediation of Energetics, Phenolics, and Polycyclic Aromatic Hydrocarbons*. Battelle Press, Columbus, OH, 2001. 313 pp.

6(4): Magar, V.S., T.M. Vogel, C.M. Aelion, and A. Leeson (Eds.), *Innovative Methods in Support of Bioremediation*. Battelle Press, Columbus, OH, 2001. 197 pp.

6(5): Leeson, A., E.A. Foote, M.K. Banks, and V.S. Magar (Eds.), *Phytoremediation, Wetlands, and Sediments*. Battelle Press, Columbus, OH, 2001. 383 pp.

6(6): Magar, V.S., F.M. von Fahnestock, and A. Leeson (Eds.), *Ex Situ Biological Treatment Technologies*. Battelle Press, Columbus, OH, 2001. 423 pp.

6(7): Magar, V.S., D.E. Fennell, J.J. Morse, B.C. Alleman, and A. Leeson (Eds.), *Anaerobic Degradation of Chlorinated Solvents*. Battelle Press, Columbus, OH, 2001. 387 pp.

6(8): Leeson, A., B.C. Alleman, P.J. Alvarez, and V.S. Magar (Eds.), *Bioaugmentation, Biobarriers, and Biogeochemistry*. Battelle Press, Columbus, OH, 2001. 255 pp.

6(9): Leeson, A., B.M. Peyton, J.L. Means, and V.S. Magar (Eds.), *Bioremediation of Inorganic Compounds*. Battelle Press, Columbus, OH, 2001. 377 pp.

6(10): Leeson, A., P.C. Johnson, R.E. Hinchee, L. Semprini, and V.S. Magar (Eds.), *In Situ Aeration and Aerobic Remediation*. Battelle Press, Columbus, OH, 2001. 391 pp.

KEYWORD INDEX

This index contains keyword terms assigned to the articles in the ten-volume proceedings of the Sixth International In Situ and On-Site Bioremediation Symposium (San Diego, California, June 4-7, 2001). Ordering information is provided on the back cover of this book.

In assigning the terms that appear in this index, no attempt was made to reference all subjects addressed. Instead, terms were assigned to each article to reflect the primary topics covered by that article. Authors' suggestions were taken into consideration and expanded or revised as necessary. The citations reference the ten volumes as follows:

6(1): Magar, V.S., J.T. Gibbs, K.T. O'Reilly, M.R. Hyman, and A. Leeson (Eds.), *Bioremediation of MTBE, Alcohols, and Ethers*. Battelle Press, Columbus, OH, 2001. 249 pp.

6(2): Leeson, A., M.E. Kelley, H.S. Rifai, and V.S. Magar (Eds.), *Natural Attenuation of Environmental Contaminants*. Battelle Press, Columbus, OH, 2001. 307 pp.

6(3): Magar, V.S., G. Johnson, S.K. Ong, and A. Leeson (Eds.), *Bioremediation of Energetics, Phenolics, and Polycyclic Aromatic Hydrocarbons*. Battelle Press, Columbus, OH, 2001. 313 pp.

6(4): Magar, V.S., T.M. Vogel, C.M. Aelion, and A. Leeson (Eds.), *Innovative Methods in Support of Bioremediation*. Battelle Press, Columbus, OH, 2001. 197 pp.

6(5): Leeson, A., E.A. Foote, M.K. Banks, and V.S. Magar (Eds.), *Phytoremediation, Wetlands, and Sediments*. Battelle Press, Columbus, OH, 2001. 383 pp.

6(6): Magar, V.S., F.M. von Fahnestock, and A. Leeson (Eds.), *Ex Situ Biological Treatment Technologies*. Battelle Press, Columbus, OH, 2001. 423 pp.

6(7): Magar, V.S., D.E. Fennell, J.J. Morse, B.C. Alleman, and A. Leeson (Eds.), *Anaerobic Degradation of Chlorinated Solvents*. Battelle Press, Columbus, OH, 2001. 387 pp.

6(8): Leeson, A., B.C. Alleman, P.J. Alvarez, and V.S. Magar (Eds.), *Bioaugmentation, Biobarriers, and Biogeochemistry*. Battelle Press, Columbus, OH, 2001. 255 pp.

6(9): Leeson, A., B.M. Peyton, J.L. Means, and V.S. Magar (Eds.), *Bioremediation of Inorganic Compounds*. Battelle Press, Columbus, OH, 2001. 377 pp.

6(10): Leeson, A., P.C. Johnson, R.E. Hinchee, L. Semprini, and V.S. Magar (Eds.), *In Situ Aeration and Aerobic Remediation*. Battelle Press, Columbus, OH, 2001. 391 pp.

A

B

E

F

G

H

I

J

K

L

M

N

O

R

S

T

U

V

W

X

Y

Z